Unruly Cities?

Steve Pile is Lecturer in the Faculty of Social Sciences at The Open University. His recent books include *The Body and the City* (1996), *Geographies of Resistance* (1997, co-edited with Michael Keith) and *Places through the Body* (1998, co-edited with Heidi J. Nast).

Christopher Brook is Lecturer in the Faculty of Social Sciences at The Open University. His recent books include *A Global World?* (1995, co-edited with James Anderson and Allan Cochrane) and *Asia Pacific in the New World Order* (1997, co-edited with Anthony McGrew).

Gerry Mooney is Staff Tutor in Social Policy at The Open University. He has published widely on issues relating to developments in social policy and in the field of urban studies. He is currently editing a collection of essays on the theme of *Class Struggles and Social Welfare* with Michael Lavalette.

UNDERSTANDING CITIES

This book is part of a series produced in association with The Open University. The complete list of books in the series is as follows:

City Worlds, edited by Doreen Massey, John Allen and Steve Pile

Unsettling Cities: Movement/Settlement, edited by John Allen, Doreen Massey and Michael Pryke

Unruly Cities? Order/Disorder, edited by Steve Pile, Christopher Brook and Gerry Mooney

The books form part of the Open University course DD304 *Understanding Cities.* Details of this and any other Open University course can be obtained from the Courses Reservations Centre, PO Box 724, The Open University, Milton Keynes, MK7 6ZS, United Kingdom: tel. (00 44) (0)1908 653231.

For availability of other course components, contact Open University Worldwide Ltd, The Berrill Building, Walton Hall, Milton Keynes, MK7 6AA, United Kingdom: tel. (00 44) (0)1908 858585, e-mail ouwenq@open.ac.uk.

Alternatively, much useful information can be obtained from the Open University's website http://www.open.ac.uk.

Unruly Cities?

Order/Disorder

edited by

Steve Pile, Christopher Brook and Gerry Mooney

London and New York

in association with

First published 1999 by Routledge; written and produced by The Open University

11 New Fetter Lane, London EC4P 4EE

Simultaneously published in the USA and Canada

by Routledge

29 West 35th Street, New York, NY 10001

The opinions expressed are not necessarily those of the Course Team or of The Open University.

Edited, designed and typeset by The Open University

Index compiled by Isobel McLean

Printed in Great Britain by T.J. International, Padstow, Cornwall

British Library Cataloguing in Publication Data

A catalogue record for this book is available from The British Library

Library of Congress Cataloging in Publication Data

A catalogue record for this book has been requested

ISBN 0-415-20073-3 (hbk)
ISBN 0-415-20074-1 (pbk)

1.1

CONTENTS

THE OPEN UNIVERSITY COURSE TEAM

John Allen *Senior Lecturer in Economic Geography*

Sally Baker *Education and Social Sciences Librarian*

Melanie Bayley *Editor*

Andrew Blowers *Professor of Social Sciences (Planning)*

Christopher Brook *Lecturer in Geography*

Deborah Bywater *Project Controller*

David Calderwood *Project Controller*

Margaret Charters *Course Secretary*

Allan Cochrane *Professor of Public Policy*

Lene Connolly *Print Buying Controller*

Michael Dawson *Course Manager*

Margaret Dickens *Print Buying Co-ordinator*

Nigel Draper *Editor*

Janis Gilbert *Graphic Artist*

Celia Hart *Picture Research Assistant*

Caitlin Harvey *Course Manager*

Steve Hinchliffe *Lecturer in Geography*

Teresa Kennard *Co-publishing Advisor*

Siân Lewis *Graphic Designer*

Michèle Marsh *Secretary*

Doreen Massey *Professor of Geography*

Eugene McLaughlin *Senior Lecturer in Criminology and Social Policy*

Gerry Mooney *Staff Tutor in Social Policy*

Eleanor Morris *Series Producer, BBC/OUPC*

John Muncie *Senior Lecturer in Criminology and Social Policy*

Ray Munns *Cartographer*

Kathy Pain *Staff Tutor in Geography*

Steve Pile *Lecturer in Geography and Course Team Chair*

Michael Pryke *Lecturer in Geography*

Jenny Robinson *Lecturer in Geography*

Kathy Wilson *Production Assistant, BBC/OUPC*

External Assessor

John Solomos *Professor of Sociology, University of Southampton*

External Contributors

Ash Amin *Author, Professor of Geography, University of Durham*

Stephen Graham *Author, Reader in the Centre for Urban Technology, University of Newcastle upon Tyne*

Kerry Hamilton *Author, Professor of Transport, University of East London*

Mark Hart *Tutor Panel, Reader in Industrial and Regional Policy, University of Ulster*

Susan Hoyle *Author, Research Associate in the Transport Studies Unit, University of East London*

Linda McDowell *Author, Director for the Graduate School of Geography and Fellow at Newnham College, University of Cambridge*

Ian Munt *Tutor Panel, Researcher, London Rivers Association*

Phil Pinch *Tutor Panel, Senior Lecturer, Geography and Housing Division, South Bank University*

Jenny Seavers *Tutor Panel, Research Fellow, Centre for Housing Policy, University of York*

Nigel Thrift *Author, Professor of Geography, University of Bristol*

Sophie Watson *Author, Professor of Urban Cultures, University of East London*

Preface

Unruly Cities? is one of a series of books, entitled *Understanding Cities*, that takes a new look at cities. The standard approach to thinking about the future of cities is to consider them as free-standing and geographically discrete places that rise and fall as a result of their strategic location, their economic viability or their political power. Typically, the history of cities is charted *from* the rise and fall of ancient cities (such as Athens and Rome), *through* the rise of mediaeval cities (such as Antwerp and Naples), *onto* the spectacular growth of cities during the industrial revolution and the age of empire (such as London and Paris), and finally *to* the sprawling 'post-modern' cities of today (normally exemplified by Los Angeles). The future of cities is then extrapolated on the basis of this historical vision – usually that urban areas will continue to sprawl across the surface of the Earth, eventually joining up to form '100-mile' or 'mega-'cities. The analysis developed in this series is, however, radically different.

In order to understand cities, we argue that it is necessary to rethink their geography. This is more than simply extending the range of cities considered beyond the London–New York–Los Angeles axis, whether from São Paulo to Sydney, from Manchester to Moscow. It also involves using a geographical imagination to understand how cities are produced, on the one hand, in a context of social relations that stretch *beyond* the city and, on the other, by the intersection of social relations *within* the city. This argument has widespread implications for our understanding of cities. These implications are comprehensively investigated in the three books that comprise this series, *City Worlds*, *Unsettling Cities* and *Unruly Cities?*. Through the series, we tease out, for example, the ways in which cities bring people from different backgrounds into close proximity; how the juxtaposition of different people and activities in cities can change and alter social interactions; how these juxtapositions can result in, or result from, urban conflicts and tensions; how different parts of cities are connected to, or disconnected from, other cities; how people network within and between cities.

Developing these arguments shows that the city cannot be thought of as having one geography and one history (and therefore one future). Instead, cities are characterized by their openness: to new possibilities, and to new interactions between people. This book series gradually reveals the difficulties and paradoxes that the unavoidable openness of cities presents, as different histories and geographies intersect and overlap. Significantly, then, it is these issues that must be understood, if people are to learn to live in our increasingly urbanized world.

A few words about the books themselves. Each book is self-contained: each chapter provides all relevant supporting readings and materials to enable the reader to come to grips with a new understanding of cities. The three textbooks are, also, a significant component of the Open University third level course, DD304 *Understanding Cities*. The book series is joined by a television series, *City Stories*, and other materials, including audio-cassettes and course guides.

The television series picks up the themes of the three books through case-study material from across the globe – visiting Sydney, Singapore, Kuala Lumpur, Moscow and Mexico City. In particular, the television series develops a line of argument about the different ways in which groups within cities locate themselves within specific networks of social relationships. Local contributors comment on and analyse their own cities, highlighting the issues of importance for their urban futures. The television series is integral to the Open University course and provides a wealth of examples to illustrate the themes and the experiences of living in cities around the world. (Details of how to obtain copies of the TV programmes are available from Open University Worldwide Ltd.: see p.ii.)

Open University courses are produced through extensive and intensive discussions amongst academic authors, a panel of experienced tutors, an external assessor, editors, designers, BBC producers, a course manager and, not least, secretarial staff (listed on p.vi). Every component of the course has been subjected to wide-ranging discussion and critical debate. At every stage, this has led to improvements in the materials, which have benefited from the academic and distance-teaching expertise accumulated at The Open University.

The end-product of this process is to produce textbooks that have a specific style. In particular, authors have sought to make these texts *interactive*. Readers are asked to become involved in the problem of understanding cities, rather than simply to digest the material. Readers are provided, too, with a wide range of readings and extracts, which they can use to enhance their own appreciation of the problems that cities confront. Certain ideas become important to this understanding of cities – and these are registered in two basic ways. First, key points are emphasized using bullet points; and, second, core ideas, that are discussed in detail, are indexed in **bold**. Meanwhile, each book's last chapter looks back over the book as a whole and draws out the key themes that have been discussed. For readers who wish to use the series as a whole to further their understanding of cities, there are a number of references backwards and forwards to chapters in other books in the series; these are easily identifiable because they are printed in **bold** type. This is particularly important since chapters are intended to build on previous discussions. By integrating the material in this way, it is hoped that readers will be able to pull out the themes that run across the series for themselves. Through this interactive and integrated approach, it is hoped that readers will develop their own understanding of cities and ultimately be able to transfer that understanding to other cities, even to other situations.

It only remains to acknowledge those who have worked hard and with such enthusiasm to produce this book series. Most obviously, these texts – and the course of which they are a part – would not be possible without the academic consultants and the tutor panel. All have helped to shape the chapters of others, while also responding constructively to suggestions and advice. The external assessor, John Solomos, provided invaluable intellectual guidance, insights, support and encouragement at every stage of the development of the books and the related course materials. The tutor panel, comprising Mark Hart, Ian Munt and Jenny Seavers, have been rigorous in their comments on the drafts as the chapters developed. Finally, Phil Pinch, who wrote the Course Guide, generously advised on the development of the course as a whole.

Producing an excellent textbook series does not end with the writing. The production process has been co-ordinated and kept on track by Deborah Bywater and David Calderwood. We have been shrewdly supported by our editors, Melanie Bayley and Nigel Draper, who scrupulously scrutinised the chapters before publication. Meanwhile, the design of the books and chapters has been thoughtfully overseen by Siân Lewis. Our cartographer and graphic artist, Ray Munns and Janis Gilbert, developed and drew the maps and diagrams. Together, they have turned the drafts into the excellent books they undoubtedly are. Finding illustrations for this course has proved, at times, a frustrating task, but Sally Baker and Celia Hart have been good humoured throughout. The production of textbooks at The Open University means a never-ending stream of requests for word-processing and the distribution of drafts, often at very short notice. With good grace and efficiency, the course team has been supported by Michèle Marsh and, in particular, the course secretary, Margaret Charters. Within the Faculty of Social Sciences, Peggotty Graham has helped to ensure that the books were produced to high standards, while – at Routledge – Sarah Lloyd has been a constant source of good advice, flexibility and encouragement.

At the centre of this process are the course managers who have worked on this series of books. We have been lucky enough to have gained from the expertise of Christina Janoszka and Mike Dawson. However, it is Caitlin Harvey who has seen the books through the most important stages. Caitlin has never failed to keep her head, even while all about her were losing theirs. Keeping everything together has meant that course team meetings have been intellectually lively and thoroughly thought-provoking. We hope that the vitality and diversity of these debates shine through in these textbooks, which offer a new way of *Understanding Cities*.

Steve Pile
on behalf of the Open University Course Team

Introduction

Cities, like nowhere else, present a series of seemingly insoluble problems. Periodically, they seem to be 'in crisis'. The problems take many forms: as ever more people live in the world's cities, services fail to provide for their basic needs; as poverty deepens, people adopt survival strategies that others consider illegal or immoral; as tensions within cities rise, people take to the streets to protest. We could go on, and on. Alternatively, cities can be seen as *the* crisis. As they sprawl across the face of the earth, cities are disordering the environment through their consumption of raw materials or through deadly, uncontainable emissions. As stock exchanges crash, bounce and crash again, people are thrown out of work in places seemingly untouched by urban financial markets, and sometimes even the world is plunged into unexpected recession. Cities, then, seem to be genuinely *unruly*, and they appear to be growing uncontrollably, exacerbating already seemingly unmanageable social tensions. However, are these dystopic stories of unstoppable (economic, social, environmental) catastrophe the only ones we can tell about cities?

This book offers a particular interpretation of the (supposedly) unruly character of urban life. We take issue with the notion that cities are entirely chaotic and without direction. Of course, it may seem that cities are often chaotic and beyond help, for they are characteristically open to wider influences, to power relations that stretch well beyond their means to control or influence – whether we are talking about the movement of people, in or out, at particular times (as in Mexico City or London), or the vagaries of global financial flows (as in Jakarta or New York). Meanwhile, inside cities, they are abuzz with flows of people, information, money, traffic and so on, that never seem to settle anywhere, but just keep moving. Yet cities cannot simply be judged unruly because of the problems that arise from having large numbers of people and things in the same place, nor because they are inextricably interwoven with cross-cutting social relationships that interact in seemingly unpredictable and unintended ways.

Despite their vastness and their perpetual flux, it is possible to see certain persistent regularities in urban social relationships. Cities are – and this is our central point – not entirely random assemblages of things and people (even if it sometimes feels like this!). Instead, we argue that the almost unimaginable intricacy of cities arises both from the specific ways in which they bring things and people together (i.e. the mixings and meetings of cities), *and also* from the ways in which cities are built to sort, sift and segregate things and people (i.e. the patternings and orderings of cities). In this book, the authors seek to understand the problems facing cities by disentangling the spatial weave that is integral to those problems. The early chapters are concerned to tease out the distinctive spatialities of cities: their 'within-ness', and significantly a 'within-ness' that emerges out of the inherent openness of cities.

It will not astonish anyone to point out that the 'within-ness' of cities is characteristically made up of identifiably different kinds of spaces. Within any city, there might be differences, say, between the inner city and the suburbs, between shanty towns and urban villages, between the financial district and the places where restaurants, cinemas, clubs and theatres are. With such a view of the 'within-ness' of cities, it is possible to assume that *all* cities have differences between neighbourhoods, streets, quarters, districts and so on. We could describe, catalogue and map these different urban spaces for each city, but this would not be very enlightening, as it would tell us very little about the social relations producing differences, be they between people or places.

What is really at issue in this book, then, is how such differentiated spaces are produced in cities. The *differentiation of urban space* is not simply about the form that differences take, but about how such differences are produced, maintained, undermined, challenged or transformed. It is one of the key arguments of this book that understanding the differentiation of urban space involves an explicit consideration both of *what* is within one urban space (and not others), and also of *how* this space is related (or not) to other places – whether these are within the same city, or whether their connections stretch far beyond the city.

We can see, for example, that inequalities between the rich and poor within any one city will often reveal themselves in stark differences between rich and poor areas, such as between the 'gold coast' and 'the slum' in 1930s Chicago, or between the rich suburbs and shanty towns of Istanbul or New Delhi. Clearly, the juxtaposition of these extremes within (many) cities has much to do with the unevenness of power relationships. However, by thinking through the spatialities of the social relationships that produce differentiated urban space, the authors of this book are able to turn up some rather unexpected issues. For example, by tracing the connections between 'gated communities' or rich suburbs – whether in Johannesburg, Los Angeles or São Paulo – and the rest of the city, it is possible to show that they are remarkably porous and open to the wider world. Similarly, 'the ghetto', which is often distinguished by its lack of connections, can be seen to be intimately connected to the rest of the city. However, these connections and disconnections need to be carefully analysed, for they neither take the same form, nor have the same social outcomes. This is also to say that connections and disconnections, that acts of connecting or disconnecting, are bound up with power relationships that are both unequal and uneven.

From this, we can begin to glimpse something else about the unruly city. It is not just that we find order in one place and disorder in another, but that there is an internal relationship, a tension maybe, between the two. Take a gated community, for example. On the one hand, the building of such segregated urban spaces (in very different circumstances) has commonly been prompted by a sense that the city is disorderly and that the only way to provide safe, habitable neighbourhoods is to 'gate out' the threatening elements of the city. On the other hand, such solutions tend to intensify the sense that the outside world is dangerous and unmanageable. More than this, the visible exclusiveness of such neighbourhoods intensifies resentment against them, and against the people within them. Nor can the gates keep out those disorders that appear within their walls. Indeed, they can also serve to prevent escape. 'Gates' and 'walls', then, are socially ambiguous, not only because they can be seen as both orderly and disorderly, but also because they intensify the relationship between order and disorder. In this simple case, therefore, we can see that one person's order might be another's disorder; this is an important point to bear in mind. But we can also see that these orders and disorders have something to do with the ways in which spatial relationships within the city are made and remade. This is one example, but we might also begin to think about other ways in which urban spaces are (dis)ordered and differentiated, particularly through the practices of planners, builders, urban social movements, and so on – as we will see in this book.

Understanding what it means to say that the city is unruly (or not) thus begins to open up questions about the differentiation of urban space, about the ways in which people are brought together – and kept apart – within cities, and about the orders and disorders that arise from the intricacies of social relationships and their integral spatialities. Such questions are not new if considered separately, but perhaps if these questions are brought together, it will be possible to offer a new way of understanding cities.

◆ ◆ ◆ ◆

The idea that cities are characterized by processes that produce differentiated urban spaces has been a key arena of debate among people trying to understand and appreciate the specific qualities of cities: their heterogeneity, their densities, their intensities and their seemingly unruly character. Somewhere out of the differentiation of urban space, questions arise about the unruliness of cities and what is to be done about urban life. In broad terms, it can be suggested that there are two opposing responses to these questions. Some believe that cities should be more orderly and they have sought to find the means through which more orderly cities can be produced, physically and socially. However, it has also been argued that 'disorder' is a fundamental feature of what makes cities worthwhile places to live in, because this underscores their vibrancy and creativity. Such a polarized understanding of cities is unhelpful, since some kinds of orderings produce stultified cities, while disorders can be menacing. Indeed, the situation is more ambiguous than this, because it depends on who is experiencing (dis)order and in what ways: order can be safe as well as boring, disorder both thrilling and threatening. This book therefore revisits the classic work of theorists who have sought to disentangle and interpret the orders and disorders of the city.

In many ways this story begins with the work of the nineteenth-century urban planner, Ebenezer Howard. Faced with the appalling urban conditions of Victorian cities in both Britain and America, Howard sought a solution to the problems of squalor, sickness and poverty. His plans for a 'good' city are important for this book because we see how ambiguous such ordered plans can be. In essence, Howard was attempting to deal with the heterogeneity and inequalities of Victorian cities by disentangling them and re-integrating them in such a way as to produce a differentiated urban space that was equitable – environmentally, socially and economically. Howard has been severely criticized for the solutions that he hit upon, as you will see. However, it is less important whether he produced the right solution than to see that these plans were directed at producing a particular form of orderly city. More than this, the arguments presented by the authors in this book lead us to consider how these plans were intended to intervene in the spatialities of the city, how they were implemented and by whom, and whom they benefited. It is not simply the case that plans for a 'better' city are intrinsically good or bad, but rather that they have good or bad effects in specific situations. Another case in point would be the work of the architect Le Corbusier.

Le Corbusier's plans for cities direct us immediately towards issues of the ordering of cities. His designs for 'cities of tomorrow' are renowned for the kind of urban space they envisaged: the city would be a vast, planned, rational machine for living in. What is of interest here is the way Le Corbusier dealt with the heterogeneity of the city. Along with Howard, he imagined a 'good' city as having a definite order, with the physical, social, environmental and economic aspects of the city being finely tuned into a complex machine, as if the city were a car engine, with specific parts for specific functions – an orderly whole directed towards the function of living. It was assumed that the unruliness of the city could, and should, be planned away through rational, planned interventions into the organization of urban space. Many people tackling urban problems across the world shared this assumption. What becomes important, therefore, are the ways in which such ideas are taken up in social practices that are directed towards ordering cities in apparently very different contexts. Such practices can sometimes be referred to as 'international best practice', since this term brings out the ways in which visions of the 'good' city are transferred between cities and contexts. Tracing out these ideas demonstrates both how it is that the built environment is produced to achieve specific social outcomes, and how it is that these outcomes are desired and sought in different circumstances, sometimes far away from the places where they are first imagined. Once again, then, we should remind ourselves that visions of the 'better' city are never produced in isolation from power relations both within and beyond the city.

If Howard and Le Corbusier are keen to provide plans for the good city as a whole and to influence the ways in which the physical structure of the city is built, others have been more concerned with the nature of social life within cities and have asked questions about what makes a good city in terms of whether they are decent places to live and thrive. One of the most influential contemporary thinkers along these lines has been the urban commentator Jane Jacobs. In her work she has engaged critically with the ideas of Le Corbusier. For Jacobs, the order that Le Corbusier wished to impose ended up constraining the vitality and creativity of a city life which she saw, in essence, as emerging out of its unruliness and disorders. We will read about her ideas at many points throughout this book. Similarly, the urbanist Richard Sennett has been keen to think about how a disorderly city life should be lived appropriately. While Sennett wishes to see the mixing and meeting of different people as an opportunity, he also recognizes that most people in the city are indifferent to the differences of others – but perhaps this is not such a bad thing either. Rather than condemning heterogeneity and attempting to disentangle it, and rather than seeking homogeneity, Jacobs and Sennett have sought to value the differences among people and to see positive aspects in encounters with people who are unfamiliar or strange. In this the figure of the stranger becomes key, as we will see. Questions arise as to whether the stranger – the unknown person who *appears* to be different – is a threat, someone to be afraid of, or whether this is someone who should be trusted or enjoyed, or if not exactly trusted, then someone who need not be feared.

In this brief review we seem to have reached an impasse. There seem to be good sides and bad sides to every story. On the one hand, it might be a very good idea to impose order on a situation, to build cities for orderliness. After all, disorderliness can be very discomforting and dangerous. On the other hand, some kinds of disorder seem to be productive of some things that are valuable about cities: their vitality, vibrancy, surprise, creativity. After all, orderliness can be very stifling and help to maintain a situation in which people are dominated by rules produced by self-serving élites. So is this impasse the fault of the theorists? Or, to put it another way, is it a problem with our account of cities? Well, no. In part, we can see, even from the brief outline above, that cities are thoroughly paradoxical places. By this, we mean that cities often bring together incommensurable activities and people, who have very different ways of living or assessments of what a good city might be like. For some, this might be part of the problem of cities: their inequities, their injustices, or alternatively their openness to overwhelming influxes of (denigrated) 'others'. For others, this may be the very best thing about cities: their heterogeneity, their opportunities for advancement, their openness to stimulating encounters with (valued) 'others'.

In this book we do not seek to resolve these opposing accounts of the city, but to draw out the ways in which they variously interpret the orders and disorders of the city. More than this, these interpretations are applied to an understanding of the differentiation of urban space. With this in mind, it can be argued that attempts to order urban social relations also have integral spatialities to them, whether they seek to segregate one group from another, or bring them together in productive ways. We can also turn this argument on its head. From the other side, it can be seen that the production of differentiated urban spaces can both order and disorder social relationships, at one and the same time. With this 'double-sided' imagination, it is possible to infer that one of the real dangers in urban analysis is the attempt to separate out order and disorder, and to privilege or value one over the other.

◆ ◆ ◆ ◆

The sub-title of this book flags the issue of order and disorder, since this is a central theme of the book. It is addressed in terms of the differentiation of urban space, as we have noted. But there is a little more to it than this. Calls for order, and claims about the significance of disorder, arise from the differentiation of urban space – whether this is about the safety of people on the streets, about the creativity and vibrancy of city life, or about the vicious segregation of one group of people from another. We might argue that interventions in urban space to produce particular effects are ambiguous, with no social outcome – for good or ill – that can be easily or definitively predicted. But this does not mean that it is impossible to say anything about the social consequences of specific interventions into the differentiation of urban space.

In this book, then, there is an intention to think through how order/disorder are produced through the relationships that differentiate urban space. We can trace these spatialities of differentiation through many aspects of urban life, whether they are social, economic or environmental. We will also see them play out differently in different cities, including New York, London, Paris, Berlin, Istanbul, New Delhi, Jakarta and Sydney. What these examples begin to suggest is that we need to think carefully about the openness both of cities and, as important, of spaces within cities, involving the ways in which they are connected up, segregated out and disconnected from one another. It is here that the unevenness and inequities of power relations can most easily be identified, and it is here that we can begin to think about new possibilities for ordering cities and for thinking about their unruliness.

Steve Pile, Christopher Brook and Gerry Mooney

CHAPTER 1
The heterogeneity of cities

by Steve Pile

1 *Introduction: living with difference*

In her novel, *Elise or The Real Life*, Claire Etcherelli writes about the relationship between a white Frenchwoman and an Arab man in Paris in the 1950s. For writer and critic Liz Heron, Etcherelli's novel is significant because it describes the tensions and contradictions that arise out of cross-cutting class, 'race' and sexual relations in the city (Heron, 1993, p.3). Indeed, it is these tensions and contradictions that make Paris life real for Elise, Etcherelli's heroine. Once in Paris, Elise takes a job on a car assembly-line. There, she begins a love affair with a co-worker, Arezki, who is Algerian. The story is set in the late 1950s, at a time when the Algerians were engaged in a bitter and brutal anti-colonial struggle – against the French. As we read the following extract (which appears in Liz Heron's anthology, *Streets of Desire*), we can glimpse how French colonial domination abroad reverberated around Paris: the heart of Empire.

EXTRACT 1.1
Claire Etcherelli: from *Elise or The Real Life*

The assembly line came to a halt and the siren went off. Mustapha brought me the gasoline cloth Arezki had given him. It was a signal. He wanted to speak to me.

I picked up my coat and left for the Port d'Italie. I felt the need to walk and talk out loud. There were gusts of wind that raised your hair on end and sliced the skin of your face, beautiful girls in warm coats who, height of injustice, were made even more beautiful by the cold and their winter clothes, Algerians walking duckfooted in spring jackets with their collars turned up; there were cops at the entrances to the Métro checking identity cards, and the windows – from the Prisunic to the most dilapidated grocer – were caught up in a fever of garlands and lights. A happy throng, well-nourished, wearing fur-lined boots and interlined coats, who spent August by the sea and wore spring clothes at Easter, a throng that paid for its leisure with the sweat of its brow, walked, sat at café tables, and looked the other way when into its territorial waters slipped ill-nourished types who wore Easter clothes in November and who, for all their brow's sweat, earned only enough for bread. These species just happened to gather in special neighbourhoods – shanty towns, run-down hotels – and, by nationalities: Algerian, Spanish, Portuguese, and, naturally, French. They also fell into other categories: alcoholics, idlers, tuberculars, degenerates. There is something to be said for the ghetto. But sometimes these types managed to sneak up on you in the Métro, in the café, and in addition, they were noisy, lost, or disgustingly drunk. And occasionally, in these caricatures of humanity, in these suffering bodies mutilated by misery, in the cold dark rooms, between the dirty laundry and the drying laundry,

> one of these dregs carried inside him – by luck or miracle – the gleam, the flame, the spark that made him suffer even more. The spirit breathed there as much as anywhere; intelligence either developed or died, crushed.
>
> These thoughts, the cold, my hair blowing around my neck, Arezki's disappearance, the Magyar's blood and the smell of the factory, the four hours on the line stretching ahead, the still unread letter from Grandmother, all this is life. How gentle it had been, the previous one, a little blurred, far from the sordid truth. It had been simple, animal, rich in dreams. I said 'one day ...' and it was enough.
>
> I am living this day, I am living the real life, involved with other human beings, and I suffer.
>
> Source: Etcherelli (undated) in Heron (ed.), 1993, pp.224–5

This story evokes the powerful emotions of a woman seeking to come to terms with the real life Paris confronts her with. What is distinctive about this real life is the way in which the city, Paris, brings together people from diverse backgrounds and concentrates them in urban space. We can also see that Paris makes some people occupy the same space – in this instance, the spaces of the car assembly-line and the cold, wet streets. In this way, the city opens up the possibility for different people to meet one another. In this story, we can see just how exhilarating and how painful these experiences can be. But there is something more in this tale than the star-crossed love affair between people from different cultures.

ACTIVITY 1.1 Read the further excerpts from Claire Etcherelli's *Elise or The Real Life* in Reading 1A at the end of this chapter. While you are reading, think particularly about what Paris – and its different neighbourhoods, such as Port d'Italie and Stalingrad – might have been like to live in. ◆

In these stories, we can see that the city of Paris juxtaposes different feelings (**Allen, 1999a**). Here we can see Arezki's fearful watching out for cops. And how the lovers' happiness transformed the streets. Further, we can also see how, when confronted by the heterogeneity of Parisian people, Elise turns away and tends to caricature them: 'alcoholics, idlers, tubercular, degenerates'. Though not indifferent to their plight, Elise cannot see them as individuals. 'Cityness' – Paris life – seems to prevent people from being seen as individuals. Instead, they are 'caricatures of humanity, in these suffering bodies mutilated by misery'. In these respects (and more), we can see that Elise's feelings arise from the ways in which cities are heterogeneous, constantly mixing differences (**Amin and Graham, 1999**). However, the ways in which people react to the heterogeneity of the city are not easy to predict. Even while Elise falls in love with Arezki, the police are arresting and 'disappearing' Algerians, and there are murderous riots. Sure, city life can lead some people to be indifferent or more tolerant, but it can also intensify racist hatred.

It is not just people of different ethnicities and sexes whom Paris brings together, though. You can see how Etcherelli records, with some bitterness, the social polarities that exist between classes in 1950s' Paris. Social polarities come in other forms too: Elise, herself, at the same time that she feels sympathy with social outcasts also feels disgusted by drunks. And these social polarities are mapped into social geography of the city. There are distinct neighbourhoods in Paris – and these are racialized: 'Algerian, Spanish, Portuguese, and naturally, French'.

Clearly, the city brings different people together and mixes them up. Paradoxically, however, the city is also a machine for *producing* differences amongst people – differences which then become the basis on which they are kept apart from one another. In this story, Algerians arrive in Paris because they have links through French imperial connections. Once they are there, they are classified as foreigners and potential terrorists – and, therefore (dangerously) *different*. And they come to live in segregated parts of Paris. Like others. Like them. The question for us, in this chapter, is how we are to understand this paradox of mixing and separation.

FIGURE 1.1 *On the steps of Montmartre, Paris, 1950s*

From our reading of Etcherelli's account of 1950s' Paris, it is possible to make some preliminary observations about the heterogeneity of cities and the materiality of city life. We can make three key points.

First, we can note from this cross-section that Paris is a city of *contrasts* – from what we have read, we could list alcoholics, idlers, tuberculars, degenerates, writers (Etcherelli), communists, car workers, the police, shoppers, the dead and so on. There are also many different activities, from lovers strolling to acts of terrorism, from leisure to work. All the people, all the events, all the experiences, all the stillnesses and movements, all the chaos, all the dangers: it's often too much to take in. So, paradoxically, cities are both exhilarating and depressing, enthralling and overwhelming. Nevertheless, these narratives point to the importance of recognizing the differences in people's experiences of city life.

Second, cities combine people whose experiences have very different *felt intensities.* The idea of 'felt intensities' immediately evokes people's subjective experiences – whether these are about the intensity of racism, or of love, or of danger. However, there is more to the idea of 'felt intensities' than this. The city is more than the backdrop for people's feelings. Instead, the city itself – cityness – provokes felt intensities. For example, as Etcherelli has her characters thread their way through different parts of Paris, she brings to life the (romantic? violent? paranoid?) intensity and (class, racial, political) diversity of the city's street life. This brings us to the next point.

Third, it is possible to discern from Etcherelli's stories something of the *intensification of social relationships* in cities. 1950s' Paris does not simply contain different histories and geographies, it also brings them together – and, indoing so, puts 'differences' into relation to one another. Differences might be ignored or exaggerated, avoided or embraced. Whatever, by concentrating and exaggerating different histories and geographies, cities intensify social relationships. And these social relations are also power relations. Their intensification, then, can aggravate social tensions: as workers and Algerians struggle for justice – supporting or clashing with each other, being supported or oppressed by other classes, other nationalities. This city, indeed, can be unruly. One consequence might be that governments impose order by diffusing the tensions that arise from the intensification of social relations. Thus, it is possible to see why a violent anti-colonial struggle in Algeria would provoke the authorities into excessive restrictions on people's – especially Arab people's – movements through the city. Or why class struggles in Paris might provoke urban planners into settling the working classes in estates on the urban periphery (see Chapter 2 of this book). In this way, we can begin to see how different spaces within the city might emerge; how it is that different histories and geographies might overlap or sit side by side or be kept apart.

Let us think about this last point.

ACTIVITY 1.2 Look back over Reading 1A. This time, look for the varied experiences that Elise and Arezki have of Paris life. Can you pick out the ways in which different groups are brought together – or kept apart – in Paris? Think, too, about the ways in which certain relationships are intensified, moderated or suppressed. ◆

Sometimes, as Elise and Arezki walk from neighbourhood to neighbourhood, different communities blend into one another and it is not at all clear where one has ended and the other begun. Sometimes, different people occupy the same place, but simply walk past one another, not even noticing that they are there. Other times, areas are sharply demarcated and some groups will have nothing to do with one another. By occupying – or avoiding – certain places, Elise and Arezki can negotiate the intensification of social relationships in certain places, either by occupying sites of pleasure or calm, or by avoiding those where racial hatred is concentrated. However, city spaces can rarely be mapped so clearly, so unambiguously.

Even if we take one of the clearest distinctions in the social geography of cities – between the inner city and the suburb – we cannot be sure that one necessarily represents order and the other disorder; one community, the other difference; one safety, the other danger. An argument that suggests that cities are heterogeneous begs questions about the relationships between order and disorder, community and difference, safety and danger that make different urban spaces different. In this way, we can glimpse the way in which urban spaces are produced through the *negotiation* of heterogeneity: whether people seek community or difference, whether they embrace the excitement of urban disorder or the calm of urban orderings. More than this, we might begin to understand how it is that the inner city and the suburb have come to be seen – *and produced* – as such different spaces within the city.

This chapter, then, seeks to re-interpret this social geography of the city by thinking about the ways in which *heterogeneity* creates distinctive urban spaces that are constituted by the ways people negotiate

- relationships with others
- the city's spatial relationships (inside and out), and
- the tensions of city life.

However, the idea that the city mixes up people from different backgrounds and intensifies social interactions – thereby becoming a place of order *and* disorder, community *and* difference, of excitement *and* danger – leaves us with the question of how exactly differences between urban spaces are produced (a question that will concern us throughout this book). Let us start our analysis, then, with the supposedly most heterogeneous and disorderly parts of the city: the inner city.

2 *From the city ...*

2.1 IS CITY LIFE RULED BY DISORDER?

In his classic concentric zone model, the urban sociologist Ernest Burgess detects an underlying order in the social pattern of the city (see Figure 1.2). And, despite its age, it is this image that continues to dominate conceptual understandings of the social geography of the city. In Zone I (titled Loop, after Chicago's district), business is transacted. Meanwhile, factories, the poor, the criminals, the sick and so on, are apparently confined to Zone II (which is in transition). As you move further from the centre of the city, so the city-dwellers become richer and richer – until you eventually arrive at a suburban commuter belt (Zone V). This mapping of the social patterns of the city suggests that each Zone is defined by its internal homogeneity and also by its difference from the surrounding areas. The city, then, is sifted and sorted in an orderly way. (Although we should remember, too, more disorderly readings of this concentric zone model: see **Pile, 1999.**)

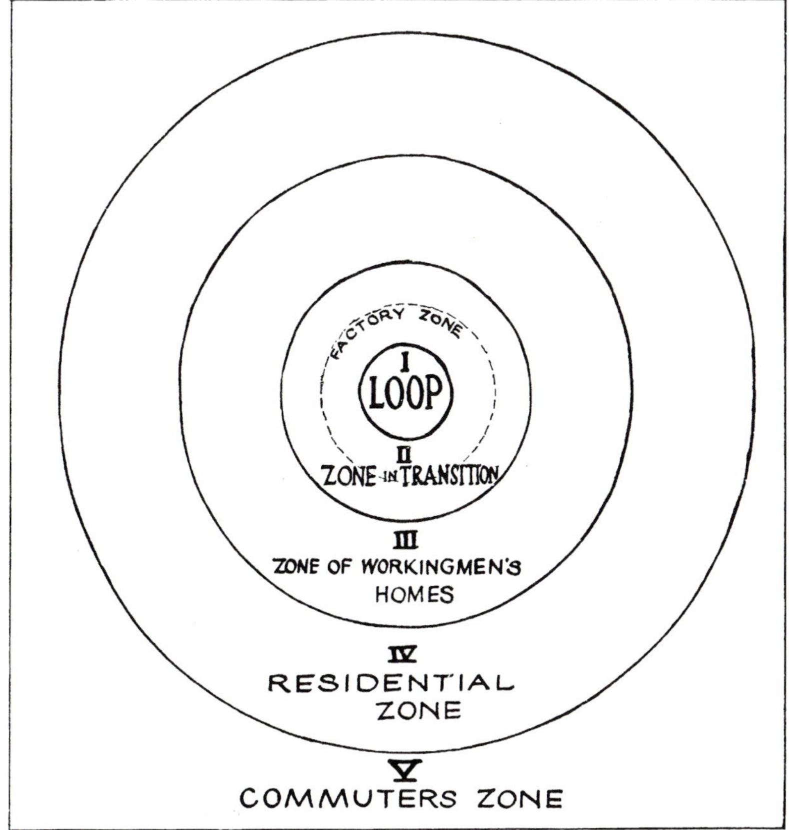

FIGURE 1.2 *The growth of the city from Burgess (1925)*

However, Burgess's own description of these zones suggests that something more chaotic is going on within them. Let's take a closer look at the inner zones of the city. In Zone I, the city's commercial, financial and political élites conduct their business. In this area, we would expect to see the most visible signs of the city's economic success. Here are to be found the skyscrapers, up-market shops, expensive restaurants and so on that represent the city's vitality and glamour. However, according to Burgess, this is also where – though less obviously – there is an adjoining street which can be described as 'hobohemia': 'a teeming Rialto of the homeless migratory man of the Middle West' (Burgess, 1925/1984, p.54). Underneath the supposed orderliness of dominant economic rhythms, people live out other kinds of lives (see **Pryke, 1999**).

While the central business district (Zone I) is not quite as orderly as it might at first appear, the next zone out (Zone II) is, according to Burgess, a highly disordered place. So disorderly, in fact, that social, moral and communal organization has broken down. For this reason, it is commonly described as 'a zone of deterioration'. The juxtaposition of these two contrasting zones is, according to Burgess, a universal feature of (US) cities. While homeless migratory men are to be found on the streets of the central business district, the surrounding neighbourhoods are 'submerged regions of poverty, degradation, and disease, and their underworlds of crime and vice' (Burgess, 1925/1984, p.55). Thus, right up alongside the wealthy central business district is an area of impoverished, dilapidated slums – the 'badlands' of the city.

For Burgess – and for other members of the Chicago School – the main reason the city becomes more unruly has to do with *who* is moving into the area. The hobos – those homeless migratory men – become a threat to city life precisely because they are homeless (and men?) and on the move. The zone of deterioration is 'the purgatory of "lost souls"' (Burgess, 1925/1984, p.56) because poor immigrants have been cast into the slums. Despite the disorderliness of this zone, Chicago School sociologists nevertheless assumed that higher ideals of community and the American Dream (natch) could pull people together and push them outwards towards a better way of life and urban living.

Whatever the Chicago School's assumptions, it is possible to suggest – by drawing on evidence introduced at the beginning of this chapter – that 1950s' Paris might be described in similar terms to 1930s' Chicago. In this light, the inner areas of cities (Zones I and II) – in fact, those parts of the city that we might think of as actually being *the* city – are clearly heterogeneous spaces. This heterogeneity, however, is more than about the juxtaposition of extreme differences between people. They are heterogeneous

- both because they bring different people into close proximity,
- and because people can meet others and mix with them, though they may not do so all the time, and they might find different ways to mix (indeed, to 'mix it' is slang for fighting!).

The social patterning of the city, then, has to do with the way in which people mix, or not. It is to these mixings that we must turn if we are to understand the heterogeneity of cities,

- partly because it is through the social relationships that bring people together, and keep them apart, that differences amongst people in the city are defined,
- and partly because these social relations of difference produce distinctive urban spaces and define the use (and abuse) of those spaces.

In the work of the Chicago School, we can note that their patterns suggest that there are sharp divisions between people (as portrayed in the concentric zones of the city). On the other hand, they recognize the proximity of contrasts, for example, between the hobo and the office worker. Meanwhile, the slum contains people with divergent backgrounds, attitudes and prospects. The slum was not irretrievably condemned to, nor endangered by, the disorder wrought by criminal activities, moral degradation, disease and social disorganization, for there was the possibility that orderly conduct might emerge through a community spirit. Disorderliness, then, was to be overcome by a sense of community. But how was this to be fostered? To begin with, it was assumed that communal spirit was fostered through contacts between people within distinct neighbourhoods.

For Robert Park, a leading member of the Chicago School, the neighbourhood is the basic building-block of the city's social organization. It is at the neighbourhood scale that people form associations that tie them into city life. Thus, for Park,

> Proximity and neighbourly contact are the basis for the simplest and most elementary form of association … in the organization of city life. Local interests and associations breed local sentiment, and, under a system which makes residence the basis for participation in the government, the neighbourhood becomes the basis of political control.
>
> (Park, 1925/1984, p.7)

We can note the importance of spatial relationships in this analysis, marked by terms such as 'proximity' and 'local'. And we should also note that there is an in-built set of ideas about political control too. Further, it was argued that, in the slum neighbourhoods of the 'zone of deterioration', local forms of association (whether sentimental or political) had broken down. The question is why this might be so.

The problem seemed to be related to the *movement* of people through a neighbourhood. Thus, Park suggested, if an area were stable enough and sufficiently homogeneous, then it could develop an orderly communal way of life. He was not making this suggestion in the abstract, however. He had a specific neighbourhood in mind: Harlem in New York. We will return to Harlem in the next section, but let us see what lessons Park draws from Harlem's experiences. First, a word of warning about the language Park uses. We have already read the term 'hobo' and we are about to read the term 'Negro'. These terms are not innocent, since they are used to describe people who were denigrated, socially marginalized and, more often, worse. With this caveat in mind, let us think about why – for Park – Harlem developed a communal life. Park extracts the following quote from James Weldon Johnson's 'The making of Harlem' (1925):

> In the history of New York the significance of the name Harlem has changed from Dutch to Irish to Jewish to Negro [*sic*]. Of these changes the last has come most swiftly. Throughout colored America, from Massachusetts to Mississippi and across the continent to Los Angeles and Seattle, its name, which as late as fifteen years ago [i.e. 1910] has scarcely been heard, now stands for the Negro Metropolis. Harlem is, indeed, the great Mecca for the sight-seer, the pleasure-seeker, the curious, the adventurous, the enterprising, the ambitious, and the talented of the Negro world; for the lure of it has reached down to every island of the Carib sea and has penetrated even into Africa.
>
> (James Weldon Johnson, cited in Park, 1925/1984, p.8)

What is of interest to Park is that the history of Harlem is characterized by successive waves of in-migration. Yet, the most recent in-migrants, African-Americans had turned Harlem into a place where talented, ambitious, pleasure-seeking black people could find success. Having *settled* in Harlem, African-Americans formed a sense of community that was both intimate and organized. This was Park's success story.

ACTIVITY 1.3 Look back over the argument presented so far in the section. Why are 'hobos' and migrants (amongst others) seen as a problem? Do you think 'order' is associated with 'community'? Or with 'stability'? ◆

For me, the Chicago School is arguing that

- a city's moral order is built up out of communal ties fostered at a neighbourhood scale,
- local loyalties can only be developed if people feel a connection to a place, and
- communal ties are only possible amongst people who are in (repeated) contact with others in that place.

The potential for local ties and loyalties is stifled where people do not feel for their neighbourhood. And they do not feel these ties when they are drifters or passing through or have every desire to leave the place. In this scheme, the movements of *strangers* become a particular problem, for their anonymity allows them to be criminal, immoral, anti-social.

However, an alternative view was suggested by Jane Jacobs in her classic study, *The Death and Life of Great American Cities* (1961). Rather than seeing strangers as a threat to orderliness, she argued that they were the very foundation of communal life in cities – indeed, urban life was impossible without them. Rather than looking to neighbourhoods to have an internal homogeneity that would guarantee their moral conduct, Jacobs suggested that streets should have a diversity of activities, which would encourage many people to use them at all hours of the day.

For Jacobs, as for Park, the continual use of streets means that intricate webs of (assumed) mutual support and trust are built up between people. However, instead of seeing the street life of poor neighbourhoods as impoverished and disorderly, Jacobs suggested that there were virtues to be found. She railed against the conventional practices of urban planners (whether based on the ideas of Ebenezer Howard, Lewis Mumford or Le Corbusier), because they could only see order in empty, clean and quiet streets. Further, she suggested that the way in which the seemingly unruly city brought strangers together contained a hidden orderliness:

> Under the seeming disorder of the city, wherever the city is working successfully, is a marvellous order for maintaining the safety of the street and the freedom of the city. It is a complex order. Its essence is the intricacy of sidewalk [street] use, bringing with it a constant succession of eyes. This order is all composed of movement and change …
>
> (Jacobs, 1961, p.60)

There is much to think about here, as we will see in Chapters 3 and 7 of this book. What is important, at this point, is the tension between order and disorder that Jacobs uncovers. City life has a complex order precisely because it continually moves people through the city, its richness and diversity a result of perpetually bringing strangers together into close proximity, into social interaction: 'sidewalk contacts are the small change from which a city's wealth of public life may grow' (Jacobs, 1961, p.83). Jacobs implies that people's reaction to strangers ought not be one of indifference (as described by **Allen, 1999a**, section 4) or fear (as seen in **McDowell, 1999**, sections 3 and 4). While she understands these reactions, Jacobs argues that strangers are constitutive of 'an intricate, almost unconscious, network of voluntary controls and standards among the people themselves, and enforced by the people themselves' (1961, p.41).

Jacobs gives two examples (1961, pp.64 and 48–9 respectively). One is when a stranger saves the arm of a boy who has fallen through a plate glass window during a scuffle. The stranger was never, she claims, seen or heard of again (see Reading 3A in this book). The other is where a man is struggling to drag off a small girl. Not realizing that the man was the child's father, many people surrounded them both to ensure that everything was all right. From these examples, Jacobs suggests that the safest streets are those that are characterized by the mixing of strangers: that is, by their heterogeneity. More than this, her argument suggests that the anonymity of strangers does not necessarily imply either sociability or unsociability/anti-sociability. What is important is that there is always someone watching what is going on, informally policing the situation.

Yet, isn't there something disquieting in this? Isn't there the potential for unruly, immoral, evil interventions by strangers? Jacobs suggests not:

> Safety on the streets by surveillance and mutual policing of one another sounds grim, but in real life it is not grim. The safety of the street works best, most casually, and with least frequent taint of hostility or suspicion precisely where people are using and most enjoying the city streets voluntarily and are least conscious, normally, that they are policing.
>
> (Jacobs, 1961, p.46)

From Jacobs' perspective, successful neighbourhoods, like Harlem in the 1920s, do not become more orderly or communal because they purge themselves of strangers. Instead, they have to 'handle' and 'make the best of' strangers:

- partly by having many 'eyes on the streets' watching other peoples' activities (Jacobs, 1961, p.64),

and,

- partly by having a continuous use of spaces.

Reading Park and Jacobs, the stranger is a somewhat paradoxical figure: the harbinger, on the one hand, of fear and danger; and, on the other, of excitement and safety. However, we need to think more carefully about *who* is moving *where*.

We have seen that the city is characterized by the internal *differentiation* of spaces,

- not simply in terms of any abstract zones that we might discern,
- but in terms of its distinctive street and neighbourhood life.

In this situation, it must now be recognized that inner urban areas are differentiated by class and by 'race'. In this respect – and to take Jacobs' own examples – it would not be the same thing for an Italian-American person to walk through the streets of the North End (an Italian neighbourhood) of Boston as it would for an African-American; nor the same for an Italian-American to walk through Harlem as for an African-American. It is not simply that strangers move through urban spaces, but that urban spaces make some people into strangers, while others are not noticed at all.

While Jacobs and Park seem to disagree about whether strangers are good or bad, and about whether diversity or homogeneity within neighbourhoods is good or bad, they are both convinced that (too much) movement and change is damaging to city life because it undermines people's sense of local ties, local affiliations and their sense of community. However, we should be more cautious than this. Not only is the interplay between movement and settlement more intricate than this (see **Allen, Massey and Pryke, eds, 1999**), but the ideal of community and stability can be equally damaging to city life. In order to understand how it is that cities bring together such ambiguous and ambivalent experiences – of community and order, of strangers and disorder – we should follow the injunctions of **Massey (1999)** and **Amin and Graham (1999)** to think about the city both *spatially* and *relationally*.

In this respect, we should note that Park's success story of communal order – Harlem – can be retold. Looking for the other geographies and histories of Harlem brings questions about the marginalization and connectivity of urban areas to light. It is to these questions that we turn in the next section.

2.2 HARLEM: ISOLATED GHETTO OR GLOBAL CITY?

For the urban geographer, Neil Smith, there are many Harlems:

> Located in northern Manhattan in New York City, Harlem is a pre-eminent national and international symbol of black culture [Figure 1.3] … The public representations of Harlem are manifold, intense, resonant, and highly imbricated with definitions of black identities. There is the Harlem of the Harlem Renaissance … or the Harlem of the 1960s – Malcolm X, Black Power, the Black Panthers. But there is also Harlem the ghetto, the everyday Harlem for more than 100,000 people, predominantly poor, working-class and black; Harlem the community, the refuge from racism, starved of services; and there is Harlem the landscape of physical dilapidation, landlord criminality, social deprivation, street crime, police brutality, drugs. Harlem as haven; Harlem as hell. The latter is real enough for local residents even as its near-monopoly of media representations of Harlem magnifies racist stereotypes of variously threatening or exotic danger.
>
> (Smith, 1996, pp.140–2)

Let us tease out some of these Harlems. Perhaps the most familiar is 'Harlem the ghetto'. In this light, Harlem is typical of inner-city neighbourhoods in western cities around the world. Even if Harlem was the way Park and Jacobs described, it certainly took a turn for the worse. Instead of city life becoming freer and more equitable, injustices and inequalities became exaggerated. Within a decade of Jacobs' warning about the impending death of American cities, there were riots in black neighbourhoods across the United States. For some, these urban riots confirmed Jacobs' thesis that city-planners had wiped out the intimacy, vitality and safety of the streets: the street had become dangerous, fearful. For others it raised new questions about the way in which class relationships and racial injustices had produced iniquitously differentiated urban spaces.

It could be argued that the prime cause of Harlem's ghetto status is the way that it has been systematically deprived of inward investment. Thus, it is because of its economic *disconnection* from the rest of the city, and from the wealth of Manhattan's central business district in particular, that Harlem has become a neighbourhood characterized by dilapidated housing, social deprivation, poor services, and the like. These inequalities are produced by *class* relations in the city. However, since Harlem is also a neighbourhood populated almost

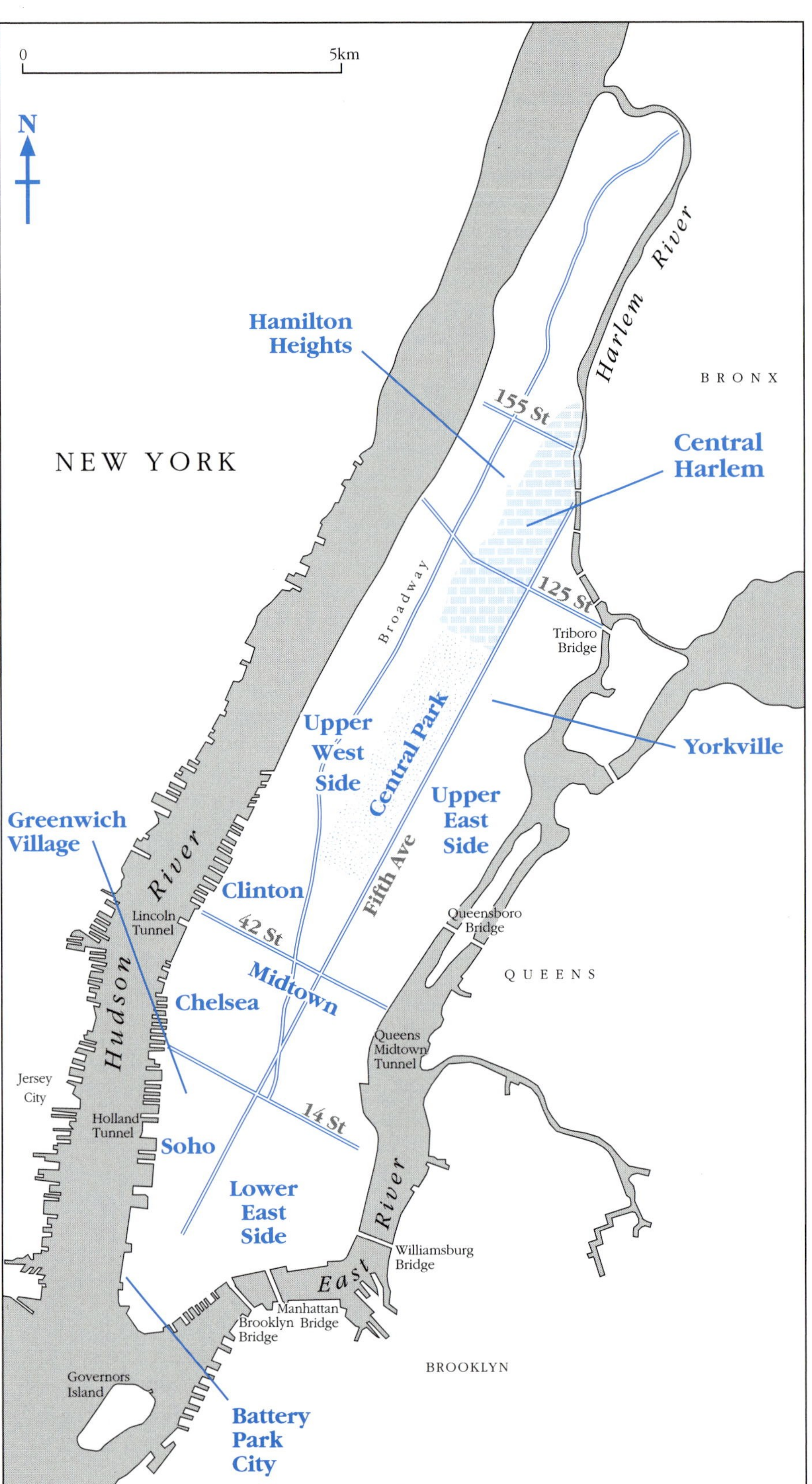

FIGURE 1.3
The location of Harlem within New York City

exclusively by African-Americans, there is also a case that its ghetto status is produced and maintained by '*race*' relations – where economic disconnection, police oppression, and government indifference are instituted by an endemic white racism. However, we might not need to choose between either class-based social marginalization or racialized social exclusion. Instead, it can be suggested that Harlem is a ghetto because of the way in which class and racial relations interact to produce Harlem as a doubly marginalized urban space (following Katznelson, 1981, and Wilson, 1987). And in suffering from both economic and social marginalization, Harlem's ghetto status is not simply confirmed, it is intensified, concentrated and exaggerated.

This inner-city Harlem, however, is only one Harlem – a disconnected Harlem; a Harlem on the edge. Harlem also articulated stories of connection, of being at the centre, and of alternative ways of living with difference. This is also an area in which people have fought against their social marginalization and exclusion – the key figures in the struggle against white supremacy are widely known: Marcus Garvey, Malcolm X, the Black Panthers. It is not just these political figures who have given Harlem such significance. Harlem has been bound up with black (urban) identities since the 1920s, when African-Americans – returning from service in the First World War and arriving from US southern states – began to move into what was then a predominantly (German) Jewish area (see Osofsky, 1971). At this time, there was great optimism about Harlem, as we have seen in section 2.1. In the 1920s, Harlem became renowned as a neighbourhood of intense artistic production.

The Harlem Renaissance was a remarkable period of cultural creativity amongst African-Americans, beginning with the end of the First World War and ending with the Stock Market crash of 1929 (though Renaissance artists continued to work after this period, they had lost the financial support of many patrons). For many commentators, this era was significant because it marks a transformation in black people's consciousness, both in Harlem and more widely (see especially Gates, 1997). The Harlem Renaissance was culturally significant, then, because

> ... black people were perceived as having finally liberated themselves from a past fraught with self-doubt and surrendered instead to an unprecedented optimism, a novel pride in all things black and a cultural confidence that stretched beyond the borders of Harlem to other black communities in the Western World.
>
> (Powell, 1997, p.16)

Artistic production in the Renaissance suggested that African-Americans need not be confined to notions of tradition and to rural lives, and that they too could be modern and lead city lives. This new black subjectivity was embodied in the idea of *The New Negro* (Locke, 1925/1992). Such a suggestion had more than an aesthetic resonance, however. For African-Americans were also claiming their *right* to participate in city life and the production of the New – to be creative

FIGURE 1.4
Loïs Mailou Jones, The Ascent of Ethiopia, *1932*

(Milwaukee Art Museum. Purchase, African-American Art Acquisition Fund, matching funds from Suzanne and Richard Pieper, with additonal support from Arthur and Dorothy Nelle Sanders)

(see Figure 1.4). More than this, through their cultural confidence, African-American artists were claiming equality.

Who were these artists? Key figures in the Renaissance were, and remain, well known. They include poets such as James Weldon Johnson (whom Park cited) and Langston Hughes; jazz musicians such as Duke Ellington and Bessie Smith; painters such as Aaron Douglas, Archibald J. Motley Jr and Winold Reiss; writers such as Jean Toomer and Zora Neale Huston; dancers such as Josephine Baker; and singers/actors such as Paul Robeson. For both Gates and Powell, these artists' endeavours were *both* a celebration and redefinition of black identity *and*

FIGURE 1.5
Archibald J. Motley Jr, Saturday Night Street Scene, *1936*

a placing of black people (back) in the city. However, Harlem was an ambiguous and, in many ways, uncomfortable place for African-Americans to be.

ACTIVITY 1.4 Look at Figure 1.5. What kinds of associations do you see amongst people? How might Harlem intensify these contacts? Finally, what do you make of the (Irish?) policeman in the foreground? (Remember to look at the background too.) ◆

Harlem certainly offered a uniquely creative atmosphere in the 1920s – intensifying relations

- by bringing black people from different backgrounds into close contact,
- by enabling those people to share their experiences, and
- by allowing them to express those experiences in new ways.

However, Harlem was also a place where African-Americans felt the shock of racism and racial exclusion. In wondering, too, about whether to become 'New' people, many felt the loss of older (and, for some, distinctly African) identities. Thus, Harlem presented people with a paradox: it was *both* a haven (for African-Americans from white supremacy) *and* a hell (of impoverishment, of social exclusion, of denigration). At the very time that the Harlem Renaissance was excitedly becoming internationally recognized, Harlem was also becoming (as we have seen) an inner-city ghetto – both black and poor.

What we can see is that Harlem was becoming more and more a *city within a city*, with its own rhythms of life and connections to the wider world – but this is a very different 'wider world' to that of Manhattan and New York. These connections were material in the sense that artists and art-works were moving through them, but Harlem was also a state of mind, the touchstone of a continually developing black consciousness. Paul Gilroy describes it this way:

> Harlem, the great black metropolis, could nurture the life of an emergent people, provide a crucible for their cosmopolitan consciousness and their historic obligations, both to their own special nature and to the wider world. Harlem would form the majestic kernel of a novel modern enterprise, bringing new life to the race after the slumbers of New World slavery and the catastrophic shock of adaptation to impoverished life in America's cities. However, as putative capital of the black world, it would have to be recognized as having significant ties to populations in other parts of the world. These groups were bonded together not by some conveniently automatic mechanisms of race consciousness but by a novel sense of blackness as a hemispheric, indeed planetary, phenomenon. Precisely that spirit was being fostered in a truly global city by the growing forces of anti-colonial nationalism and anti-imperial communism as well as the foolish, petty and irrational workings of the system of white supremacy itself.
>
> (Gilroy, 1997, p.105)

Gilroy is arguing that Harlem became important because it acted as a point of translation and sharing between people who had similar histories and experiences – of slavery, racial injustice and poverty. Harlem, then, is produced by its connections to 'the black world' *and* it also (re)produces those connections by being at the centre of them. Its influence over, and place in, these networks made Harlem a powerful symbol (following **Allen, 1999b**). To be sure, Harlem's connections were not only with the black world: Josephine Baker became famous in Paris as a dancer, while jazz clubs sprang up in cities throughout the western world. Harlem demonstrates that different parts of the city can have dramatically different connections to the wider world: almost completely isolated from New York as a whole (especially as white ethnic groups fled to the suburbs and ceased even to invest in the area), Harlem was nevertheless a 'global' city within a 'global' city. And it still is.

ACTIVITY 1.5 Read Sharon Zukin's description of Harlem's 125th Street in Reading 1B. What kinds of connections can you see between 1990s' Harlem and the wider world? What kinds of conflicts are there in Harlem? Why have they arisen? ◆

Conflicts over the African Market in Harlem demonstrate, at least, that Harlem has maintained its links with the black world. It is also important to note that, in these contestations over the market, the meanings of 'black identity' and the connections to a 'black world' are neither static nor unchanging, nor are they uncontested or incontrovertible. Nevertheless, Harlem has a continued significance in struggles to establish black political claims. It is partly for this reason that when Nelson Mandela, the imprisoned black leader of South Africa's African National Congress, was released from prison, he soon made a 'pilgrimage' to Harlem. This pilgrimage becomes less surprising once Harlem's *connections* to South Africa *are traced.* Thousands of miles away and some three decades after the Renaissance of the 1920s, Harlem was to have a significant impact on the cultural production of a neighbourhood in Johannesburg: Sophiatown (or Sophia).

If Harlem came to symbolize black cultural creativity for African-Americans, then Sophiatown has a similar significance for black South Africans. In the 1950s Sophiatown experienced a Renaissance that had similarities to Harlem's. For the cultural critic Rob Nixon, these similarities arose from connections that were both material and experiential. For black South Africans, the racist marginalization of African-Americans in US cities had clear resonances with their own. As white supremacism intensified in South Africa in the 1950s, the apartheid state sought to exclude black South Africans from the city or, temporarily, to confine them to ghettos (issues which will be picked up again in Chapter 4). In order to justify this social exclusion, white élites argued that black people did not belong in cities: black people, they declared, were 'naturally' unsuited, since they were 'tribal', 'rural' and incapable of adapting. On the back of these racist logics, black South Africans were forcibly and violently removed from the cities and relocated in homelands, and those who were left were confined to state-designated townships.

Sophiatown, however, threatened these white South African rhetorics that justified and necessitated the segregation of different 'racial' and ethnic groups. To begin with, Sophiatown was racially and ethnically heterogeneous. Like Harlem, too, Sophiatown contained a mix of different classes, from the very poor to doctors, lawyers and so on. This produced a neighbourhood with a distinct feel, like Harlem, where people from widely divergent backgrounds and interests – but with similar experiences of racial injustice – mixed and met:

> Sophia, or Kofifi as the locals knew it, lingers in South African argot as shorthand for 50s cultural brio – for the journalists and fiction writers, the shebeen queens presiding over speakeasies, and the jazz artists and gangsters who revered style, especially style that borrowed amply from Hollywood.
>
> (Nixon, 1994, p.12)

This 'cultural brio' was widely influenced by American popular culture, refashioning everything from the style and slang of 1930s' Hollywood gangster movies to the rhythms of American jazz (see Coplan, 1985). However, the writings of the Harlem Renaissance also had a significant impact. In particular, Alain Locke's anthology, *The New Negro*, suggested that black South Africans, too, had a right to occupy the increasingly white South African city. Using these arguments, writers such as Bloke Modisane, Nat Nakasa, Lewis Nkosi and Can Themba fought against apartheid's racist images, which associated black people with tradition, primitivism and rural life. If such links are intangible or tacit, more material links were also made – such as that between the poet Langston Hughes and the editor of Sophiatown's *Drum* magazine, Es'kia Mphahlele – through correspondence and publication. Even so, Harlem's influence was limited to scattered, chance discoveries by Sophiatown artists.

However, the increasing vitality and renown of Sophiatown's cultural brio – especially the gangster aspects – increasingly played into white racist fears. In part, these have to do with Sophiatown's heterogeneity. From one side, white South Africans were being drawn to the neighbourhood's illicit delights: its gangster culture, its speakeasies, its night-life. Meanwhile, black South Africans delighted in crossing over into white areas to make mischief. Sophiatown was becoming uncontainable; this city within a city was establishing too many connections with the white city beyond. As problematic for the apartheid state was that many Sophiatown dwellers had property rights. Very unusually, Sophiatown residents owned land – indeed, it was for this reason that professional black South Africans were drawn to live there. This was a circumstance that the apartheid state could not tolerate and from 1955 to 1960 Sophiatown residents were gradually evicted. The diverse and creative Sophiatown community was scattered and dispersed,

> ... compelling most of the black homeowners to become renters in a distant, soulless, segregated township called, with typical euphemism, Meadowlands. On the spot where they killed Sophia, a white suburb was raised up and christened Triomf – Afrikaans for Triumph.
>
> (Nixon, 1994, p.11)

The fates of Harlem and Sophiatown have been different. However, it is possible to draw some conclusions from the similarities in their experiences. These experiences are characteristically ambiguous, paradoxical even.

- Harlem and Sophiatown were both havens and hells;
- both were cities within the city;
- both established wider connections, yet both were marginal, disconnected;
- both came to symbolize black people's place in the city, yet both were sites where black subjectivities were being contested and transformed.

In both Harlem and Sophiatown, the cosmopolitanism of the city produced periods of cultural creativity and confidence. According to Gilroy (and the

Renaissance artists themselves), these cultural responses cannot be interpreted as the expression of some innate black (or other) identity. Instead, we can see these Renaissances as being produced out of the interactions between the differing histories and geographies of the people within the city – and the possibilities that these neighbourhoods offered for cultural responses to urban life. Both Harlem and Sophiatown intensified these social interactions by providing a context in which similar experiences could be shared, expressed and communicated more widely. In Harlem and Sophiatown, cultural brio resulted

- both from the way people were brought into, mixed within, settled in, and made connections beyond, the city (contradicting Park's assumption that communal life is best fostered by localism),
- and from both the *felt intensities* of people's experiences and the *intensification of social relations* through the production of these distinctive urban neighbourhoods (that is, of differentiated urban spaces).

However, it is clear too, that Harlem's and Sophiatown's wider connections and cultural brio were not to everyone's liking (to put it mildly). We have noted that white ethnic groups fled Harlem in the 1920s and also that the apartheid regime responded to Sophiatown's reputation by obliterating it. In both cases, we can see that the heterogeneity of these neighbourhoods was too much for many.

We have seen that Harlem and Sophiatown were both havens and hells. However, for many people, the inner city is *only* associated with the hell of urban life: the dirt, noise and smell, the fear of strangers, the crowdedness and congestion, the dangers and disease, its rat-race pace. So, they attempted to make places where the city's intensities could be diffused, perhaps also defused; a place we can call suburbia. And it is to the suburbs that we now turn in our examination of the heterogeneity of the city.

3 ...to sweet suburbia

3.1 SUBURBAN ESCAPISM: HAVEN OR HELL?

At the outset, we should note that the circumstances under which suburbanization takes place vary markedly: for example, we have seen (white) people fleeing Harlem, and the apartheid state forcing black South Africans out of Sophiatown (we will develop these issues further in Chapter 2). Nevertheless, for many, suburbia offered an escape from the tensions of city life, while still allowing people to be connected enough to the city to gain from its advantages (see Fishman, 1987). The suburbs were meant to achieve this

- partly by housing a stable – preferably homogeneous – community,
- partly by separating home from work and so providing a suitable environment for domestic life and leisure, and
- partly by providing a moral landscape in which people took care of their property and of their own (good) conduct.

Thus, suburbs were based on the assumption that communal life and orderly conduct were best fostered in neighbourhoods that were settled and domesticated (an assumption that closely mirrors Park's views about successful urban areas). Taken together, the suburbs promised a calmer, safer, more prosperous way of life – in direct contrast to life in the city. Roger Silverstone puts it this way:

> ... the experience of modernity is grounded in the life of the city, the visible vitality of the street, jumble, jungle, a vitality destroyed and denied by the sweeping redevelopments of freeways, malls and suburbs. The tension between old and new, the creative and destructive impulses, the paradox of ordered disorder, of an accessible, securable safety amidst the tense but creative struggle for the soul of the city – the capacity to make oneself at home in the maelstrom – this is what marked the essence of urban space and modern times. Yet, for millions, and mostly by choice, the city was too much to bear. It was a place to leave. And for these millions, throughout the modern period, the experience of modernity was the experience not of the street, but of the road, not of the sidewalk but the lawn, and not the jarring and unpredictable visibility of public spaces and public transport, but the enclosed private worlds of fences, parlours and automobiles. Public. Private. Paradise. Prison. Palpable danger was replaced by hidden dread. Batavia, St John's Wood, Levittown, Crestwood Heights, Crawley New Town, Ramsey Street, Watts.
>
> (Silverstone, 1997, pp.4–5)

At the beginning of this passage, Silverstone is implicitly attacking those views of the city that were discussed in section 2.1 of this chapter. In particular, he takes to task the idea that everybody's experiences of the city are grounded in street life, in public space and in public transport. Instead, he sees suburbanized cities as being dominated by the use of the road, private spaces and the private car (see **Hamilton and Hoyle, 1999**). Though these are 'private' experiences, we should question whether the suburbs can entirely escape the intensities of 'public' city life. Let's take the 'ambiguous' stranger, for example. The suburbs are not stranger-free: anyone from the milkman (the focus for many anxious jokes) to door-to-door sellers to confidence tricksters can knock at the door. Silverstone argues that it is only the experience of the stranger that has changed, from palpable danger to hidden dread: the stranger is a more fearful figure for being rarer, less familiar (see **Allen, 1999a**, section 4).

Even in the suburbs, then, there is 'the paradox of ordered disorder': the stranger is still an ambiguous figure, but the social interactions have changed subtly. Here, strangers are 'managed' differently: by twitching curtains, by neighbourhood watch, rather than by eyes on the street (see Chapter 3). People are not the only strangers that get 'managed' in the suburbs. Nature, too, is a stranger (as **Hinchliffe, 1999** suggests). In this sense, suburbia is an odd city-nature formation; odd because it seems to be neither city nor nature. Nature is represented by the domestication of plants, children and the household pet: no longer feral. In this light, suburbia is a particular compromise between the city and the countryside (see also Chapter 6): a kind of utopia, where even nature is ordered.

However, for many commentators, this unbounded enthusiasm for order has its down side. The suburbs have been caricatured as a boring, uniform, isolated, domestic, oppressive hell – full of identical people doing identical things, whether they are putting out gnomes, cooking the Sunday roast, or catching the (delayed) 8.15 to work.

There is, it seems, a pervasive anti-*sub*urbanism. Thus, when Mumford complained about urban sprawl in 1930s' America, he identified the unplanned development of land-hungry, unimaginative suburban housing as the core of the problem (see **Pile, 1999,** section 1.2). In most accounts of the city, the suburbs are beneath the city (literally, *sub* urban): a place where the richness and diversity of city life does not exist; and, a place where nature is manicured, artificial and has a price-tag on it. However, we must question this interpretation of the suburbs. Do they really eliminate heterogeneity? Have they really managed to dispel/disperse the tensions of the inner city? Are they quite so orderly? The cultural critic Roger Silverstone would think not. He attempts an alternative description of the suburbs by describing a walk through Bromley, a London suburb.

ACTIVITY 1.6 Read Silverstone's description of Bromley in Reading 1C. What do you think Bromley might be like to live in? Bear in mind Etcherelli's stories about Paris and Zukin's observations on Harlem. ◆

Silverstone is suggesting that – against the stereotypical view of the suburb as comfortable, uniform and bland – the suburbs are as heterogeneous as the city proper. To begin with, he describes a busy street, full of women and children, and adolescent and elderly people. It isn't just busy, or prosperous; there is also evidence of economic marginalization and crime. It would appear that suburbs juxtapose success and failure, like the city. Indeed, Silverstone takes the time to note the diversity of images and objects that clutter the scene – clutter that would not be out of place in any inner-city neighbourhood from Marrakesh to Moscow. But, for me, once he leaves Bromley's high street, the place becomes less intense, less chaotic – in fact, less urban. Against his wishes, Silverstone begins to describe a street scene that is very different from those described by Etcherelli and Zukin. Instead of romantic walks, instead of racial tension, instead of police raids, instead of inter-communal conflict and violence, there is a 1959 pink Cadillac. Sure, there is diversity in the suburbs, but it is not being brought together – *intensified* – in the same way as it was in Port d'Italie or Harlem.

Of course, the pink Cadillac shows that suburbanites have intense feelings about things. This is not at issue. It is not felt intensity that Bromley lacks. Instead, the argument is that Bromley – and similar suburbs around the world – is an attempt to manage, and disperse, the intensification that arises out of the concentration of social relationships in the city. Bromley is heterogeneous, but the heterogeneity – the mixings and meetings – is of quite a different pitch to that of the city. Simply, Bromley is not as marginal as Harlem (despite the unemployment and racism). It is not as creative as Sophiatown (despite the emergence of some punk rock bands in the mid 1970s). Nor as stylish as Paris (despite the affluence). And obviously so. Because it is not meant to be like that. The point is this: suburbanization is an attempt

- to eliminate or marginalize the tensions of urban life, and
- to separate out different geographies and histories
- and also to *de-intensify* urban social interactions.

However, this does leave a problem for some. If the suburb is meant to dissipate tensions, to produce spaces and times of calm in the midst of the maelstrom of urban living (such as 'the garden' and 'the weekend'), to diminish the intensity and pace of city life, then this will have material consequences for the people who live there. Let us look, particularly, at gender relations.

3.2 IDEAL HOMES: PARADISE OR PRISON?

Undoubtedly, many suburbs were built for a domestic life – to make the most of leisure time, the weekend and the family. Indeed, new suburbs in western cities were deliberately designed to encourage the most desirable attitudes in people. Plans used straight, regular avenues, with houses that had space on every side and large fences, with the intention of ensuring that people focus their attention on their private, family lives (see Chambers, 1997).

Suburbanization, then, has an in-built set of utopian ideals about what *should* characterize domestic life and relations between men, women and their families. Silverstone puts it this way:

> Suburban culture is a gendered culture ... The suburban home has been built around an ideology and a reality of women's domestication, oppressed by the insistent demands of the household, denied access to the varied spaces and times, the iteration of public and private that marks the male suburban experience and which creates, for them, the crucial distinctions between work and leisure, weekday and weekend. In particular, postwar suburbanization was buttressed by a concerted effort by public policy and media images to resocialize women into the home, and into the bosom of the nuclear bourgeois family.
>
> (Silverstone, 1997, p.7)

The suburbs were a different kind of urban space, made of ideal homes built for women and children first – but is this paradise or a prison? The design critic, Helen Grace (1997), has taken a closer look at the internal architecture of display homes in a recent suburban development in western Sydney, Australia. She found that the rooms in the 'ideal home' were distributed and decorated in particular ways. The homes had a 'Master Bedroom' (quote) which, she suggested, had a bed made for reading in, rather than anything else. There was a girl's room, in soft pastel colours, of course. And a nursery ready for that newborn baby. But there was no boy's room. No teenage clutter. No loud stereo, or dirty clothing. No ragged furniture, or unclean surfaces. In fact, it was the model of suburban bliss. Ideal, but not real. Like Silverstone, she sees this as a 'feminization' of the suburban home. And, like Silverstone, she refuses to see this feminization of the home as necessarily a bad thing for women. Instead, she argues strongly that it is men who are de-privileged in the home. Thus, suburban homes were places where women could take control of their lives, rather than their being inescapable prisons. The question for us is how women negotiated, and (re)produced, suburban spaces.

ACTIVITY 1.7 Read Deborah Chambers' description of women's experiences of suburbanization in Western Australia in Reading 1D. What was the impact on women's lives? And what did they do about it? ◆

Suburbanization in Sydney initially left many women isolated. They commonly found themselves amongst strangers, disconnecting them from the support of friends and relatives. Often they were further away from amenities and facilities, such as shops, health centres and schools. To make matters worse, public transport in suburbs was planned around the (man's) journey to work and not around (women's) domestic needs (as we will see in Chapter 5). Nevertheless, Chambers demonstrates that women found many ways in which to create support networks for themselves. In this way, women were active in the constitution of the suburbs and suburban communal life (see Chambers, 1997).

FIGURE 1.6
Suburban Tupperware Party, c. 1956

Alison Clarke has built up a detailed history of one such network, established through the now famous (or notorious) Tupperware Party.

ACTIVITY 1.8 Take a look at Figure 1.6. What do you notice? ◆

There are several things of note in this image. I will pick out just a few. First, we can see the range of Tupperware products in the foreground. Behind that, we can see the woman hosting the Tupperware Party waving at the group of women arriving. And, in the background, we can see that this is a suburb, because there is a modern bungalow, complete with tidy lawn and a (pink?) family car (a Cadillac?) in the driveway. Let us develop these observations as a way of describing the social relations in the suburbs.

For some people, Tupperware is a bit of a joke. It is emblematic both of a trivial plastic culture and of the artificiality of suburban life. But we should not be so dismissive. Let's begin with Tupperware itself. Tupperware was invented by an American, Earl Silas Tupper, in the mid 1930s. Thanks to his detailed observations of, and participation in, domestic life – involving himself in everything from jazz (from Harlem?) to cooking – he understood the need for a range of non-toxic, odourless and lightweight polyethylene household products. Significantly, these products were meant to be clean, functional, simple and, consequently, aesthetically pleasing. In this sense, Tupperware represented the condensation of American ideals about domestic life: good, clean, fun. The first items were produced in 1939 by the Tupper Corporation. However, sales of Tupperware took off – and peaked – in the 1950s. In this period, Tupperware successfully became one of *the* icons of suburbia. How? The answer lies in the way in which *women* built up networks amongst themselves.

Initially, Tupperware was sold in hardware stores. However, sales were poor. Impressed by the direct sales successes of other companies, Tupper decided to employ Brownie Wise's Patio Parties company to sell Tupperware. Famously, Tupperware was sold exclusively through Tupperware Parties, involving the creation, maintenance and extension of networks by women. The success of these networks was marked. By 1951 Brownie Wise held the rights to sell Tupperware and she changed the name of her company to Tupperware Home Parties Incorporated: 'By 1954 over twenty thousand [American] women belonged to the Tupperware party network as dealers, distributors and managers' (Clarke, 1997, p.138). And this figure does not include those who attended the parties.

Tupperware Party networks were built up by 'dealers'. They would contact friends, relatives, local women's groups and social organizations in the area (like those described in Reading 1D) and their members would be invited to host and attend Tupperware Parties. Each party would be used to recruit more women, to extend the network still further. During the parties, women would examine products, would be treated to demonstrations – and, for us most importantly, they would socialize. Interestingly, women would be invited to play party games. Clarke notes that some of these games were quite unruly: for example, women were invited to write adverts for a newspaper that sold their husbands (Clarke, 1997, p.142): 'For sale, one ...'. (Don't try this at home!) What becomes important for us in thinking about this situation is

- that women were creating social and economic *networks* at one and the same time and
- that, through these networks, women were *both* leading new lives – as 'dealers', 'distributors' and 'managers' – that were outside of the family *and* making an independent income.

Nevertheless, the Tupperware Party network relied on women's centrality in domestic life. Thus, Bonnie Wise dissuaded women from identifying with 'corporate' images and ideals, instead asking them to think of the network as her 'family'. More critically, women in the Tupperware organization were being exploited, since they were given no social or insurance benefits (Clarke, 1997, p.142). However, this should not stop us from recognizing the paradox in the situation. For sure, women were still centred on their domestic responsibilities. But they were also behaving in ways that are more often associated with men's business activities – and gaining self-respect and new social skills as a result.

By creating new relationships between women, these networks changed the form and content of suburban life: for sociability was the key to developing and extending the sales network. In this sense, we should think again about Figure 1.6. We see here a group of white women, dressed in stylish clothes, being greeted by a young white woman. However, Clarke notes that key figures in the Tupperware sales networks were more socially marginal women than this image of white middle-class respectability suggests. From the outset, single, lone-parent and divorced women were able to use Tupperware Parties to gain benefits that other employment opportunities did not allow, as were black American women. Wise recognized the importance of black and Hispanic American women's networks (Clarke, 1997, p.144).

The image of suburban life as dull, boring, domestic and simply emasculated makes it seem as if there is nothing 'else' going on. However, by tracing the spatial relationships amongst women, it has been possible to bring to light other social orders in the suburbs. Women's lives in 1950s' suburbia were clearly disordered by the imposition of the new suburban order. Consequently, they sought new forms of association, new forms of communal ties – such as Tupperware or bartering. Yet, these sat uneasily with their supposed 'domestication'– as women learnt new skills, new forms of self-respect. In fact, women were actively making new (social, economic) connections, new kinds of lives for themselves.

It is too simple, then, to argue that the suburbs eliminated social tensions, say, between men and women, since an orderly domestic life proved quite disorderly too. It is also too simple to suggest that suburbs are essentially characterized by homogeneity, as women (amongst others) struggled to negotiate the uneven consequences of suburbanization. Paradoxically, moreover, suburbs intensified certain social tensions (as women juggled 'home' and 'work'), and continued to bring strangers together, though differently (as women networked). Nevertheless, suburbs can be seen as an attempt to gain respite from many of the city's hells.

By tracing spatial connections *within* the city, we have been able to disrupt the idea that the suburbs were characterized either by women's confinement or by ideal homes (even while people may have experienced them this way). The question is whether following other connections, this time *between* cities, allows us to rethink other presumptions about the suburbs – their insularity and their parochialism. Let us think again about the historical connections that eventually enabled suburbs such as Bromley, London, to be envisaged.

3.3 WHOSE SUBURBS?

We can tell a story of the expansion of cities, as it were, from the inside out. Let's see. By the end of the nineteenth century, London was the largest city on Earth, partly as it gathered in the huddled masses thrown off the land, and partly as urban dwellers had families that swelled the numbers still further. As London expanded, the bourgeoisie sought to escape the perils of life in a city that was becoming increasingly impoverished, as the poor crowded into the slums of the East End. A rapidly expanding middle class sought desirable areas further and further to the west and south of the city – including Bromley. The suburbanization of Bromley followed the sale of Bickley Park in 1841. Soon, grand villas had been built. Once the railway had arrived in 1858, Bromley became even more popular. However, the railway company resisted cheapening fares so as to exclude workers from making the short journey to London. Nevertheless, Bromley could not withstand the pressures for further development for long and 'it quickly succumbed to more speculative, underplanned and less ostentatious development' (Silverstone, 1997, p.3).

Bromley was not unique; nevertheless in other places people were dreaming of utopian solutions to the horror of Victorian cities. One such thinker was Ebenezer Howard. His experiences – of Leeds, Chicago, farming in Nebraska and London – led him to conceive of a 'garden city' that would provide people with a perfect blend of town and nature (these ideas will be followed up in Chapters 5, 6 and 7). Howard's designs proved extremely popular. Between 1877 and 1897 the first garden suburb was planned, laid out and built at Bedford Park, near Turnham Green tube station in London (see Bayley, 1975). Encouraged by an exhibition about garden cities in Manchester in 1903, the idea quickly spread across Britain – with developments in Letchworth (1905), Hampstead (1906) and, outside Manchester, Wythenshawe garden suburb was laid out for 100,000 working-class people in 1919 (see Chapter 5). An intellectual or bureaucratic élite was seeing in these developments a way of producing liveable and peaceable sub-urban neighbourhoods – again, by overcoming the intensification of social relations in urban environments.

In this story of suburbanization, we can see two kinds of spatial relation: *on the one hand*, there are the connections that bring people into the city; *on the other hand*, there are connections being made out from the city, as people move out, but still maintain their links with, the city. However, this story is, as it were, told from the centre of the city outwards. Perhaps there are other stories to tell. What if we were to stand on the very fringes of the city and look in. Let us start our story of London's suburbanization somewhere else: in India – in Madras and Calcutta.

According to the historian John Archer (1997), the British had begun to build suburbs in the colonial cities of Madras and Calcutta from the seventeenth century onwards. The 'colonial suburbs', however, took a particular form.

Keen to escape the cramped conditions in the European districts, rich merchants began to build homes outside the colonial city. It was not just a desire for more space that prompted this move. They also wanted to ensure that their lives and the lives of Indians were as segregated as possible. For this reason, they built lavish homes in unoccupied areas of jungle (i.e. on uncultivated land), even building roads that deliberately avoided Indian villages. Although the merchants would still travel into the city to conduct their business, family life was protected from the surrounding environment – partly by using fences and trees to ensure privacy and partly by simulating English conditions within the house's grounds.

These development were not, to be sure, exactly suburbs. However, we can see certain features that underlie later thinking behind the suburban ideal:

- first, the separation between home and work
- second, a separation between the public sphere of commerce, business and the city and the private sphere of the family
- third, the colonizers were building up alternative networks and different ways of engaging with city space. Most Indians' experiences of the city would be very different, even while some rich Indians also began to build country houses, and
- fourth, these houses represented a particular relationship between 'nature' and 'city dwelling': as jungle was turned into lawn.

As Calcutta and Madras expanded, the country house was transformed into a suburban compound. Europeans, believing that tropical diseases were associated with poor air circulation, built homes surrounded by open spaces, and then enclosed those open spaces with high walls. By the early eighteenth century, merchants returning to England from India had begun to build similar dwellings in the vicinity of Richmond, in the Thames Valley. Suburbanization in London was taking on a logic that had begun in India:

> Located at a sufficient distance from London that the proprietors could feel wholly disconnected from the city – though necessarily they remained politically and economically quite connected – they were close enough to London that one could commute back and forth on a weekly, weekend, or daily basis.
>
> (Archer, 1997, p.40)

And now we can see another feature of suburbanization. It is not just the production of a particular kind of space that marks the suburbs, it is also the making of a different sense of time: with the suburbs came the weekend – and the rhythm of the five-day working-week. By the late nineteenth century another Indian form – that was to become an iconic feature of suburbs all over the English-speaking world (and beyond) – was imported into London: the bungalow. According to the art historian Anthony King (1984), the first bungalows were built in 1891. The bungalow's popularity, however, could not be guaranteed. Nevertheless, the bungalow's use of space, although strange to British architects, also made sense for the growing middle classes of the time:

> The image was of a dwelling characterized by social and spatial apartness; separate not only from other people and other dwellings, but also separate from the city itself, a house surrounded on all sides by space.
>
> (King, 1997, p.60)

This social and spatial separateness was highly desirable for a middle class seeking to separate itself from the city and to guarantee a private space in which family life could be protected and fostered (as we have seen). It is now possible to argue that these suburban pioneers were deploying an ideal developed one and a half centuries earlier and thousands of miles away.

> By the end of the nineteenth century, London was unquestionably the political, social and economic heart of Empire. Its pre-eminent position as banking, finance, and services centre in the international economy had centred a substantial professional and service class, some of whom (along with 'returnees' from the colonies), in adapting a scaled-down version of the country house, were already making their social and architectural presence felt in the Home Counties.
>
> (King, 1997, p.65)

And not only in the Home Counties: bungalows were built in the Leeds suburb of Bramley in 1905, and bungalows were included in the plans for Letchworth Garden City. Meanwhile, bungalows were making their first appearance in Australia in what was to become know as 'Federation style' architecture. And, as we have seen in Figure 1.6, bungalows are now a familiar feature in US suburbs. So, our story about the suburbs has another component. We could imagine other solutions to the problems set by the horrors of Victorian urbanization. However, it made perfect sense – given the particular configuration of class, gender and 'race' (following colonial logics) – for the middle classes to flee the city, yet stay close enough to the city that they could still remain connected to it.

- The suburbs, then, like the inner city, are heterogeneous: that is, they are produced out of the mixings of different histories and different geographies (Driver and Gilbert, 1998). In this instance, we have traced suburbanization back to India. But there will be other connections, and disconnections, too.
- Thus, heterogeneous histories and geographies underwrite the production of differentiated urban spaces.

This is to say more than cities have different spaces in them. It is to suggest that differentiated urban spaces are produced through the negotiation of heterogeneity, one way or another. For there is a stubborn *difference* between the suburbs and the inner city in the examples we have seen. And it is on this point that we can draw some conclusions.

4 Conclusion

We seem to have ended the last section much as we began, on the idea that differentiated urban spaces are produced out of the ways in which people negotiate the intensification of social relations in the city. This idea has proved to be quite intricate. To begin with, it involves a sense of the differentiation of urban space. We saw one attempt to conceptualize this differentiation of urban space in Burgess's concentric zone model (see Figure 1.2). We saw that he produced a model in which there was a clear separation – in social *and* geographical terms – between the inner and the outer rings of the city. The concentric zones, rather than joining the city together, seem to cut it up socially and geographically. Throughout this book, we will see that this understanding of the differentiation of urban space leaves much to be desired.

For now, we can take another look at Figure 1.2. Take a pencil and mark where you think a 'Harlem' or a 'Sophiatown' would be – and then trace connections within and beyond this neighbourhood. Now do the same for a suburban neighborhood. Enjoy defacing this diagram, because it shows that urban neighbourhoods (inner or suburban) are not so easily characterized by their isolation, stability and internal homogeneity. Let's take Harlem. It was not simply disconnected or connected, but connected and disconnected in particular ways to other places. It was a heterogeneous place, in the sense that it was bringing together people and mixing them up, but on the other hand it is also a homogeneous space: black and poor. Meanwhile, the suburbs might look connected, but the experiences of women showed that this is not necessarily so. Women, however, refused to be 'ghettoized' in the suburbs, forming their own networks within and beyond suburbia. Nor are suburbs simply homogeneous, as we have seen.

Apparently, we are left with a paradox: inner cities and suburbs seem to be exactly the same – disconnected and connected, homogeneous and heterogeneous. However, we have also demonstrated that the social relations producing these connections and disconnections were often unequal (see also Allen, 1995). Thus we can draw some initial conclusions about the production of different spaces within the city. In this chapter, we have seen that the differentiation of urban spaces

- emerges out of the uneven and unequal configuration of connections – and disconnections – within and beyond the city, and it
- emerges out of people's multiform responses to the production of differences between people in the city (i.e. out of heterogeneity).

Moreover, we have been able to demonstrate that the differentiation of urban spaces also

- emerges out of social relations within – and beyond – the city, primarily through the interplay of class, gender and 'race', though we should not forget other relations, such as age, sexuality, nationality and so on. This is also to acknowledge that some people are better able to negotiate, influence and/or control their circumstances than others.

In effect, this suggests that

- the production of markedly different (and relatively persistent) urban spaces arises out of *cross-cutting* social relations operating within and beyond the city, but it also implies that
- this tells us very little until we trace these social relations *spatially*.

In this way, we can demonstrate how the differentiation of urban spaces is produced. However, we have achieved more than this in this chapter. By following social relations spatially, it can be shown how people control and manipulate urban space – in the attempt sometimes to produce orderly spaces, other times to produce alternative orderings of space, occasionally to disorder space. For example, we have seen that the concentration and intensification of social relations in particular spaces led many people to seek to escape them, by imagining – then building – ideal spaces that were calmer, safer, more orderly. Or so they believed. If attempts at producing and maintaining urban order have been continual, then this is partly because different people in cities have differing views of what is unruly about cities (historically, geographically). Of course, accounts of the unruliness of cities have material consequences for how people (are allowed to) live in cities. It is to these that we turn in the next chapter.

References

Allen, J. (1995) 'Global worlds' in Allen, J. and Massey, D. (eds) *Geographical Worlds*, Oxford, Oxford University Press/The Open University.

Allen, J. (1999a) 'Worlds in the city' in Massey, D. *et al.* (eds).

Allen, J. (1999b) 'Cities of power and influence: settled formations' in Allen, J. *et al.* (eds).

Allen, J., Massey, D. and Pryke, M. (eds) (1999) *Unsettling Cities: Movement/ Settlement*, London, Routledge/The Open University (Book 2 in this series).

Amin, A. and Graham, S. (1999) 'Cities of connection and disconnection' in Allen, J. *et al.* (eds).

Archer, J. (1997) 'Colonial suburbs in south Asia, 1700–1850, and the spaces of modernity' in Silverstone, R. (ed.), pp.26–54.

Bayley, S. (1975) Unit 23: 'The Garden City' in *A305 History of Architecture and Design 1890–1939*, Milton Keynes, The Open University.

Burgess, E.W. (1925/1984) 'The growth of the city: an introduction to a research project' in Park, R.E. *et al.*, pp.47–62.

Chambers, D. (1997) 'A stake in the country: women's experiences of suburban development' in Silverstone, R. (ed.), pp.86–107.

Clarke, A.J. (1997) 'Tupperware: suburbia, sociality and mass consumption' in Silverstone, R. (ed.), pp.132–60.

Coplan, D.B. (1985) *In Township Tonight! South Africa's Black City Music and Theatre*, London, Longman.

Driver, F. and Gilbert, D. (1998) 'Heart of Empire? Landscape, space and performance in imperial London', *Environment and Planning D: Society and Space*, vol.16, pp.11–28.

Etcherelli, C. (undated) *Elise or the Real Life*, trans. by J.P. Wilson and B. Michaels, in Heron (ed.) (1993).

Fishman, R. (1987) *Bourgeois Utopias: The Rise and Fall of Suburbia*, New York, Basic Books.

Gates, H.L. (1997) 'Harlem on our minds', *Critical Inquiry*, issue 24, pp.1–12.

Gilroy, P. (1993) *The Black Atlantic: Modernity and Double Consciousness*, London, Verso.

Gilroy, P. (1997) 'Modern tones' in Skipworth, J. (ed.), pp.102–9.

Grace, H. (1997) 'Icon House: towards a suburban topophilia' in Grace, H., Hage, G., Johnson, L., Langsworth, L. and Symonds, M., *Home/World: Space, Community and Marginality in Sydney's West*, Annandale, New South Wales, Pluto Press Australia, pp.154–95.

Hamilton, K. and Hoyle, S. (1999) 'Moving cities: transport connections' in Allen, J. *et al.* (eds).

Heron, L. (ed.) (1993) *Streets of Desire: Women's Fictions of the Twentieth-century City*, London, Virago.

Hinchliffe, S. (1999) 'Cities and natures: intimate strangers' in Allen, J. *et al.* (eds).

Jacobs, J. (1961) *The Death and Life of Great American Cities*, New York, Random House.

Katznelson, I. (1981) *City Trenches: Urban Politics and the Patterning of Class in the United States*, Chicago, IL, University of Chicago Press.

King, A.D. (1984) *The Bungalow: The Production of a Global Culture*, London, Routledge and Kegan Paul.

King, A.D. (1997) 'Excavating the multicultural suburb: hidden histories of the bungalow' in Silverstone, R. (ed.), pp.55–85.

Locke, A. (ed.) (1925/1992) *The New Negro: An Interpretation*, New York, Atheneum Press.

Massey, D. (1999) 'On space and the city' in Massey, D. *et al.* (eds).

Massey, D., Allen, J. and Pile, S. (eds) (1999) *City Worlds*, London, Routledge/ The Open University (Book 1 in this series).

McDowell, L. (1999) 'City life and difference: negotiating diversity' in Allen, J. *et al.* (eds).

Mumford, L. (1937) 'What is a city?' in LeGates, R.T. and Stout, F. (eds) (1996) *The City Reader*, London, Routledge, pp.184–9.

Nixon, R. (1994) *Homelands, Harlem and Hollywood: South African Culture and the World Beyond*, New York, Routledge.

Osofsky, F. (1971) *Harlem: The Making of a Ghetto*, New York, Harper and Row.

Park, R.E. (1925/1984) 'The City: suggestions for investigation of human behaviour in the urban environment' in Park, R.E. *et al.*, pp.1–46.

Park, R.E. and Burgess, E.W. with McKenzie, R.D. and Wirth, L. (1925/1984) *The City: Suggestions for Investigation of Human Behavior in the Urban Environment*, Midway Reprint, Chicago, IL, University of Chicago Press.

Pile, S. (1999) 'What is a city?' in Massey, D. *et al.* (eds).

Powell, R (1997) 'Re/Birth of a Nation' in Skipworth, J. (ed.), pp.14–33.

Pryke, M. (1999) 'City rhythms: neo-liberalism and the developing world' in Allen, J. *et al.* (eds).

Silverstone, R. (1997) 'Introduction' in Silverstone, R. (ed.), pp.1–25.

Silverstone, R. (ed.) (1997) *Visions of Suburbia*, London, Routledge.

Skipworth, J. (ed.) (1997) *Rhapsodies in Black: Art and the Harlem Renaissance*, London, The Hayward Gallery, and Berkeley, CA, The Institute of International Visual Arts and the University of California Press.

Smith, N. (1996) *The New Urban Frontier: Gentrification and the Revanchist City*, London, Routledge.

Wilson, W.J. (1987) *The Truly Disadvantaged: The Inner City, The Underclass and Public Policy*, London, University of Chicago Press.

Zukin, S. (1995) *The Cultures of Cities*, Oxford, Basil Blackwell.

READING 1A
Claire Etcherelli: from *Elise or the Real Life*

Extract two

'Listen. Take the Métro to Stalingrad. All right? Get off, take a seat, and wait for me on the platform. While you wait, read a paper folded in front of your face. If any people from here get off, they won't recognize you.'

I followed his directions; he joined me on the platform at Stalingrad where I had buried my face in the front pages of my paper. This made him laugh. He tapped on the paper and said we'd go on to Ternes.

'It's near the Étoile. I think it's a good place.'

Arezki had dressed carefully. He was wearing a white shirt, a tie hidden by his scarf, and his brown suit, shiny with wear, was spotless.

At last, I was seeing Paris by night, the Paris of postcards and calendars.

'You like it?'

Arezki was having fun. He suggested that we walk up to l'Étoile and then come back on the opposite sidewalk. It would be easy to lose ourselves and become a part of the scene. To feel one had a place in this beautiful city, to be integrated …

We spent some time discussing the Magyar's accident. We were both cold. Arezki glanced toward the cafés as we walked. He must be worrying about their being expensive, I thought. Three days until payday. He must be almost broke, too.

As we turned back toward Les Ternes, he said: 'You're cold,' and we went into a café whose sidewalk terrace was heated. But he preferred the interior, picked out two places and ordered two teas. The process was always the same. Our neighbours studied us in silence for several seconds and it was easy to guess their thoughts. I tried to say to myself: 'So what? It's Paris, the city of outlaws, fugitives from all over the world. This is 1957. Am I going to come apart because of a few stares? We are a scandal in this lovely neighbourhood. Are these people responsible?'

'… But where are the police? Look at that guy sitting right next to you in a nice place where you've made a date with a nice girl you're going to take home in the car you've parked nearby … and there's an Arab with a French girl! She's French and working class for sure, you can tell right off. We're fighting a war with those guys … Where are the police? No, we don't want to make them suffer; we're human. There are camps, places they can be assigned to. Clean up Paris. Maybe this one has a gun in his pocket. They all have.'

Every one of their stares said that.

Extract three

When Arezki joined me at Stalingrad, he stated that we wouldn't go to Les Ternes anymore. It wasn't a good neighbourhood.

'We're going to … the Trocadéro.'

We went to the Trocadéro. We even returned two days later. We walked in the gardens where the freezing fog raised protective walls around us.

We went to the Opéra and circled the building several times.

We crossed the bridges.

We lost ourselves on the streets around Saint-Paul.

We walked up the boulevards toward Saint-Augustin.

Starting at Vaugirard, we ended up at the Porte d'Auteuil.

The Rue de Rivoli we did in both directions.

And the Boulevard Voltaire, and the Boulevard du Temple, and the little streets behind the Palais-Royale. And La Trinité and the Rue Lafayette.

We never returned to the same neighbourhood. The smallest incident, a gathering of people, the shadow of a police car, someone who seemed to be following us, and our walk was abruptly ended. We had to part, to go home separately. These interrupted evenings, our conversations cut short, and the anxiety – never knowing, leaving him behind, waiting until the next day to find out if anything serious had happened – these bound me to him in that well-known way where the more fleeting the thing, the dearer it is.

He saw police everywhere. I thought he exaggerated. I protested a bit when he'd say:

'Look. See that guy in front of the window. He's a cop. You don't believe me? I tell you it's so.'

'So what? What does it matter?'

We continued our walk.

There were lots of police raids. Arezki dreaded them.

'But you don't break any laws.'

'You think that satisfies them?'

And the next night, we changed neighbourhoods. I asked no questions. Time passed, we met almost every day. I tried to address him as 'tu', for he became angry one night at my continual 'vous'. I loved to hear him talk. His tongue made a soft little roll when he pronounced his 'r's'. We passed from serious to gay, we made fun of our friends on the assembly line. I told him about Lucien's youth, I often talked of Grandmother. She had become familiar to him; he knew her faults, her expressions, her manias. Mustapha, Grandmother, Lucien – these people who made up our company helped us to discover each other. Out of shyness, we made use of them to talk about ourselves.

One evening, we were walking in the gardens of the Trocadéro. We found a hole in the shadows and Arezki kissed me violently. With my new ideas, I thought, this is it, now he's going to take me to his room. But nothing happened. Our understanding was miraculous: anyone else would have been more impatient, more audacious. If he wasn't, it was because to the difficult circumstances that already hampered us had been added the calculated pleasure of our moving forward together.

We observed each other for a long time with growing tenderness. In front of others, we feigned indifference, that game where the smallest gesture, a blink of the eye, an inflection of the voice, takes on intense meaning.

Each time we separated, Arezki swore me to secrecy, which annoyed me a little. Actually, it suited me perfectly.

Rain, sleet, we walked. Paris was an enormous ambush through which we moved with ludicrous precautions. Our love heightened the background of our wanderings. Nothing was ugly. The rain polished the pavement and the lone light of an alley made a prism of the shimmering stones. The squares had a provincial charm and the broken-down sheds took on the look of old abandoned windmills. Our happiness transformed Paris.

Extract four

'Here's Paris.' The cloth tears. The countryside and the soft wind in the trees, anticipating the summer to come, prolonged still further the funeral ceremony and its capacity to appease. But here begins the city's overflow. A clock marks the time. The streets are rectilinear and without mystery. The horizon now is a fragment of sky between the many buildings. It is decidedly blue. It is going to be hot, and the women are wearing dresses without collars or sleeves. Some Arabs are digging a sidewalk. Once we've passed the viaduct at Auteuil, the traffic grows heavier. This is Paris. Delivery trucks, trailer trucks, buses, it's the start of a day. From the Porte de Versailles, we move slowly and I examine the people on the sidewalk to my right, as if they could answer my questions. It's because here, in the noise of the city, in its colours and mixtures, I've found Arezki again.

Now the buildings of the 'Cité Universitaire'. The red brick of their walls reminds me of the English colleges, the way they looked in my brother's school books. Between two pavilions, a garden gives to the whole a quality of fullness. Lecture halls, rooms from which it must be possible to see the distant roses amidst the green … because of that, because of the old stones and a few students walking toward the boulevard, I tell myself that Arezki risks nothing. Further along, coming out of the Moroccan pavilion, a boy yawns, his collar open. He stretches his free arm. And even if Arezki didn't come back, I'd rouse Paris. There are lawyers, newspapers. A man's life, that matters here. A few would rise up to cry out, protest, make demands. The 28th of May was not a dream.

At the Porte de Gentilly, the road goes gently downhill. The concrete of the stadium steps is blinding in the sun. On a sign, I read 'Poterne des Peupliers'. It reminds me of gallows. Articles 76 and 78: 'Attack on the internal and external security …' They won't let go all that fast.

We pass a monument made of white stones: 'TO FRENCH MOTHERS'. The homage, the veneration, they come later, when it's too late. The slope flattens out toward the Place d'Italie. I know it too well; I barely look to the left toward this old whore of a factory where I read the inscription 'Automobiles; Wood-working machines'. I feel as if the unnerving noise of the assembly line were reaching out for me. I smell the warm metal.

When we begin the descent towards Charenton, the vibrating motions of the car – the boulevard is being repaired – throw me from hope back to anguish. And memories are mixed in as we pass the square of La Limagne. Arezki used to say 'de la Limace'. He also said: 'Le Mont de Pitié', and I loved this last word.

On the Pont National, at the sight of the water, I think about the bodies that float under it. Bodies that are thrown in on nights of big riots, in the paroxysm of hatred; the bodies of the weak who have talked too much and whom death punishes. Out of place in this area, L'Auberge du Régal watches those pass whom no red light stops.

On the Boulevard Poniatowski, buildings rise to circle Paris with their pre-war ugliness. Unfriendly houses with rough façades, dull stones, shapeless doorways, large interior courts no sun could ever reach; there lives the workers' aristocracy aspiring to the bourgeoisie. Crushed and constricted by indifference, by new ideas, what price the life of an Arab here? The love of order oozes from these buildings. He's been sent away, sent back into the war. I could cry, but who would hear me? If he's alive, where is he? If he's dead, where is his body? Who will tell me? You've taken his life, yes, but what have you done with his body? At the Porte de Vincennes, the boulevard comes to an end and a vast housing project takes over: new apartment buildings with terraces shaded by blue and orange awnings. They suggest hot afternoons where you drink from frosted glasses while listening to a record. Who will think of Arezki?

Henri slows down still more. We're behind a truck that belches its exhaust. Montreuil is at my right and the Rue d'Avron opposite. The stalls of les Halles challenge a painter's palette. The rows of fruit, the pyramids of vegetables tear the fabric of my hopes. In front of the mounds of garden produce, thousands of ants act as a rampart before the displays.

On the hill between Bagnolet and Les Lilas, the car struggles between two buses. A road gang at the Porte de Ménilmontant is taking time off for a drink. Tomorrow, one of them won't be back and fifty will appear to pick up his shovel. There are so many, there are too many, inexhaustible reserves, forever replenished.

After Les Lilas, on the curve going down toward the Pré-Saint-Gervais, you see before you Aubervilliers, pale in the heat haze. On the barren esplanade, a curious solitary church attracts me. But now Henri is driving very fast and it's only after the Porte de Pantin that we reach the slums of that other Paris that comes to Paris only for the 28th of May. Not dangerous, easy to control, easy to satisfy. We enter the tunnel under the Porte de la Villette. I have a presentiment that I will never see Arezki again.

Source: from Heron, ed., 1993, pp.225–31

READING 1B
Sharon Zukin: '125th Street'

... The eight crosstown blocks of 125th Street that make up the heavily traveled, commercial center of Harlem remain surrounded by ghettos. From Eighth Avenue east to Madison Avenue, 125th Street is an almost entirely black shopping street, with a large African presence, in the center of dilapidated, renovated, and abandoned apartment houses, with a mainly black, partly Latino, population.

From the late 1970s through the mid 1980s, four major public-private commercial projects were announced. Through most of the 1990s, however, none of them broke ground ... 125th Street lacks even a memory of large, corporate-owned department stores. Its narrative is that of a low-income shopping center, with fast-food franchises, empty land, and local and national chains of low-price stores.

But this is Harlem. Memories, like the street itself, are long and deep. The Apollo Theater, declared a New York City landmark in 1983, is the only black theater left in New York that can claim performances by all the great African-American artists of jazz, bebop, and rock and roll. Like some of the old stores, it had a "whites only" policy until 1934 ...

...

A couple of local department stores still bear the names of the German Jewish merchants who opened

View of 125th Street, 1943. Blumstein's Department Store, opened in 1896 and target of a community boycott during the Great Depression, is on the south (left) side of the street. The Apollo, Victoria, and Harlem theaters ... are on the north (right) side.

125th Street, 1995. Blumstein's has been divided into smaller stores. Does anyone remember the "whites-only" policy?

them around 1900, when Harlem was still a predominantly white community. The owner of one of these stores sold it in 1930 rather than admit black customers (Osofsky, 1971, p.121); the building now houses medical services. Other stores, such as Blumstein's, resisted the hiring of black employees, but gave in to community boycotts in the Great Depression. Wertheimer's is still in business on 125th Street, as it is in Jamaica, Queens, and downtown Brooklyn. But most of the white storeowners gradually disappeared after the riots of 1964 and 1968. Reminders of those times – when "Soul Brother" painted on a window identified black-owned stores – are now iconic business emblems. Stores bear names derived from African history and geography and their windows are emblazened with African statues, posters of Marcus Garvey, the Rev. Martin Luther King, Jr., and Malcolm X. Posters along the street urge people to "Buy Black. Boycott non-black-owned businesses. Pool resources, create jobs." The Studio Museum, the city's main African-American art museum, and the National Black Theater are both on 125th Street. If you are white, walking along 125th Street is a constant reminder that the Other is you.

125th Street as a crucible of African-American identity: sidewalk vendors display their wares while Nation of Islam militants spread the word.

Also if you are Korean. From September 1988 to December 1989, several years before a notorious black boycott of a Korean-owned grocery store on Church Avenue in Brooklyn, Harlem residents boycotted Koko's, a Korean-owned grocery store on 125th Street, where a Korean employee had mistakenly accused a black shopper of theft and assaulted him with a knife. The storeowner criticized bias against her by the community and the police: "I think if a white policeman had come on Sunday morning, my employee would not have been arrested" (Picard and Cates, 1990, p.11; emphasis added).

Despite periodic rumors of economic revival, many store-fronts, and second and third floors, are empty. Rents do not approach the level of 14th Street in Manhattan or Jamaica Avenue in Queens, streets with similar stores and many black shoppers. "It is difficult being a business person here," says the [black] owner of a full-service office equipment store, the only one on 125th Street. "…I think we have to work harder to stay in business than people in other areas do" (Kennedy, 1992) …

Problems of business owners on 125th Street indicate the special circle tightly drawn around race and class in American cities. Harlem was never developed in terms of "good" jobs for community residents, and its economy has historically lacked a financial and wholesale base to support the development of 125th Street as a profitable retail shopping center (Vietorisz and Harrison, 1970). Even

Riots in Harlem in July 1964 protested against the police and white-owned stores. Here, the shopwindow announces, "THIS IS A BLACK STORE."

On 125th Street today, a "Harlem USA Boutique" in a McDonald's franchise sells Afrocentric books and toys.

with large increases in black-owned businesses from the mid 1950s on, blacks for many years owned smaller stores than whites, with fewer employees outside their own families, smaller inventories, less insurance, and less access to credit (Caplovitz, 1973)
...

...

Yet plans to revitalize 125th Street aim at making it attractive to middle-class black shoppers, most of whom would have to be drawn into Harlem to shop. Here revitalization implies a huge contradiction, for 125th Street already is the metropolitan region's premiere shopping street for Afrocentric goods and ethnic foods from the African diaspora. Ironically, while the black middle class has greater resources than ever before to live in more expensive neighborhoods and shop in the suburbs, 125th Street is one of the places where a new African-American ethnicity is being formed.

These issues shaped a conflict over the removal of street vendors from a large, informal "African Market," that grew in recent years at the corner of Malcolm X Boulevard and Martin Luther King, Jr., Boulevard (125th Street). The vendors were expelled by police in 1994. Until then, they congregated on the very same vacant lot where the Harlem International Trade Center was announced in 1979. At that time, local black politicians, led by Rep. Charles Rangel, envisioned an import-export bank that would aid a trilateral trade between the United States, Africa, and the Caribbean, but the project was never funded by the federal government. Instead, trade with Africa soon took a different form. During the 1980s, as political repression and economic crisis in their countries grew more severe, an increasing number of West Africans emigrated and wound up selling goods on the streets in major cities of the world, such as London, Paris, Florence, and New York. Some were connected to established, though informal, distribution networks; others brought over goods made by family members and neighbors. The African trade, then, that was finally set up on 125th Street was in mud cloth and kente cloth; in vests, robes, and jackets; and in sculpture and jewelry. Vendors came mainly from Mali, Senegal, Ghana, and Liberia.

While the African Market seems to have been recognized as a coherent, though unofficial shopping space around 1986, the number of vendors and shoppers increased dramatically in the early 1990s. At least on the vendors' side, self-employment and political oppression contributed to the latest wave of Afrocentrism.

Before Mayor Rudolph Giuliani sent police officers to clear the peddlers from 125th Street, in 1994, we spoke with Yadda, a Liberian woman who was described as the Queen of the Market. She said the market reminds her of Africa. "The market is an open place, not controlled; you sell what you want to sell. Each vendor has a designated spot, and all the merchants look out for one another. People know better than to steal from us." Yadda claimed the vendors had permission from HUDC [the Harlem Urban Development Corporation] to sell at this site. But the peddlers on 125th Street – and it is estimated that were 500 to 1,000 of them – had no legal status. Although they belonged to at least one merchants' association, most were not licensed by the city government. Even when fully licensed by either the New York City Health Department, to sell food, or the Department of Consumer Affairs, to sell other goods, street vendors are not supposed to sell on such a busy street. Moreover, while some of them rented storage space in the basements of stores on 125th Street, the storeowners ceaselessly lobbied the mayor and local officials to force the peddlers to go elsewhere. (Despite the presence of Koreans and a few remaining whites, most of the small shopkeepers are black, as are nearly all local elected officials.) So, from time to time, over a 10 to 15 year period, the peddlers were harassed by the police ...

...

Just when the African Market began to draw more shoppers, in the early 1990s, the Dinkins administration was pressured into enforcing the local laws against peddlers, including those on 125th Street. Merchants on many shopping streets opposed the proliferation of street vendors, the result of a confluence of immigration and recession. As they did on 125th Street in the 1970s, storeowners complained about the crowded pavements and litter street vendors left behind. They claimed they lost business to peddlers selling cheap knock-offs and bootleg goods. They contrasted themselves, as honest merchants, with unlicensed vendors who did not pay taxes ...

Ghetto shopping center or Third World bazaar?: African Market on 125th Street, Harlem, 1994.

...

Although the street vendors were forcibly removed from 125th Street, the issues they pose loom over the public spaces of most cities. "Good" and "bad" representations of vendors reflect different notions of public order in the streets, as well as a different sense of social class, ethnicity, and public culture. On the one hand, as an HUDC official says, "Peddling is bad economic development policy. These people don't pay taxes. According to the police, 80 percent of the goods are stolen or illegally produced, with false brand names or bootleg videos and cassettes. ... The peddlers drive away more people than they attract. It leads to street congestion. People get disgusted and they don't come back." Moreover, he claims that the African peddlers on 125th Street were dumped there from all over the city: "They drove the Senegalese from Fifth Avenue and now they are here."

On the other hand, the street vendors of 125th Street were almost all black. (On two Saturdays, the busiest shopping day, Danny [Kessler] saw only two vendors out of 200 to 300 who were not black.) They sold all kinds of music created in black communities, from jazz and reggae to rap, blues, soul, and slow jams. At the height of a Saturday market, a black torso displayed leopard skin underwear at a sidewalk table. A poster at a jewelry stand announced it was black owned. Food peddlers sold cakes and pies, Southern fried chicken, beef sausages, and West Indian-style fried whiting. This was, in all senses, a bazaar. People stopped and talked with the peddlers. Conversations ranged from haggling over prices to street gossip, politics, and discussions of business prospects. When no customer was at hand, the vendors called out descriptions of their products and prices. Most vendors were men, and they called out compliments to women shoppers, telling them how great they would look in the products they sold. Several vendors dealt in posters and smaller framed photographs and prints, including pictures of black families, jazz collages by the artist Romare Bearden, photographs of black leaders, and village scenes in Africa. It could fairly be said – allowing for differences of gender, social class, and national origin – that the street vendors represented ethnicity on 125th Street.

But this is not the representation desired by local political leaders and HUDC. They want black culture *and* economic revitalization. They want 125th Street to look like shopping streets in other neighborhoods. Mart 125 is their model of retail shopping – accessible to the street but not outdoors, a midpoint between bazaar and shopping mall. Indeed, Mart 125 has two floors of small stalls, each 60 or 120 feet square, that are fully rented and draw a lively crowd of shoppers. Their goods are not easily distinguishable from those sold on the street. Restaurants on the second floor serve Caribbean food, soul food, and health food. On the ground floor are two bookstores, one owned by the Nation of Islam. An African tie-dye stall sells fabrics and clothing. African fabric stalls have sewing machines where the proprietor sews while waiting for customers. One of two hat stalls offers name brands next to "personalized" hats on which buyers can design their own messages. But while the goods may be similar, the mart is organized differently from the street. HUDC provides a security staff and a mart manager, and runs education courses and small business training seminars for the merchants who rent stalls.

HUDC also supports the expansion of Harlem as a tourist attraction. Since the 1980s, a number of tour companies have brought busloads of tourists to view 125th Street and eat at Sylvia's, a well-known restaurant with a varied clientele. German, Italian, and Japanese tourists visit the jazz clubs at night. A street photographer with a Polaroid plies his trade, taking pictures of visitors with jazz musicians, attaching himself to tourists with the same tenacity as his confrères in Paris or Rome. A specific form of black culture, as well as the distance from a hegemonic downtown, continues to distinguish 125th Street …

Clearly, 125th Street is a complex representation of race, class, and ethnicity. With the newest source of its Afrocentrism in post-1985 immigration from Africa, it negotiates and identity that is at once global and local. With older sources of Afrocentrism in religious and Black Nationalist literature, the street is typically African American, separatist, aggressively Other. Moreover, 125th Street is not only a site but a means of reproducing difference, exclusion, and "ghetto culture." It is half in the mainstream economy, with both corporate products and their bootleg imitations, and half bazaar. The African Market has to be seen as a form of shopping that both challenges and affirms a ghetto culture. It remains to be seen whether the shopping street will be changed by tourism and the corporate identities of new retail stores or remain submerged in poverty.

References

Caplovitz, D. (1973) *The Merchants of Harlem: A Study of Small Business in a Black Community*, Beverly Hills, CA, Sage.

Kennedy, S.G. (1992) 'New momentum builds on 125th Street', *New York Times*, November 8.

Osofsky, F. (1971) *Harlem: The Making of a Ghetto*, New York, Harper Row.

Picard, J. and Cates, S. (1990) 'Clash at uptown Korean market', *Village Voice*, June 5.

Vietorisz, T. and Harrison, B. (1970) *The Economic Development of Harlem*, New York, Praeger.

Source: Zukin, 1995, pp.230–6, 238–42, 245–7

Mart 125, an indoor market with official approval.

READING 1C
Roger Silverstone: 'Introduction to *Visions of Suburbia*'

...

Seeing Bromley from the forecourt of Bromley South Station. Through the dust of a high summer's day the High Street curves away right, up the gentle hill towards the new pedestrian precinct and The Glades, the pastelled, decoed, glass elevatored, fully let shopping mall at the north end of town. Lunchtime. School holidays. The street is busy. Young wives and their tiny children. Bare midriffed girls and shaded boys. The elderly on the occasional bench. Chain store jostles with department store, thrift with fast food. Everything seen on television can be bought.

Ahead, a ravaged shopping precinct, by-passed by the glossy mall, stands mostly empty. Next to it a cut-price supermarket. On a wall a sign, faded with age and inattention, reads: *Graffiti Is Vandalism, An Offence For Which You Can Be Fined Up To £300 And Be Given A 3 Months Imprisonment Or More Under the Criminal Damage Act 1971 And Also Quite Often Pay For The Repair. A New Video System Is Now Being Installed Within This Area.* Deep in the precinct's bowels but heralded by a transfixing collection of assorted junk and pot plants, 'Caligula', Richard Niazi's amazing folly of a restaurant (live opera on Sunday and Monday), a cornucopia of an eating house, whose frontage is decorated with pages torn from the magazine *Understanding Science* as well as the *Collected Works of Walter Scott* (including *The Bride of Lammermoor*) and made over to loom like a *fin de siècle* prostitute's boudoir, stuffed to bursting with silks and satins, lampshades, feathers, pearls, toys, birdcages, bottles, dead clocks and china birds, the flotsam and jetsam of suburban fantasy, the whole misassembled as if in a car boot sale in Marrakesh. Outside, next to a cactus growing out of some cast-off industrial machinery and a toy koala bear tied to a tree with a piece of turquoise polyester stands a television with its power wire lying flaccidly on the concrete; an L-shaped pipe sits proudly above the blank screen.

Look left and the road forks one way down Mason's Hill; the other leads to Westmoreland Road. St Mark's Church (nineteenth-century Gothic) can just be seen three-quarters hidden behind the social security and tax offices on the corner of Sandford Road. Sandford Road itself is resolutely residential. Large detached suburban villas, only one of which masquerades as a dental surgery, stand comfortably side by side. No one the same, each pitched and chimneyed, pebble-dashed and porched, paved and glazed in its own way, neither conceived nor inhabited identically, bearing the scars of alternate neglect and regentrification, their bay and bedroom windows open for the heat but masked by flapping nets, their interiors private and protected. Sandford Road leads to the Bromley Lawn Tennis and Squash Club (founded in 1880) (for the middle and upper middle classes) and the Bromley allotments (for the lower middle). Walk along its shaded length, past the 1970s town houses of Streamside Close, and listen to the sounds of suburbia. The groan of a low-flying light plane, the regular explosion of an airliner on its way from Heathrow to destinations east, the distant hum of traffic and the occasional rattle of the commuter train – the noises of passage – creating a kind of passacaglia with the sparrows, pigeons, blackbirds and the voices of children and neighbours. The houses get smaller and semi-attach. Gardens are, mostly, carefully groomed; slab-paved space created at the front for the second car. Not quite new Fords, Vauxhalls, Rovers and Nissans stand in driveways and against the kerb. Late Victorian gives way to an uneven blend of Edwardian, midwar and 1960s styling.

A footpath leads to the main Hayes Road, past Ravensbourne School (grant-maintained). On a lamp post at the corner and with due cognizance of earlier notices, find the following: *Jodie woz here 1994 won't come back again.*

Hayes Road leads to Hayes one way, and to Orpington, Chislehurst and Sevenoaks the other. Opposite the school are what look like open fields, horses grazing, and then a string of more suburban houses, detached this time, but grubby from their main-road-side location. Suddenly as if planted entirely for your benefit, in the driveway of number 25A (the Meadows), spot a 1959 pink Cadillac, registration FSK 279. In the front garden a well, a pump and a gas lamp, a stone otter and a stone hedgehog. On the wall of the mock tudor residence a security alarm. The house next door is called The Squirrels.

Suburbia lives.

Source: Silverstone, 1997, pp.1–3

READING 1D
Deborah Chambers: 'Women's experiences and negotiations of suburban planning'

Although the railway service from western Sydney to the city was regular, there were very few bus services interconnecting the western suburbs. The public transport system was not designed to give women access to local facilities. Like the telephone service, it was designed for men and their businesses. Yet it was women in particular who urgently required public transport, since most either did not drive or did not have access to the family car – which was typically used by the husband. Most women walked, with small children in tow, to the shops and other local amenities. Support networks were necessarily limited to small areas, intensifying the sense of community within them. As the towns grew in size, and cars became more cheaply available in the 1960s, women began to drive regularly to the shops which led to a fragmentation of the community of women. Walking everywhere guaranteed that people met one another regularly with the small community. Shopping later became a more impersonal activity, with less likelihood of meeting neighbours on the way to town.

As the suburbs expanded in the 1960s and 1970s, and housing settlements were built farther and farther away from the town centres, so the women of the newer suburbs experienced an increased sense of social isolation. Public transport and amenities were not expanded to meet the increasing demands of the new residents, particularly in the poorer housing commission areas. Women and children living on Housing Commission estates were so cut off from the central activities of the town that the inhabitants were inevitably treated as marginal, as a species foreign to the community that they were supposedly part of:

> I think the biggest upheaval was when they built the housing estate at Mount Druitt. It caused a lot of resentment around this area because, you know, the old established people didn't like the idea of 'no hopers', as they called them, coming into this area because they felt they were worse than the migrants you know. That's the way they felt. You know, because they were down and out. (Margaret)

Pat recalls the same event through her memories of direct involvement as a foster parent and a prominent member of community services, and was actively involved in local politics, which indicates the level of active intervention of women in the improvement of amenities in the urbanization of the region:

> Also, of course, they brought in huge floods of people, like 30,000 people. At one stage the population of Mount Druitt was 90,000, 60,000 of whom were under the age of 15. So it was like having a giant kindergarten suddenly plonked in the middle of the area, with no facilities, no postal, no telephone, and people were just suiciding like crazy out there … And also they put people together, like Serbs and Croatians, who traditionally hadn't spoken to each other for thousands of years and in fact were very violent to one another. They put all the old people in certain streets. All the people with nine children got the five-bedroomed houses in other streets. So you could never get away from your problem. If you were old and couldn't open your window because it was stuck, neither could the person next door, because they were old as well. If you had nine children and were going round the bend, you couldn't get away because the people next door had nine as well. So it was really terrible planning, without facilities … (Pat)

…

Both the women who were brought up in the region, and those who migrated to the region in the 1940s and 1950s were highly critical of the manner in which the suburbs were planned, but some reluctantly interpreted the changes as part of 'progress', a catchphrase of the period:

> Well it's a shame really because it was pretty, you know, down there – the river flats and looking towards the mountains. I thought it was very pretty you know. But it's gradually been sold. Particularly out at Cranebrook … there's a terrible lot of houses out there now. Oh, I did see part of it one day. Some friends drove me around from the Kingswood end. But oh, in one way I think it's a shame because Penrith was quiet and it was more of a rural atmosphere when we came here. Now it's commercialized – it's really spoilt. But that's progress I guess! There's nothing you can do about it. (Eugiene)
>
> …

Women's work in neo-suburbia: the home as a productive and consuming unit

During Australia's postwar period, the Factory Acts, unions and government promoted middle-class values to encourage and force women to move out of the jobs they had held during the war, re-enter the home and take up motherhood and housekeeping as full-time careers. Radio programmes and popular women's magazines such as *The Australian Women's Weekly* emphasized this middle-class ideology of domesticity. Not only were home and employment divided along gender lines, but the ideology of motherhood was also linked to ideas about the correct manner in which to raise children and care for husbands in the interests of the 'nation'. Not only did most husbands actively deter wives from working but married women were barred from public service careers such as nursing and teaching until the late 1960s:

> No. That really wasn't the done thing in those days, you know. I used to say I'd go and get a job but Doug said 'Oh my wife's not going to work.' It's amazing how quickly it's changed hasn't it? It's just the thing to work now isn't it? (Audrey)

The ideology of homemaker and increased consumption were intertwined in the sense that, on the one hand husbands and employers collaborated to deter married women from entering the workforce, and on the other hand husbands bought domestic appliances 'for their wives' from manufacturers who advertised their goods as household necessities.

The women interviewed were, not surprisingly, much more concerned about the hot water and toilet facilities than domestic appliances. Some homes in Blacktown area did not have sewerage connected until the 1980s, but others received the facility in the 1960s, depending on the zoning. Most of the women interviewed had experienced outside toilets until at least the 1960s.

...

Husbands' low 'family' wages forced wives to be highly resourceful. The home continued as a productive unit but also gradually extended into a consuming unit during this neo-suburban transitional phase between the rural and urban community. Women sewed clothes for themselves and their children from the material of old clothes or from cheap cloth bought locally. Many commented that they never bought jumpers but knitted everything, even socks for the children. They grew their own vegetables and kept hens and cows for the domestic supply of eggs and milk as butter was rationed up to 1954. They made their own cakes, jams, pickles and preserves and some women proudly claimed that they did not need to shop for vegetables and fruit at all. Those who could not afford, for example, the doctors' fees engaged in bartering by offering their home-grown produce as remuneration. Bartering between neighbours was a common activity:

> The Yugoslav people that were our neighbours, she'd send me down tomatoes 'cause that's ... and I'd send her up eggs, and down would come more tomatoes and up would go more eggs. (Audrey)

Women's support networks, which were so vital in country towns before the Second World War, continued to be conveyed through kinship, friendship and bartering structures thereafter. Although bartering extended through to the 1960s it became increasingly difficult for these older families to sustain hens and fruit and vegetable gardens as they aged. Their offspring and the newer residents relied increasingly on retail produce as so many women began to engage in paid employment by the late 1960s.

It is clear that, in some respects, shopping was actually perceived to be easier before the superstores arrived in the towns. The now commonplace traumas associated with the search for parking spaces and waiting in shop queues in the crowded town centres of Blacktown and Penrith were frequently portrayed as experiences inferior to that of the convenient weekly deliveries made by horse and cart by the local grocer, baker, milkman, butcher, wood fuel merchant and even dry cleaning service in the 1950s. Despite the regular deliveries, some women took leisurely walks to the shops every day with their children. and some with their mothers. The experience is generally recalled as an enjoyable one, except when the weather was scorching hot. Shopping was perceived as a social activity which took hours, not only because the women generally walked to the shops, pushing prams, but because they stopped to talk to virtually everybody they met in the streets: they simply 'knew everybody' in the township.

...

Suburban community networks

The women of the emerging suburban communities were a significant group that initiated and established community and neighbourhood networks of a social and voluntary nature at the crucial stages of suburban development. The community networks in Blacktown and Penrith were considerable in number and highly active in supporting charities that aimed to serve the needs of the elderly, the sick and the poor. Almost all the women interviewed were active in several associations or voluntary organizations both as young women and continuing as older members of the community.

Voluntary work was centred on the church groups and the Country Women's Association. The latter was initiated during the 1930s to cater for the needs of women and families on the land within farming communities and adapted over the years to cater for the suburban women. The Red Cross was established in Blacktown in the 1920s. The Women's Guild, the Inner Wheel, the Quota Club (service club), the Torch Bearers for Legacy, the Hospital Auxiliaries, the bush fire brigade ladies auxiliaries, parents and citizens associations all held regular meetings and fetes. These organizations served the purpose of creating neighbourhood networks and raising money for regular dances, card nights, social afternoons, and carnivals as well as supporting older widowed women and those who lost their homes in bush fires. Events were regularly advertised in the local press. Meetings were often held in women's homes as there were few meeting halls to call upon.

Source: Chambers, 1997, pp.92–7, 99

CHAPTER 2
Urban 'disorders'

by Gerry Mooney

1 Introduction

I know nothing more imposing than the view which the Thames offers during the ascent from the sea to London Bridge … all this is so vast, so impressive, that a man cannot collect himself, but is lost in the marvel of England's greatness before he sets foot upon English soil.

But the sacrifice which all this has cost has become apparent later. After roaming the streets of the capital a day or two, making headway with difficulty through the human turmoil and the endless lines of vehicles, after visiting the slums of the metropolis, one realizes for the first time that these Londoners have been forced to sacrifice the best qualities of their human nature, to bring to pass all the marvels of civilization which crowd their city … The very turmoil of the streets has something repulsive, something against which human nature rebels. The hundreds of thousands of all classes and ranks crowding past each other, are they not all human beings with the same qualities and powers, and with the same interest in being happy? … And still they crowd by one another as though they had nothing in common, nothing to do with one another, and their only agreement is a tacit one, that each keep to his own side of the pavement, so as not to delay the opposing streams of the crowd … The brutal indifference, the unfeeling isolation of each in his private interest becomes the more repellent and offensive, the more these individuals are crowded together, within a limited space. And, however much one may be aware that this isolation of the individual, this narrow self-seeking, is the fundamental principle of our society everywhere, it is nowhere so shamelessly barefaced, so self-conscious as just here in the crowding of the great city …

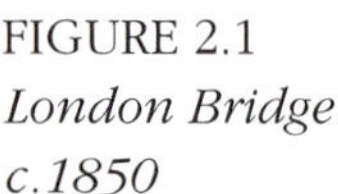

FIGURE 2.1 *London Bridge c.1850*

> What is true of London, is true of Manchester, Birmingham, Leeds, is true of all great towns. Everywhere barbarous indifference, hard egotism on one hand, and nameless misery on the other ... and [one] can only wonder that the whole crazy fabric still hangs together.
>
> (Engels, 1844/1987, pp.68–9)

Writing in the 1840s Friedrich Engels appears to capture something of the intensity and 'disorderliness' of urban life: the 'turmoil' and congestion; the concentration and mixing of large numbers of people from a variety of social backgrounds. Engels does not see this as a positive development but one which brings its own 'culture', a pronounced 'brutality' and a visible anonymity. Brought together yet pulled apart, where difference and indifference become sharply counterposed. In the preceding chapter Steve Pile explores the ways in which the differentiation of urban space is the product of cross-cutting social relations, both within and beyond the city. Cities are heterogeneous places characterized by social differentiation, places where differences are often *intensified.* This is reflected in the spatial stories he presents from Paris, Johannesburg and New York. Throughout, you will have noticed that he touched on issues of order and disorder in the city. In this chapter, we will examine how some of the ways of thinking about different spaces in the city brings into sharp focus the tension between order and disorder.

In exploring urban 'disorders' the first problem we encounter is with the notion of 'disorder' itself. Disorder only emerges out of a vision of order: one cannot have order without some sense of disorder. Both concepts are highly ambiguous reflecting particular interpretations of urban life and wider social arrangements. In other words, ideas of order and disorder are embedded in particular world views and as such they both reflect, and reproduce, relations of inequality and power. In discussing the theme of disorder in the city, therefore,

- the questions 'whose order?', 'whose disorder?' must occupy centre stage.

How are we to view urban 'disorder': from the ghetto or from the multinational boardroom or local authority chamber? Is the experience of global financial flows, where stock market collapse in one city has major consequences for markets in other cities, any more 'orderly' than the commonly depicted 'disorderly' behaviour of the urban 'underclass' or the shanty-town dweller? (**Pryke, 1999**). Such questions demonstrate that order/disorder in the contemporary city are real, in that they do not simply exist as free-floating representations but as social and spatial divisions and as serious concerns about congestion, pollution, violence, poverty, inequality and a long list of other 'urban ills'.

The tension between order and disorder is reflected in the main aims of this chapter, which are:

- To explore the ways in which the intensification of social relations within the city produces *both* orders *and* disorders.
- To open up our understanding of cities by problematizing ideas of order and disorder.

- To foreground different interpretations of urban (dis)order.

As you read through this chapter you will encounter ideas of 'urban order/ disorder' in a variety of contexts: from suburban 'sprawl' in Los Angeles, the redevelopment of Istanbul in the 1980s and 1990s, to the survival strategies of shanty-town dwellers. Together with other examples from several North American and European cities, the main concerns are

- to disrupt common-place assumptions about urban disorder, and
- to highlight some of the distinctive senses of (dis)order in different urban settings.

Moreover, the chapter explores the tension between order and disorder. The case studies explored here tend to focus on the 'seamier' side of urban life. However, it is hoped that in the stories of the urban 'underclass' and shanty-town dwellers you will develop an appreciation of the myriad of survival strategies which have been developed by the disadvantaged and also of the innovatory forms of governance and participation 'from below' which are emerging in many such places: that there is a richness and diversity in city life which often results in conflict.

Recognizing urban diversity and difference is one thing, but we must also recognize that the mere fact of urban differentiation does not preclude the existence of dominant representations and ideas of urban order and disorder. Before preceding to the next section stop for a minute and think about the ways in which urban order and disorder have been imaged and represented.

2 *Imaging order and disorder*

"HELP!"

ACTIVITY 2.1 Study carefully the photographs and images in Figure 2.2. Which do you think represents urban 'order' and 'disorder' most effectively? What factors do you consider to have influenced your choice here? ◆

FIGURE 2.2

Toxteth, Liverpool, 1981

Suburban estate, Newcastle upon Tyne

Shanty town, Rio de Janeiro

Park Hill estate, Sheffield

The first point to stress is that there is no clear-cut, unproblematic demarcation of order and disorder. Certain images and pictures *appear* to depict urban disorders more adequately: images of riots, shanty towns or run-down housing developments come immediately to mind but they leave me with a sense of unease. 'Riots' are often a legitimate response to oppression and the ordering of the city in particular ways which privilege certain groups and marginalizes others; 'disorderly' shanty towns and run-down housing estates are the product of the 'normal' operation of the housing market and the capitalist economy which seeks to secure the economic and political fortunes of dominant groups.

In approaching the issue of urban disorder then, we seldom arrive without some baggage, without some sense of what order 'looks like', what disorder 'is' and where 'it is to be found'. Thus images of the shanty town and urban violence come to the fore in constructions of a volatile and dangerous urban 'cocktail'.

For urban theorist Manuel Castells, this world is

> made up of multiple black holes of social exclusion throughout the planet. The Fourth World comprises large areas of the globe, such as much of Sub-Saharan Africa, and impoverished rural areas of Latin America and Asia. But it is also present in literally every country, and every city, in this new geography of social exclusion. It is formed of American inner-city ghettos, Spanish enclaves of mass youth unemployment, French banlieues warehousing North Africans, Japanese Yoseba quarters, and Asian mega-cities' shanty towns. And it is populated by millions of homeless, incarcerated, prostituted, criminalized, brutalized, stigmatized, sick, and illiterate persons. They are the majority in some areas, the minority in others, and a tiny minority in a few privileged contexts. But, everywhere, they are growing in number, and increasing in visibility.
>
> (Castells, 1998, pp.164–5)

While Castells' depiction of a 'Fourth World' may not be shared by all observers of contemporary city life, his concerns about growing 'urban problems' are widely and increasingly voiced. For many commentators the landscape of many of the world's cities is one which is characterized by an increasing divide between rich and poor; increasing levels of poverty and hunger; growing levels of homelessness, disease and mortality; by crime, social unrest and rioting; and by overcrowding, pollution and permanent traffic jams (MacGregor and Lipow, 1995). But while these issues may occupy centre stage in representations of urban disorder, we should also consider the ways in which attempts to order may have unintended outcomes.

Consider the following examples: suburban shopping centres may provide better shopping opportunities for some social groups, but they often result in empty retail premises in other parts of the city, encourage greater road use and congestion at other times, as well as limiting access for those without private transport. In the process they impose new patterns and rhythms (and stresses!) of shopping in a world of consumption far removed from the local corner shop. What value are huge shopping-centre car parks when the shopping centres are closed? Out of town sports stadiums may be also more accessible for those with cars, but for many football fans they represent an alien world distant from the days of football grounds located alongside densely packed working-class housing in central areas of cities, the close proximity of which often added a distinctiveness to the character of the club itself. Urban motorway networks may impose a certain kind of order in the city but they sit uneasily alongside housing estates and contribute to real concerns about the impact of pollution on the residents of such estates. No doubt you could generate many more examples, but within each of those presented here, the tension between order and disorder is apparent, highlighting not only that order itself can be viewed in different ways but that the bringing together and separating out of different activities and groups of people is itself potentially 'disordering'.

In focusing on the tension between the two, as products of the way social relations, groups and places are juxtaposed and (dis)connected, as the outcome of urban segregation, differentiation, social conflict and unequal power relations, then it is possible to conceive of the city as a place where attempts to create order produce disorder. Further, what is often portrayed as the disorderly practices characteristic of ghettos, shanty towns and inner cities, may be little more than organized relationships for those who live there. In other words, we need to be conscious of the varying (and often conflicting) sources of urban (dis)order: from above in the form of major urban 'redevelopment' programmes, the activities of multinational firms, and market-led plans for the use of urban space; or forms of (dis)order which emerge through the day-to-day activities and social relations of disadvantaged groups in the city.

ACTIVITY 2.2 Read the story which follows on the legacies of colonialism in Kenya. As you read through this think about the use of ideas of order and disorder. What was it that was considered 'disorderly' and how were attempts made to 'order' it and by whom?

During the period from 1934 to 1947, Mombasa posed numerous problems for colonial administrators. This Kenyan city experienced successive waves of strikes and industrial agitation by dock workers. For the British this industrial unrest appeared initially as the result of the casualized nature of much of the labour market. In the early 1930s the District Commissioner commented that 'large numbers of unemployed natives on whom no check can be kept must necessarily be a danger and such a life is undoubtedly demoralizing'. This was reinforced by the Town Clerk who feared that 'criminals and deserters' would invade the city in greater numbers (Cooper, 1987, p.35). But the British were soon to realize that the 'disorders', as the strikes were termed, did not stem from the so-called 'dangerous classes', but represented the growing militancy of an urban working class in many African cities at the time, and an emerging labour movement and trade union organization. That new management methods and new work regimes were to be introduced which effectively fragmented worker solidarity should not allow us to ignore the enormity of the struggles. As Frederick Cooper argues, 'the experience of disorder was a blow to a conception of African society that saw urban workers as so many interchangeable units of labor power and rural society as the repository of the true Africa, naturally tranquil and immutable' (Cooper, 1987, p.248). ◆

For me, this story emphasizes the conflicting senses of urban spatial organization, and that space is used and organized in disparate ways in different cultures. In Mombasa, worker resistance and unionization was viewed with alarm by colonial administrators. Their response to the growing militancy of workers was to re-order housing provision in a way which disordered groups of workers and which segregated groups along 'racial' lines, thereby imposing a particular spatial order (see also Chapter 4).

That the colonists were able to import and largely impose such forms of urban life should not blind us to the struggles which surrounded such episodes. Nor were existing forms of urban life and settlement completely eradicated in the process. The example sharply brings into focus the central theme of this chapter: whose order?, whose disorder?

3 *Whose disorder?*

In the introduction we noted that the idea of urban disorder has become central to many accounts of the contemporary city. There are several interrelated developments which have contributed to this: growing social and spatial polarization (frequently labelled as the 'divided city' or 'dual city') and the 'ghettoization' of areas of cities around the world; growing levels of crime and social unrest; and heightened concerns about urban 'sprawl', congestion, pollution and smog. While we can point to growing levels of inequality and material poverty in many cities (Castells, 1998; **Massey, 1999**), such processes never quite work out in the same way in different places. Take the issue of population, for example: fears about the future of Western cities, and about disorder within them, often invoke concerns regarding population loss – the so-called 'doughnut-effect'. In some North American and European cities, selective population loss is leaving cities with an increasingly impoverished, welfare-dependent, and 'socially excluded' population. But in many of the cities of Asia, Africa and South America, the concern is one of rapidly expanding populations. So while there may be a shared concern with the issue of 'population', this is played out in very distinct forms in different cities, but in ways which tend to focus on the disorderly effects of population movements.

While the concern with 'urban disorders' has increased in the latter stages of the twentieth century, the unruliness of cities has long been a feature of social commentary. Particular places and groups of people have at different times been constructed as disorderly. Some cities (or locales within them) have long been constructed and presented as chaotic and 'dangerous'. In his depiction of the mid-nineteenth century disorder of capitalist urbanization, Engels was not alone in choosing Manchester as a prime example of this, reflected in sharp class segregation and a potentially 'uncontrollable' city:

> The town itself is peculiarly built, so that a person may live in it for years, and go in and out daily without coming into contact with a working people's quarter or even with workers, that is, so long as he confines himself to his business or to pleasure walks. This arises chiefly from the fact, that by unconscious tacit agreement, as well as with outspoken conscious determination, the working people's quarters are sharply separated from the sections of the city reserved for the middle class … Manchester contains, at its heart, a rather extended commercial district, perhaps half a mile long and about as broad, and consisting almost wholly of offices and warehouses. Nearly the whole district is abandoned by dwellers, and is lonely and deserted at night … This district is cut through by certain main thoroughfares upon which the vast traffic concentrates, and in which the ground level is lined with brilliant shops … these shops bear some relation to the districts which lie behind them, and are more elegant in the commercial and residential quarters than when they hide grimy working

> men's dwellings; but they suffice to conceal from the eyes of the wealthy men and women of strong stomachs and weak nerves the misery and grime which form the complement to their wealth.
>
> (Engels, 1844/1987, pp.85–6)

Engels' depiction of Manchester again draws attention to the intensity of social relations in the city, the close proximity yet sharp juxtaposition and separation of different social classes. Social segregation takes on a particular spatial form with clearly demarcated boundaries between the city of the working classes and the city of the business classes: the *different worlds within cities* (**Allen, 1999**).

We can refer to the differing worlds of the city in another way, as Chapter 1 did, as a contrast between central areas of cities and their suburban hinterlands, though we will now see that this takes very distinct forms in different cities around the world.

Arthur Krim discusses Los Angeles' portrayal within American society as a *terra incognita*. Its largely negative imagery, he argues, has been constructed throughout the course of the twentieth century, and in part derives from its depiction as a city which lacked an obvious 'order': a city characterized by urban–suburban 'sprawl'. In the 1920s and 1930s, Los Angeles was different, in terms of its spatial organization, from other American cities, such as Chicago, in that it lacked a visible central core and the high-rise office-block skyline which was a feature of many other US cities. Indeed, during the 1930s Los Angeles was depicted less as a city, more as 'six suburbs in search of a city' (Krim, 1992, p.123). For Krim, Los Angles' distinctive urban order was reflected in widespread suburbanization. The main agent promoting this process, along with the housing market, was the rapid development of highways and the expansion of car ownership (see **Hamilton and Hoyle, 1999**). During the years of the post-Second World War economic boom this served only to link the city, in the eyes of outsiders in other US cities, with vehicle congestion in a 'spaghetti-like tangle of super-highways'. More significantly, it associated the city with smog and pollution:

> The sprawling city in many places is dammed and parcelled into islands, isolated from each other by torrents of cars that can neither be slowed nor penetrated. There is no focus, no place where the body and eye can come to rest, no point where people might converge and enjoy the amenities of life, if there is any. The same blight yearly creeps into outer areas, through once magnificent canyons and mountains. The worst consequence is that the air itself has now become polluted by exhaust gases to such extent that respiration and the act of vision become painful processes.
>
> (Conduit, 1961, quoted in Krim, 1992, p.125)

Krim argues that the suburban imagery of Los Angeles was an 'invented tradition'. But it is an imagery which has helped to construct it as the 'other' city within the USA and, indeed, in much of the contemporary Western world. For this Los Angeles can thank one of its own prestigious industries – movie making. Our understanding of urban order and disorder is heavily influenced by the media, significantly through the imagery of television and films. Disorderly, dystopian images are to the fore in numerous Hollywood film productions: *The Asphalt Jungle, The Bonfire of the Vanities, Escape from New York/L.A., Assault on Precinct 13, Fort Apache, the Bronx, Chinatown* and *Blade Runner*, to name but a few. While several American cities are represented here, Los Angeles plays a symbolic role in imagery of urban disorders, matched only by the imagery of the 'Third World shanty town'.

From the discussion above we can see that the dominant representation of urban 'disorder' can relate to the particular spatial form that a city takes. In the case of Los Angeles the expansion of car ownership and the development of expressways contributed significantly to urban–suburban 'sprawl'.

In exploring the notion of urban disorders we can detect a number of recurring themes:

- That dominant notions of urban disorders are often imbued with a political significance, with 'immorality'/social conflict/a lack of formal rules and/or government often depicted as a source of disorder.
- That which is considered to be disorderly – or the source of disorder – often lies in close proximity to processes, activities and places which are presented as ordered. But practices and a language of distancing is often utilized to construct such places and social groups as 'alien', 'foreign', as the 'other'. In other words, these groups and the worlds which they inhabit, are actively 'disconnected' through the process of constructing them as disorderly (see **Allen, 1999**; **Massey, 1999**).

The 'other' city

From Engels' account of Victorian Manchester one is left with the clear impression that certain parts of the city are disconnected or marginalized from the sites of business and commerce. In this section we focus on the ways in which some areas of cities are constructed as marginal or 'peripheral'. But how is the idea of the 'periphery' to be understood? Before we proceed think about the following claim by Geoffrey Mulgan.

ACTIVITY 2.3

> The centres of two cities are often for practical purposes closer to each other than to their own peripheries.
>
> (Mulgan, 1991, p.3)

What kinds of connection does Mulgan imply are most important between cities? What does this suggest about 'peripheries'? ◆

Implicit here is the suggestion that global interconnections between business centres work to marginalize and disconnect other social groups and activities. But perhaps like me you also have a strong sense of spatial segregation and the distancing of certain places within the city. In this respect Mulgan's comments have strong echoes of Engels' impression of Manchester. However, the idea of 'periphery' is not simply spatial but reflects the spatial constitution of lived social relations. Thus we need to distinguish between a *locational periphery* and a *social periphery*. While these tend to be mutually reinforcing, they are not homologous.

In this section we will explore the idea of periphery in two interrelated ways. First, to refer to particular spatial locations on the 'periphery' of large cities or urban areas, for example, Glasgow's 'peripheral' housing estates, the *banlieues* of the French urban periphery, or the satellite housing developments which have sprung up around Istanbul during the past decade. Second, there is another way in which the 'periphery' or the 'peripheral' can be understood: peripheral in the sense of 'marginal' groups of people, or marginal places. In this respect we can refer to the multiple, cross-cutting geographies of the city: the existence of places which appear geographically central to the city – such as the 'inner city' or urban council estates – but which are socially and politically marginalized. In other words, within cities there are:

- places which are physically close, but
- which are relationally distant and relatively disconnected from the 'rest' of the city and its main activities (see **Amin and Graham, 1999**).

Let us look at some examples: first, Istanbul.

4.1 'MODERNIZING' ISTANBUL: ORDERING THE DISORDERLY

The story which follows is largely based on studies conducted by Kevin Robins and Asu Aksoy in Istanbul in the 1990s (Robins and Aksoy, 1995, 1996; Aksoy and Robins, 1997) and by Ayse Oncü (Oncü, 1997). Istanbul is being re-imagined and re-ordered by élite groups in the city, who seek its future as a 'global city'. Arguably it is better placed than many other cities to achieve this, conveniently located between Europe and Asia and adjacent to the markets and tourist destinations opening up in Eastern Europe following the collapse of the Soviet Bloc. Turkey itself has sought to become more open to global markets over the past decade, but in looking to the West for a new global identity, it remains culturally attached to the East through the continuing legacies of Islam and its location within the Islamic world.

For those groups seeking to transform Istanbul into a global city, 'modernizing' the urban landscape is regarded as a strategic objective. This involves new transport and communications infrastructure – motorways, hotels, conference centres, tourist attractions, shopping malls, executive housing and other 'essential' facilities deemed necessary for a global city. As Robins and Aksoy highlight, the existing urban form, largely reflecting Islamic culture with its strictly gendered spaces (see **Allen, 1999**, section 2.3), was considered to be an obstacle to plans to transform the city.

From the late nineteenth century there have been a series of attempts to turn Istanbul into a 'European' city. In particular these have aspired to 'modernize' the city along European/Western lines by creating 'regularity' in the order of the city. In the most recent phase of urban re-ordering this has involved, in addition to the developments already highlighted above, land use separation with the city being functionally zoned. This has involved the planned segregation of working and trading areas and the creation of specific residential zones, a process which was alien to Istanbul and other Islamic cities.

But the 'disorder' which is being 'ordered' through this redevelopment has been central to the culture and fabric of Istanbul. Robins and Aksoy claim that this modernization (which is interpreted largely as 'Westernization') conflicts with the texture of urban culture in cities like Istanbul which tend to be characterized by anti-(European) planning attitudes. The urban landscape was characterized by a patchwork-like form with a dense mixing of housing and non-residential activities. There was little segregation of housing and industry and the haphazard scattering of small businesses and enterprises across the city had historically been regarded by the city's residents as entirely 'civil' and orderly.

In their studies Aksoy and Robins highlight the costs to urban life of this redevelopment: motorways lead to increasing levels of pollution; and the development of towering office-blocks and hotels has seriously disrupted the previously low-rise skyline, juxtaposing high-rise office-blocks with low-level traditional forms of housing. But there are also other, perhaps more significant costs of this redevelopment – the unintended outcomes of the plans to transform the city.

Istanbul has long been the locus of cultural encounter and interaction, reflecting its strategic position bridging East and West. The intensity and pace of this has increased significantly during the 1980s and 1990s, with the city growing rapidly through migration from more remote parts of Turkey, from the Black Sea area and Central and South-East Anatolia, with Kurds, rural villagers and other peoples either pulled to the city through the promise of better job and housing opportunities or to escape oppression elsewhere. This immigration has seen the population of Istanbul increase dramatically in a relatively short time from 5.8 million in 1985 to 7.3 million in 1990. Many of these incoming groups have sought refuge and housing in the large squatter settlements – the *gecekondus* – on the periphery of Istanbul. The housing in such areas, like the large settlement at Sultanbeyli (see Figure 2.3), was illegally built, and it has been estimated that by the early 1990s approximately 65 per cent of all housing in Istanbul was unauthorized.

For those groups and agencies driving the city's redevelopment, the existence of the *gecekondus* posed a serious threat to their efforts to transform the city. These settlements were represented as the 'other Istanbul', their populations depicted as 'backward', 'rural', 'ignorant', 'uncultured', and incapable of assimilation into the 'new Istanbul'. The settlements were constructed as locales of 'terrorists', 'mafia gangs' and other criminal groups. The overall picture generated was that the *gecekondus* were places of chaos and unruliness.

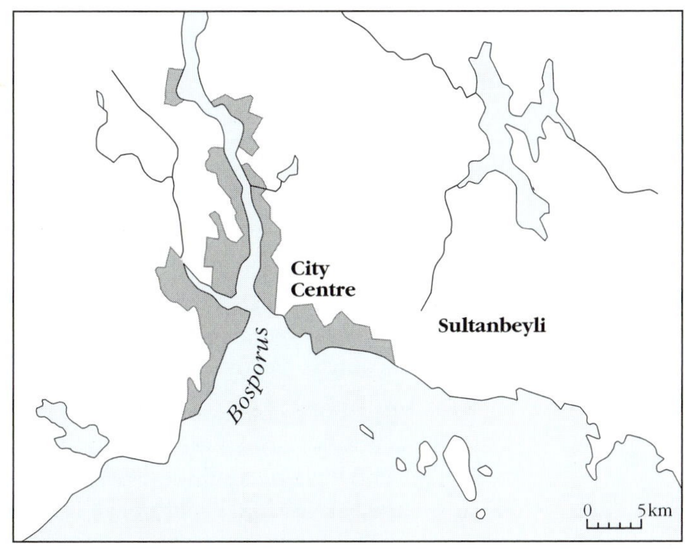

FIGURE 2.3 *Map of Istanbul*

While this construction of Istanbul's peripheral areas invokes an imagery and language used many times over in numerous urban settings around the world, there was an additional and highly potent factor at work here which marks out the experience of Istanbul as a distinctive one. The *gecekondus* were viewed with increasing hostility by Istanbul's burgeoning middle and business classes who, in aspiring to more 'Westernized' lifestyles, were adopting an increasingly secular outlook on life. But in the *gecekondus*, Islamic ideas and values were a significant feature of settlement life, particularly among recent groups of migrants. Thus the settlements were 'othered' in ways which relied on religious discourses. Further, inward migration is leading to increasing tension between Turks and Kurdish peoples who now live in growing numbers in close proximity in Istanbul itself.

Robins and Aksoy argue that urban life in Istanbul has become increasingly 'stressful' for both rich and poor alike and while this is experienced in very different ways, there is a widening sense of alienation amidst the 'turmoil of urban life' (Robins and Aksoy, 1995, p.229). Many of the affluent sections of the population have sought to escape to large-scale suburban housing projects, typified by Bahceshir ('garden city') (Oncü, 1997). These privately-run developments come complete with their own security guards. Elsewhere, other suburban middle-class housing estates, such as Bujuk Cekmece which is modelled on a 'modern American village', have 'villa-style' housing with recreational and entertainment facilities, schools and medical centres in 'natural' surroundings. Oncü (1997, p.63) notes that housing apartments in such estates are typically given names which denote 'back to nature', such as 'Lilac', 'Jasmine' and 'Honeysuckle', immediately distancing them from the 'pollution' of other areas of Istanbul. A glossy brochure advertising such housing states:

> Istanbul ... The legendary city of splendid architecture which has inspired songs, poems, books ... The gate across continents and the cradle of ancient civilizations ... But unfortunately a city which has lost much of its former beauty to become a metropolis of ten million today ...
>
> Istanbul's pollution has become oppressive ... It contaminates not only the air, water, soil of the city, but its traffic, its people and its culture ... Those who have to continue working in this polluted environment are moving away to escape its influence in their living spaces. They are searching for clean, happy, peaceful settings ...

> And in Istanbul's hinterland new towns are emerging to answer this need ...
>
> But GARDENCITY is very special among them ...
>
> - Luxury villas in gardens with $500m^2$
> - Only 20 minutes to Istanbul
> - Swimming pool, tennis courts, sports club, children's park, Country Club
> - Entries and exits guarded by special security systems
>
> (quoted in Oncü, 1997, p.56)

Oncü argues that the domestic ideal of the private home, the symbol of middle-class identity worldwide, has been imported into Istanbul but reconfigured in a way which distances the middle class from the chaos and 'pollution' (chemical, social and cultural) of other parts of the city. Such middle-class 'dreamlands' stand in sharp contrast to the new housing developments being built by public authorities for the mass of Istanbul's working class.

While the growth of Istanbul has brought increasing numbers of people together in one city, as the expanding middle and business classes seek workers and domestic servants, it also works (or is actively organized) to keep them apart. There is, for Robins and Aksoy, a growing divide, an increasing polarization between *two Istanbuls*; the Istanbul of the emerging business élites and the Istanbul of those who are excluded from the story of economic growth. The class and economic aspects of this divide are reinforced by religious, cultural and ethnic differences. Thus a *new order has emerged* (or is in the making) – but an order which is seen as *increasingly disorderly*. The culture which prevails in these peripheral settlements has been labelled as disruptive and chaotic by the new middle class and local politicians. But it is also evident that within these areas, communal support and collective attitudes are highly valued, that there are strong values of mutual aid reinforced, in part, by a mixture of necessity and Islamic beliefs.

However, while these illegal settlements and the peoples they house are regarded as marginal,

- this periphery has also become increasingly central to the story of modern Istanbul.

The *gecekondus* are no longer isolated, peripheral ghettos. They have become a major political issue and are seen increasingly as a threat to the fortunes of the 'new' city. Left-of-centre political parties, such as the Social Democratic Republican Party, have sought to construct mass public sector housing estates to replace the illegal settlements. On the outer edges of Istanbul new estates have been developed in an effort to 'civilize' the shanty-town dwellers. In another twist, however, these new estates have been largely rejected by *gecekondu* dwellers, preferring the complex networks of social relations, obligations and support which have been built up in the illegal settlements. As Robins and Aksoy indicate, these settlements were ordered in ways which allowed their

FIGURE 2.4 *Illegal housing in Istanbul*

populations, or at least sections of them, some degree of control over their lives – in contrast to the managed estates with small-sized houses and little else.

The experience of Istanbul provides yet another sense of urban disorder – the existence of a particular culturally-influenced spatial organization and differentiation which is viewed as an obstacle to 'progress'. But from attempts to order the disorderly cityscape of Istanbul, new forms of order and disorder have been created. Islamic-inspired political groups, hostile to Western-style modernizing forces, argue that only by restoring Istanbul's Islamic culture and identity can this new disorder be overcome. But this new order/disorder has its roots in a variety of wider social and economic processes, occurring at overlapping geographical levels: through the attempts by a city to project itself as a global centre; through regional economic and political change; through immigration and a 'clash' of cultures; and through the existence of marked social division and inequality in Istanbul itself. A mix of state-sponsored planning and market forces have also conspired to produce the very disorder they were seeking to overcome, through modernization.

Before this section closes you might wish to reflect on the increasing pressures which cities such as Istanbul face as a consequence of increasing exposure to global processes and influences. Ironically, as it is projected as a global city, Istanbul runs the risk of becoming a pawn, rather than a real 'player', in global developments.

If Istanbul is becoming a pawn to global developments, in other cities specific areas have become pawns to local developments – such as the attempts by local governments to order the city, on the basis of their imagined perfect urban neighbourhoods.

4.2 BRITAIN'S 'DANGEROUS' PLACES

The inner city has long been central to accounts of urban disorder in many British cities, and numerous policies and strategies developed in an effort to manage it. However, in the last two decades, this has steadily been replaced by a new socio-spatial construct – the 'deviant council estate'. From being the solution to the intensity of the post-Second World War problems of slum housing and overcrowding, many of these estates are now widely regarded as no better than the slums they replaced: in other words they represent the unintended outcomes of past attempts to re-order and re-imagine urban areas.

The 'problem' housing estate is by no means a recent invention, but the council estate has become increasingly a by-word for crime and disorder, demonized as a symbol of modern urban social problems. In Scotland's main urban centres, for example, the council estate, especially 'peripheral estates', has now been delineated and constructed as a key spatial problem for the public authorities. Elsewhere in urban Britain there are similar concerns with the disorders of the council estate, reflected in media and government concern with teenage delinquency, gang 'warfare', and welfare dependency. Beatrix Campbell focuses on such estates in her account of Britain's 'dangerous places' (Campbell, 1993). The estates she considers, Meadowell in Newcastle upon Tyne, Blackbird Leys in Oxford and Ely in Cardiff, all experienced urban disorder (rioting) in the summer of 1991. Like estates in other cities, these particular locales had long been labelled as 'problem' areas – locales of 'social disorganization', as Louis

FIGURE 2.5 *Newspaper headlines depicting the 'other' Britain*

Wirth would put it (Wirth, 1938; see **Pile, 1999**). Such localities are portrayed as home to a criminally-minded and welfare-dependent 'underclass', comprising single-parents, 'welfare-junkies' and a volatile mix of other groups with pathological traits and behaviours – increasingly 'othered' as the 'enemy within', whose existence undermined 'morality' and 'decent values'.

Estates such as the Meadowell in Newcastle upon Tyne have been used as exemplars of the 'problem estate' throughout the 1980s and 1990s, though Campbell is guilty of suggesting that the estate has always been characterized by criminality as a central experience for which little or no supporting evidence is provided. Order appears to have been much more normal than disorder. Campbell interprets rioting in Meadowell as an expression of 'lawless masculinity' among the male youth of the estate – as 'incendiary young bucks' confronted 'the Boys in Blue' (Campbell, 1993, p.3). These young men were excluded from regular employment and lacked 'proper family roles' and 'responsibilities', with few 'appropriate role models' to learn from. They 'took over the streets' and 'terrorized' their own communities. For Campbell, these riots signified a new spatial ordering of urban 'dangerousness':

> Crime was *spatialised* in the Nineties. The collective gaze was directed at localities rather than, for example, the grandiose corporate frauds which vexed, and ultimately exhausted, the judicial system. The 'symbolic locations' shifted from the *frisson* of chaos and cosmopolitanism in the inner city – the *interior* of the celebrated metropolis – to the edge of the city, archipelago, out there, anywhere. These were places that were part of a mass landscape in Britain, *estates* were everywhere. But in the Nineties estates came to mean crime.
>
> (Campbell, 1993, p.317)

FIGURE 2.6 *Manningham District, Bradford, June 1995*

Campbell highlights the conflict between different groups of people engaged in rather different activities: where car-related crime is the source of much of the tension between the youth of the estate and the police; where there is excitement in the intensity of the rivalry between the youth and the police, at least on the part of the youth. But implicit also are the contrasting voices and the very distinct interpretations of life in such estates. Council estates, such as those discussed by Campbell, have become local and national symbols of urban 'breakdown'. This often reflects a grim reality of large-scale poverty, unemployment, poor housing and crime. But such estates are all too frequently portrayed as intense, uniform locales of conflict and unrest, housing homogenous groups of 'socially excluded' people. Little thought is given to their internal social and spatial differentiation and heterogeneity, to the strategies which residents adopt to 'cope', and to the residents' day-to-day struggles with public agencies. For this reason the inner city and the council estate can be maintained as symbol of social decay, multiple deprivation and social disorganization (remember Activity 2.1?). There is something else in the images, however. Campbell argues that there was a clear 'racial' dimension in the rioting in some council estates – but it is with the inner-city discourse we find a more explicitly racialized component. We can understand this further with an examination of the different images of Manchester.

4.3 MANCHESTER: 'GUNCHESTER' OR 'MADCHESTER'?

Manchester's Moss Side estate rose to national prominence in the aftermath of urban riots in 1981. It was widely regarded as typifying the 'inner city': population loss, environmental decay, crime, unemployment, poverty and social exclusion (see Mellor, 1985). But Moss Side was also, argue Taylor *et al.* (1997, pp.205–9), '*The* locale in the whole of the North of England which most closely approximated the mythologized black ghettos of the United States'. By the 1990s the area had achieved, suggests Fraser (1996, p.44), 'an iconic status nationally as the symbol of "dangerous Britain" and has become more or less a social exclusion zone vis-à-vis the city-region of Manchester itself'. In the early 1990s Manchester, or 'Gunchester', to borrow one label conjured up and indiscriminately applied by the media, was achieving fame as a centre of 'gang-land' killings and 'drug wars', much of it centred on estates such as Moss Side. There were widespread claims that Moss Side was a centre of 'Jamaican gangsterism' – confirming its imagined status as a racialized ghetto estate, as 'England's L.A.'! But as we shall see, estates such as Moss Side are not the only urban locales which are stigmatized in such ways.

Constructing Manchester as Gunchester uneasily sat alongside attempts by public agencies to construct a new role for the City of Manchester, re-imagining it as *Madchester*: a centre of cultural industries based on a vibrant local music scene in nightclubs. The 'youth scene' in particular was a crucial aspect of Madchester which both reflected and generated 'alternative' cultures, revolving

This man is the face of modern Manchester: a veteran of the city's increasingly brutal gang wars who boasts without remorse of his role as a drug dealer and 'field marshal'. GORDON BURN hears

A MOSS SIDE STORY

FIGURE 2.7

around the sometimes illicit pleasures of soft drug use, diverse expressions of sexuality and new musical forms; where large numbers of young people cruised the streets at night in search of 'buzz' and excitement. Madchester was a source of excitement and exhilaration for some, but a source of potential danger for other residents of the three 'Man'-chesters.

Thus while there was a divergence between the imagery and realities of Gunchester and Madchester, paradoxically the boundaries between the two overlapped considerably. The hidden economies of crime and drugs permeated both Manchesters, with legitimate and illegitimate activities blurring and impossible to distinguish. Each fed the other, albeit in ways which were unplanned and uncoordinated, through practices and activities which were frequently condemned, but all of which contributed to the excitement and exhilaration of the new Manchester as a 'happening' place.

However, both the Gunchester and Madchester labels represent 'cartoon images';

Estates overwhelmed by drug wars, where young gunmen fire on the police and threaten a summer of violence

Salford: A bleak but sometimes deceptive picture of boarded-up windows, burnt-out buildings, youths on mountain bikes with anti-police attitudes — a lawless place gone wild./Photograph by Ged Murray.

Manchester disunited

Last week fire engines on a Salford estate were stoned and bullets hit a police van. Gang leaders apparently immune from prosecution rule

THE crop-haired youth in the grey sweatshirt swaggered across Salford's Ordsall estate, ...

estate's young men. He has achieved a status of almost mythical invulnerability in the ...

Fila trainers and the expensive shell-suits and they want to ... the bandwagon. It's a ...

underworld was the attention that the leaders of the gang were coming under from the ...

FIGURE 2.8

FIGURE 2.9 *Map of Greater Manchester*

reflecting dominant representations of the city. There are always other images and stories in circulation, but which tend to be marginalized by and through the dominant stories.

In both Istanbul and Manchester we can see tensions emerging from differences in urban space and an uneasy tension between 'centre' and 'periphery'. Housing estates in both cities were the product of particular attempts to order the city in specific ways. There are, of course, very unique sets of conditions which structure each of these but there are strong linkages in both the depictions of particular areas of Istanbul and Manchester as unruly/disorderly and through the attempts to socially engineer the city through mass housing construction.

From the 1930s through to the 1970s, such mass housing estates were to be among the main recipients of people cleared from older slum and overcrowded housing in central areas of Britain's main cities. One of the largest redevelopment programmes in post-Second World War Europe took place in Glasgow. Let us briefly consider attempts to re-order Glasgow in the 1940s and 1950s.

FIGURE 2.10 *Madchester: a centre of cultural industries?*

4.4 MUNICIPALIZING SUBURBIA: RE-ORDERING POST-WAR GLASGOW

By the end of the Second World War, Glasgow was synonymous with poor housing and other urban problems. It was arguably the most overcrowded city in Europe, with one-seventh of the entire Scottish population crammed into three square miles of its central area. Over half the housing stock lacked the most basic amenities, much of it in such a poor state of repair that demolition was the only option. Waiting lists for houses in the city were increasing rapidly and landlords and housing companies were unable to provide enough houses at a cost which was within the reach of the vast majority of Glaswegians. Combined with the effects of pollution from industrial areas, the absence of space for new housing, and factories in central districts there was a distinctive sense of urban 'congestion'.

In an effort to 'modernize' and spatially re-order post-war Clydeside, The Clyde Valley Regional Plan of 1946 proposed massive population and industrial decentralization from central Glasgow in a process which became known as 'overspill' (see also **Pile, 1999**, section 1.2). New Towns were to be built throughout central Scotland to receive many of those cleared from the slums and 'new' industries were directed away from Glasgow. 200,000 new council houses were also constructed on Glasgow's outer edges. Four large 'peripheral' housing estates – at Castlemilk, Drumchapel, Easterhouse and Pollok (see Figure 2.11) – were built to house around 200,000 people.

The Clyde Valley Regional Plan created a new spatial order on post-war Clydeside: planned decentralization and a new ring of suburban council housing developments around the edges of Glasgow. A new layer of social and

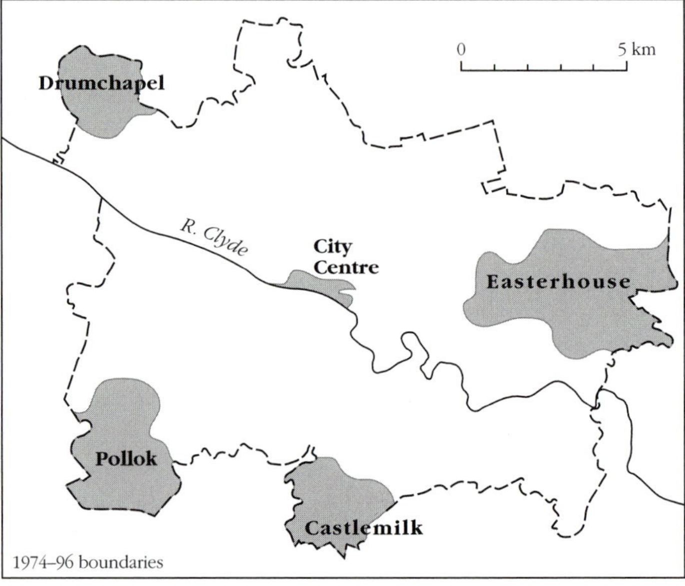

FIGURE 2.11 *Map of Glasgow*

spatial differentiation was superimposed on previous layers of class segregation which had been formed during the previous century. As in Istanbul, motorways would cut a swathe through working-class areas of the city, in the process dispersing populations to more peripheral areas.

It is a difficult task to disentangle the urban disorders which were a feature of Glasgow and other large British cities in the immediate post-war period. In post-war Glasgow these were generally interpreted as the legacy of decades of 'unplanned' urbanization and industrialization and of the inability of the market to provide for the majority of the city's residents. But the new housing estates were to have contradictory effects and unplanned consequences.

The peripheral estates which were to become such a marked feature of the (sub)urban landscape of post-war Britain may have provided better quality housing than the old slums they replaced, but they furnished little else. These estates, which were generally premised upon assumptions of the nuclear family and a gendered division of labour between home and paid employment, ironically disrupted many of the family and friendship networks which had been a marked feature of the older tenemental and slum housing areas where there was a greater sense of belonging amidst the tight-knit communities (Meegan, 1989). While the 'traditional working-class community' has often been mythologized and romanticized it was none the less 'real' in the sense of attachment. The absence of 'community', and feelings of isolation and anonymity which was characteristic of many of these new estates, was compounded in many cases by their relative geographic isolation from the main centres of employment and entertainment, a problem further magnified by poor public transport provision.

Within a relatively short time since their construction, from being the solution to housing problems in post-war Britain, these estates were increasingly labelled as 'problem' areas by public and private agencies. They were characterized by poverty, deprivation and enjoyed a bad reputation as areas of 'lawlessness', particularly in relation to 'juvenile delinquency' and 'youth disorder', and in the context of Glasgow's 'regeneration' in the 1980s and 1990s were portrayed as obstacles to continuing redevelopment: as the 'other' side to the 'new' Glasgow (see Figure 2.12).

Across Europe suburban housing estates are increasingly viewed as revealing similar trends towards poverty, polarization and 'social breakdown' to those in Britain (Power, 1997). In different ways **Allen (1999)** and **McDowell (1999)** discuss François Maspero's journey on the suburban express train across Paris. One of the destinations is Aulnay-sous-Bois, a 1970s housing estate built to house workers at a new Citroën car plant. Aulnay is a *banlieue* – literally, 'suburb'. But in France during the 1990s, the term *banlieue* has become a by-word for ethnic conflict, turmoil and disadvantage (Maspero, 1994). Like Aulnay, many of these suburban estates were built in the 1950s and 1960s in an attempt to modernize greater Paris, as part of a strategy to manage the city more effectively and to provide housing for the new workers needed for the city's

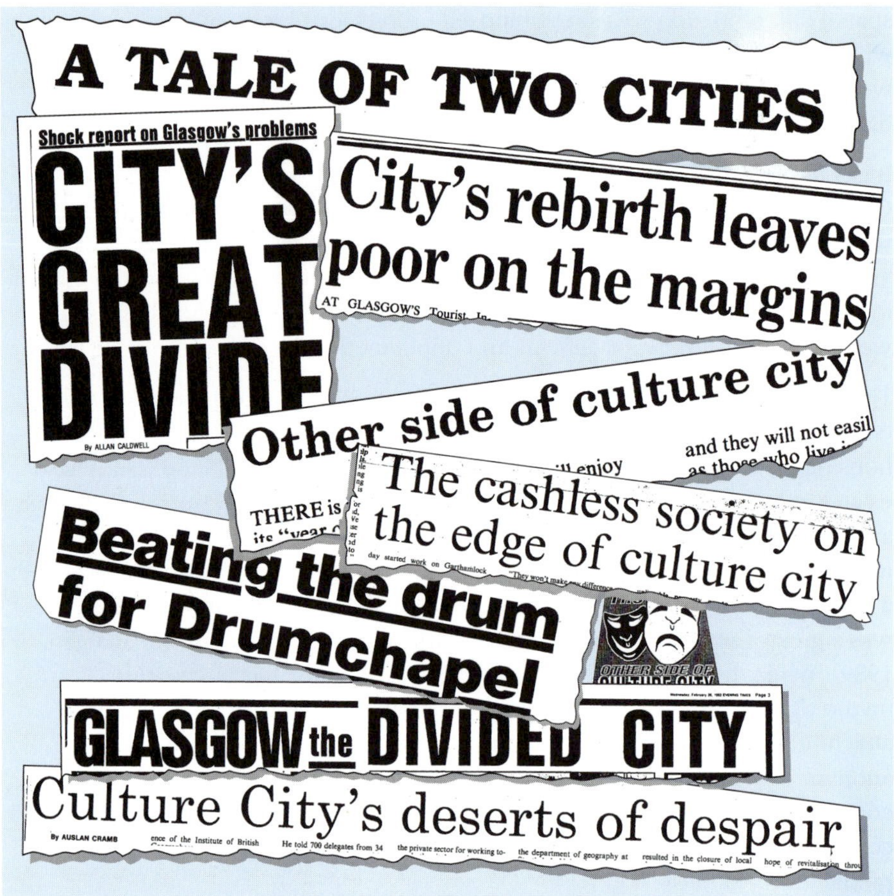

FIGURE 2.12 *Glasgow: 'dual city'?*

post-industrial transformation. But the French outer estates are distinct from those in Britain in a significant way: there has been a steady racialization of the low-income urban periphery of many French cities during the past two decades, with such estates housing increasing numbers of new immigrants and experiencing deepening levels of unemployment, poverty, racism and urban unrest.

Urban redevelopment programmes and other state-sponsored policies in central Paris are pushing larger numbers of low-income families into these estates, where cheaper public sector accommodation is on offer. White (1998) has argued that the outcome of this process is an increasing tension between Paris and its suburbs. The embourgeoisement of central Paris and the proletarianization of much of the suburban area takes place hand-in-hand: there is an interdependency but also increasing isolation and strain in the relationship between both places. Those living outside the suburbs, in central Paris itself for example, tend to stigmatize and group together all suburban estates negatively as 'ghettoized' areas, thus denying legitimacy to the views of the residents themselves.

"Personally, I blame it all on Le Corbusier."

The label *banlieue* has come to be used by the media and policy makers to depict 'problem' areas across France, and to focus in particular on the immigrant populations within them. These populations are depicted increasingly as a key source of urban disorder, neglecting the ways in which youth from different ethnic backgrounds have frequently come together in conflict with the police. This has prompted claims that there has been a ghettoization or an 'Americanization' of the French city in the period since the 1980s (Wacquant, 1997a).

In comparing the Parisian *Red Belt*, reflecting a history of Communist Party dominance in the suburbs of Paris, with Chicago's *Black Belt* of segregated ghettos, Löic Wacquant argues that the French *banlieues* and American ghettos, while sharing a similar position at the bottom of their respective urban hierarchies, represent what he terms distinct 'socio-spatial constellations'; that there is a different 'stitching' together of class, race and place on both sides of the Atlantic, with race being the primary agent in spatial segregation in the USA (Wacquant, 1996, p.122). The experience of the Parisian suburbs also contrasts with the imagined orderly life of suburban England and the USA where the suburb has historically been viewed as a place of tranquillity and rest, as a refuge from the disorder gripping inner cities (see Chapter 1). Suburban enclaves for the affluent and middle class have long been a feature of the British urban landscape. But the examples from both Glasgow and Paris highlight that the outer edges of the large cities are also constructed as places of disorder.

The estates explored in this section have been planned, built and managed by the state. While some of the worst forms of housing are the outcome of state attempts to provide for large sections of the population unable to occupy housing provided through the market, many of these housing developments were implemented for well intentioned reasons (albeit in ways which were often paternalistic) and their 'problems' the unintended consequences of utopian plans for housing estates. However, large-scale redevelopment and 'modernization' projects often reflect a top-down approach to organizing the city: one in which the state, or other public bodies, in conjunction with a host of private interests, plan and order for less powerful groups within society. Few if any of these estates were the product of involvement of those who would live in them.

The *banlieues*, 'peripheral' estates, and estates such as Meadowell can be understood as

- the product of attempts to de-intensify urban life and to impose a new sense of urbanity in different settings throughout post-war Europe. The result of urban differentiation and segregation, they have contributed to new forms of urban spatial difference, standing in sharp contrast to the locales of the affluent and middle classes.

Thus the examples explored above demonstrate the differing ways through which wider social relations become spatially constituted.

While the populations of the areas considered here experience largely similar problems of poverty, unemployment and marginalization, place and context matter and the experience is never quite the same. In many of the cities highlighted thus far there appears to be growing social inequality and social polarization. Thus

- the existence, in close proximity, of rich and poor areas, reflect a particular ordering and combination of social groups, but it is also a source of tension and disorder. The juxtaposition of places of affluence and extreme poverty in the urban setting reflects the highly unequal distribution of social and economic resources. This is both 'disorder', and also the product of a specific ordering by the economics of the 'free' market.

5 'Disorderly people: disorderly places'

One of the themes which runs through this chapter is the idea that there exists within major urban areas groups of people who are often spatially segregated, and marginal to the life and dominant rhythms of the city. Think back to the comments by Engels on nineteenth-century British cities, for example. To borrow a term from Castells, such groups contain 'structurally irrelevant people' (Castells, 1996). By this Castells suggests that there are sections of populations who are actively disconnected from 'mainstream' urban society (see also **Massey, 1999**). As you read through this section keep this idea of *mainstream* in focus. What does it mean? How is it defined and by whom? In this section we focus on two groups which are constructed as 'social outcasts' from mainstream society: first, the urban 'underclass' in Western cities and, second, the population of Third World shanty towns. While both occupy distinct urban worlds in very different kinds of cities, they share a similar position within these cities, constructed as socially disorganized groups and places, places which suffer from *territorial stigmatization*.

Further, while condemned and constructed as deviant, 'there is a tendency to exoticize the ghetto and its residents ... to highlight the most extreme and unusual aspects of ghetto life as seen from outside and above ... from the standpoint of the dominant' (Wacquant, 1997b, p.342). Particularly in relation to the urban underclass in the USA there is, in what can rightly be termed the 'underclass industry', a clear focus on the pathologies of people living in ghetto areas, in ways which are racialized and gendered. The main lesson to be learned here is that

- in studying the shanty town and urban ghetto we need to avoid what Edward Said has referred to as a new 'urban Orientalism'. In other words the shanty town and ghetto

  > should be studied not merely as pathological departures from what is good and right, but as part of the total complex of human activities and enterprises. In addition, they should be looked at as orders of things where we can see the social processes going on, the same social processes, perhaps, that are to be found in the legitimate institutions.
  >
  > (Hughes, 1980, p.99)

5.1 THE URBAN 'UNDERCLASS'

ACTIVITY 2.4 Read the following two quotations carefully. What kinds of ideas about urban life are mobilized within them? In what ways might these provide contradictory expressions of particular parts of American cities?

> Behind [the ghetto's] crumbling walls lives a large group of people who are more intractable, more socially alien and more hostile than almost anyone had imagined. They are the unreachables: the American underclass.
>
> (cover from *Time* magazine in the 1970s, quoted in Weir, 1993, p.100)

> Ghettos, as intrinsic to the identity of the United States as New England villages, vast national parks, and leafy suburbs, nevertheless remain unique in their social and physical isolation from the nation's mainstream. Discarded and dangerous places, they are rarely visited by outsiders … Ghettos are pervaded by abandonment and ruin; they openly display crude defenses and abound in institutions and facilities that are rejected by 'normal' neighborhoods.
>
> (Vergara, 1995, p.2) ◆

For me these brief extracts reveal a paradox about order/disorder and urban space. A clash of different forms of urban ordering and different senses of order. The external portrayal of the ghetto diverges unmistakably from the internal sense of ghetto life, its survival mechanisms and networks. This is not to deny that ghettos are also frequently places of violence and dire living conditions, but that ghettos cannot be dismissed simply or solely as places of disorder and have to be interpreted as locales with differing layers, characterized by a multi-dimensionality which cannot be understood through a focus on the 'exotic'.

FIGURE 2.13 *'Underclass': from the USA to Britain?*

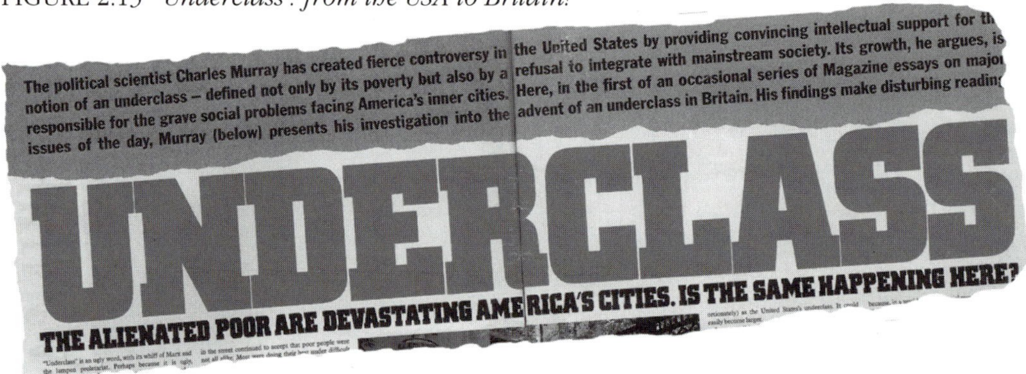

The political scientist Charles Murray has created fierce controversy in the United States by providing convincing intellectual support for th notion of an underclass – defined not only by its poverty but also by a refusal to integrate with mainstream society. Its growth, he argues, is responsible for the grave social problems facing America's inner cities. Here, in the first of an occasional series of Magazine essays on majo issues of the day, Murray (below) presents his investigation into the advent of an underclass in Britain. His findings make disturbing readin

UNDERCLASS

THE ALIENATED POOR ARE DEVASTATING AMERICA'S CITIES. IS THE SAME HAPPENING HERE?

Fear of the urban 'mob' or the 'dangerous classes' has long been a source of potent representations and sentiments of urban disorder (as we will see in Chapter 3). The language may have changed over time, but the imagery and the role that these groups play in discourses of urban disorder has constantly remained in dispute. The underclass perspective has been used almost as a code for urban disorder, particularly in American cities. The notion of an underclass is largely Western and while it has been mobilized on both sides of the Atlantic as a metaphor for urban and wider social problems, albeit in different ways, its use has been predominantly driven by US discourses of social life. This is not the place to enter into a long discussion of the history of the underclass idea, or the myriad of definitions and competing perspectives which underpin its use – but take your pick from the listing which follows.

Key characteristics of the 'underclass'

dysfunctional behaviour

deviant behaviour

lawlessness

anomie

alienation

social exclusion

chronically poor

drug using

welfare dependency

violent criminals

socially isolated

illegitimacy

moral breakdown/deficiency

differential values

culture of poverty

irrational behaviour

uncivilized behaviour

The ghetto underclass have been depicted as one of the key sources of urban disorder. In this sense they are constructed not only as pathological and deviant, but also as 'contagious' and 'corrosive of the social fabric' (Devine and Wright, 1993, p.81).

In such accounts the idea that ghettos can offer feelings of security, ways of maintaining other lifestyles and ethnic entrepreneurship, together with the organizational basis for activity within wider society, is neglected. Further, ghettos allow the development of subcultures which are often vital to struggles for survival. As Wacquant again comments, ghettos can be understood in very different ways from their 'official' categorization and representation:

> Yet intensive, ground-level scrutiny based on direct observation – as opposed to measurements effected from a distance by survey bureaucracies utterly unfit to probe and scrutinize the life of marginalized populations – immediately reveals that, far from being disorganized, the ghetto is *organized according to different principles*, in response to a *unique set of structural and strategic constraints* that bear on the racialized enclaves of the city as on no other segment of America's territory.
>
> (Wacquant, 1997b, p.346)

Within the ghettos there are complex and intricate networks of relationships, with their own unwritten codes and rules. Take crack dealers for example. They operate within the underground economy in ways which often mimic the formal economy. Phillippe Bourgois argues that

> the vast majority of the residents of East Harlem are honest and hard-working. It is a struggle to live there, and they struggle honorably. Nevertheless, to many, especially the young, the underground economy beckons seductively as the ultimate 'equal opportunity employer'. The rate of unemployment for Harlem youth is at least twice the citywide rate ...
>
> Most of the people I have met are proud that they are not being exploited by 'the White Man'. All of them have, at one time or another, held the jobs – delivery boys, supermarket baggers, hospital orderlies – that are objectively recognized as among the least desirable in American society. They see the illegal, underground economy as not only offering superior wages, but also a more dignified workplace.
>
> (Bourgois, 1989, quoted in MacGregor and Lipow, 1995, p.13)

In opposition to claims that the underclass do not share or identify with wider social values and/or are badly socialized, Bourgois continues that 'On the contrary, ambitious, energetic, inner city youths are attracted to the underground economy precisely because they believe in the rags-to-riches American dream'.

The idea that anyone could be attracted to ghetto life and can derive a sense of belonging and security from it does not feature in dominant representations of the underclass and the ghetto. But Bourgois's account emphasizes once again the tension between different senses of urban life and contrasting forms of social organization. We will pick this theme up again in a different context in the following section which considers conflicting stories and representations of the shanty town.

5.2 THE 'DISORDERLY' SHANTY TOWN

ACTIVITY 2.5 Read Benedita da Silva's story (Reading 2A). How does it challenge some of the dominant ways in which shanty towns have been imagined and portrayed? What do you think about what it was like to live in the shanty town she refers to? ◆

Benedita da Silva's story reflects a powerful personal commitment to the shanty town and it provides a clear response to the question 'why would anyone want to live in a shanty town?' There is a strong sense of attachment to community here, a sense of belonging. As in the American ghetto, da Silva's portrayal of the shanty town is markedly different from the dominant representation of such places.

The Third World shanty town plays a particular role in the story of urban disorders. While the high-rise office towers, hotels and shopping malls spread around numerous Third World cities are symbolic signs of international capitalism – as 'landscapes of power' – the shanty town occupies a very distinct urban social space. From the *favelas* of Rio de Janeiro, the *pueblos jouenes* of Lima, the *bustees* of Calcutta, the *bilonvilles* of Dakar to the *corralones* in El Salvador, shanty towns and 'illegal' housing/squatter settlements occupy a distinctive position in dominant depictions of urban chaos, and not just from those in the West. It is often those groups who live in closest proximity to the shanty towns who are most vocal in their denunciation of such places and the people who live within them. Jeremy Seabrook shows that in Bombay the Dharavi district, which conforms to much of the stereotypical portrayal of shanty towns, is depicted within India itself as the 'largest slum in Asia', a 'blot on the landscape of India's richest city', 'a refuge for anti-social elements' (Seabrook, 1996, p.49).

Once again it is important to understand these representations as social constructions. Shanty towns are depicted as marginal places in Third World cities. But Janice Perlman in her study of the *favelas* of Rio de Janeiro argues that marginality is 'both a myth and a description of social reality' (Perlman, 1976, p.242). By myth Perlman refers to the stereotypical portrayal of *favelas* as a political weapon which the ruling classes use to justify their policies, which increase the stigmatization, exclusion and marginalization of shanty-town dwellers.

The *favelas* are seen as disorderly locales of deviant social groups, 'disease-ridden places with a seething, frustrated mass of people'. Perlman refers to an official report on shanty towns in Rio de Janeiro in the mid 1960s:

> Families arrive from the interior, pure, and united – whether legally or not – in stable unions. The disintegration begins in the favela as a consequence of the promiscuity, the bad examples, and the financial difficulties there. Children witness the sexual act. Young girls are seduced and abandoned; they get pregnant but don't feel any shame ... Liquor and drugs serve to dull the disappointments, humiliations, and food deficiencies of favela life. The nights belong to the criminals.
>
> (quoted in Perlman, 1976, p.93)

Among the other characteristics which the same report attributes to the *favelas* are

> irregular agglomerations of sub-proletarians with no professional capacities, low living standards, illiteracy, messianism, promiscuity, alcoholism, the habit of going barefoot, superstition and spiritualism, lack of healthy recreation, refuge for criminals and marginal types, and spreader of parasites and contagious diseases.
>
> (quoted in Perlman, 1976, p.93)

What we have here is pronounced class bias: the moralizing and stigmatization reflect particular ideologies which work to reproduce the status quo. There is, however, an alternative vision which sees the shanty town in a markedly different way, as the product of wider social, economic and political arrangements that stretch beyond the boundaries of the shanty town – in much the same way as Harlem and Sophiatown (see Chapter 1). Shanty towns can exist as self-organized communities with well developed networks of social and political organizations, ties that stretch back to places of origin in rural localities. In contrast to images of poverty, inactivity and disintegration, these can be places of energy and intense activity where there is hard work, mutual trust and complex internal social affiliations, all of which contribute to survival amidst some of the worst living conditions in the modern world. But this also contributes to the reproduction of Rio de Janeiro itself and Brazilian society in general. The 'disorderly' shanty towns are part of the 'ordering' of the city as a whole.

Perlman argues that shanty towns are central to Brazilian life in a number of ways. One of the best examples of this is their contribution to the cultural identity of Brazil. The samba, spiritist cults and colourful slang all originate in the *favelas* but contribute much to the vitality and spirit of Rio. Likewise with the carnivals for which Rio is world famous. The order for which cities such as Rio de Janeiro are renowned worldwide, is predicated and dependent upon the disorder of the shanty town.

Thus the *favelas* play important economic, social and cultural roles within Brazilian society, in ways which challenge their 'marginality'.

ACTIVITY 2.6 The readings from Lea Jellinek (Reading 2B), Ray Bromley (Reading 2C), and Larissa Lomnitz (Reading 2D) highlight survival strategies in three distinct cities in very different parts of the world: Jakarta, Cali in Colombia, and in Cerrada del Cóndor in Mexico.

In what ways do these studies disrupt the dominant portrayals of 'Third World cities' and their shanty towns? ◆

The readings illustrate many different things about shanty towns and other similar places depicted as marginal. Cities contain distinct but related worlds which are packed with different activities and groups working to different rhythms. As Bromley highlights in his study of Cali, and Jellinek explores in Jakarta, there is an intensity in the range of street occupations. We get a feel for this through his discussion of the constant movement and 'bustle' of street level activity, but it also emerges in the context of conflict between legal retailers, government officials and the illegality of much of the street work. While working the streets and the illegal economy is necessary for survival, these activities and the people concerned are interconnected in different ways to the wider organization of social and economic life in these cities, but few if any of these groups and places depicted in the readings are tied in to networks of power and influence (**Massey, 1999**). However, 'working the streets' covers a broad spectrum of activities, some of which are deemed by some/many to be anti-social/immoral and/or parasitic, and contribute little to the social reproduction of the shanty town.

In important ways the worlds of the urban underclass and the shanty town become a 'problem' for dominant social groups when they 'spill over' into other worlds of the city: when forms of living and endeavours which are considered deviant or illegal are 'out of their place'. Contained within the ghettos and the *favelas* such subcultures may be considered disorderly, but in ways which reflect some sense of ordering of the city. However, the potential for the wider disordering of city life remains.

Each of the readings depict in their own ways, and in their particular contexts, the tension between order and disorder in the city, as the product of the intensity and intensification of social relations. Different groups and activities are (dis)connected in different ways to the dominant social and economic networks which characterize the city.

The existence of places of extreme wealth and affluence and the ghetto and the shanty town represent the coming together of disparate groups of people, working to sometimes very different rhythms (**Allen, 1999**) in segregated spaces within the city. Here we have, once again, the intensification of difference, underpinned by unequal power, class and social relations.

6 Conclusion

In this chapter we have explored the notions of urban order/disorder in a variety of contexts. Cities are places of wealth, poverty, opportunity, oppression, pollution, excitement, danger and struggles for survival. They are places where disparate groups and ventures exist in tension. Cities are also places where different histories and stories come together, but in ways which privileges some narratives and relegates others. All too often the social, cultural and political legitimacy of those who live in the places explored here is denied.

I have highlighted the existence of competing senses of order and disorder: conflicting ways in which urban space is organized, perceived and used by different groups of people. This chapter highlights that:

- There exist competing, sometimes conflicting, senses of urban order and disorder: conflicting ways in which urban space is organized, interpreted and used by different social groups in ways which frequently conflict.
- Disorders can emerge from the intensity of urban life – the coming together and juxtaposition of different groups and practices.
- Order and disorder can have very different meanings for these social groups.

There are strong echoes of the past which remain forceful in representations of urban disorder which are dominant today through, for example, notions of 'dangerous classes' such as the 'urban underclass' and through the continuing use of spatial metaphors such as 'out of place'. Thus issues of order and disorder in the city have long histories (as we will see in the following chapters). While there are historical continuities, however, notions of urban order and disorder are best understood

- as a dynamic and shifting matrix of competing meanings and understandings,
- an elusive rhetorical formulation that mobilizes a range of representations about the conditions of cities around the world, but which also reflect both the power and social relations which are prevalent in cities.

Thus,

- the construction of a variety of social groups and places as marginal, peripheral, deviant, socially disorganized or dangerous has material consequences in terms of policy and strategies – actions which might be contested and resisted by populations of these places, and

- those places, people and practices which are constructed as, and portrayed as, marginal/peripheral (and disorderly) *are often symbolically central to the dominant representations of urban life* and its problems.

I have stressed that (dis)orders cannot be understood in isolation from inequality, social divisions, and power relations. Urban landscapes are etched with highly uneven social and power relations and this is reflected in social–spatial differentiation and segregation in the urban context, a theme which is explored in the next two chapters.

References

Aksoy, A. and Robins, K. (1997) 'Modernism and the millennium', *City*, 8 December, pp.21–36.

Allen, J. (1999) 'Worlds within cities' in Massey, D. *et al.* (eds).

Allen, J., Massey, D. and Pryke, M. (eds) (1999) *Unsettling Cities*, London, Routledge/The Open University (Book 2 in this series).

Amin, A. and Graham, S. (1999) 'Cities of connection and disconnection' in Allen, J. *et al.* (eds).

Branford, S. and Kucinski, B. (1995) *Brazil: Carnival of the Oppressed*, London, Latin America Bureau.

Bromley, R. (1997) 'Working in the streets of Cali, Colombia: survival strategy, necessity, or unavoidable evil?' in Gugler, J. (ed.).

Campbell, B. (1993) *Goliath: Britain's Dangerous Places*, London, Virago.

Castells, M. (1996) *The Rise of the Network Society*, Oxford, Blackwell.

Castells, M. (1998) *End of Millennium,* Oxford, Blackwell.

Cooper, F. (1987) *On the African Waterfront: Urban Disorder and the Transformation of Work in Colonial Mombasa*, New Haven, Yale University Press.

Devine, J.A. and Wright, J.D. (1993) *The Greatest of Evils: Urban Poverty and the American Underclass*, New York, De Gruyert.

Engels, F. (1844/1987) *The Condition of the Working Class in England*, Harmondsworth, Penguin.

Fraser, P. (1996) 'Social and spatial relationships and the "problem" inner city: Moss Side in Manchester', *Critical Social Policy*, vol.49, November, pp.43–65.

Gugler, J. (ed.) (1997) *Cities in the Developing World: Issues, Theory, and Policy*, Oxford, Oxford University Press.

Hamilton, K. and Hoyle, S. (1999) 'Moving cities: transport connections' in Allen, J. *et al.* (eds).

Hughes, E.C. (1980) *The Sociological Eye*, New Brunswick, Transaction.

Jellinek, L. (1997) 'Displaced by modernity: the saga of a Jakarta street-trader's family from the 1940s to the 1990s' in Gugler, J. (ed.).

Krim, A. (1992) 'Los Angeles and the anti-tradition of the suburban city', *Journal of Historical Geography*, vol.18, no.1, pp.121–38.

Lomnitz, L. (1997) 'The social and economic organization of a Mexican shanty-town' in Gugler, J. (ed.).

McDowell, L. (1999) 'City life and difference: negotiating diversity' in Allen, J. *et al.* (eds).

MacGregor, S. and Lipow, A. (eds) (1995) *The Other City*, New Jersey, Humanities Press.

Maspero, F. (1994) *Roissy Express*, London, Verso.

Massey, D. (1999) 'On space and the city' in Massey, D. *et al*. (eds).

Massey, D., Allen, J. and Pile, S. (eds) (1999) *City Worlds*, London, Routledge/The Open University (Book 1 in this series).

Meegan, R. (1989) 'Paradise postponed: the growth and decline of Merseyside's outer estate' in Cooke, P. (ed.) *Localities*, London, Unwin Hyman.

Mellor, R. (1985) 'Manchester's inner city: the management of the periphery' in Newby, H., Bujra, J., Littlewood, P., Rees, G. and Rees, T.L. (eds) *Restructuring Capital*, London, Macmillan.

Mulgan, G. (1991) *Communications and Control*, Oxford, Polity.

Oncü, A. (1997) 'The myth of the "ideal home" travels across cultural borders to Istanbul' in Oncü, A. and Weyland, P. (eds) *Space, Culture and Power*, London, Zed Books.

Perlman, J.E. (1976) *The Myth of Marginality*, Berkeley, University of California Press.

Pile, S. (1999) 'What is a city?' in Massey, D. *et al*. (eds).

Power, A. (1997) *Estates on the Edge*, London, Macmillan.

Pryke, M. (1999) 'City rhythms: neo-liberalism and the developing world' in Allen, J. *et al*. (eds).

Robins, K. and Aksoy, A. (1995) 'Istanbul rising', *European Urban and Regional Studies*, vol.2, no.3, pp.223–35.

Robins, K. and Aksoy, A. (1996) 'Istanbul between civilisation and discontent', *City*, 5–6 November, pp.6–33.

Seabrook, J. (1996) *In the Cities of the South*, London, Verso.

Taylor, I., Evans, K. and Fraser, P. (1997) *A Tale of Two Cities*, London, Routledge.

Vergara, C.J. (1995) *The New American Ghetto*, New Jersey, Rutgers University Press.

Wacquant, L.D. (1996) 'The rise of advanced marginality: notes on its nature and implications', *Acta Sociologica*, vol.39, pp.121–39.

Wacquant, L.D. (1997a) 'America as realized social dystopia: the politics of urban disintegration', *International Journal of Contemporary Sociology*, vol.34, no.1, pp.39–49.

Wacquant, L.D. (1997b) 'Three pernicious premises in the study of the American ghetto', *International Journal of Urban and Regional Research*, vol.20, pp.341–53.

Weir, M. (1993) 'From equal opportunity to "the new social contract". Race and the politics of the American "underclass"' in Cross, M. and Keith, M. (eds) *Racism, the City and the State*, London, Routledge.

White, P. (1998) 'Ideologies, social exclusion and spatial segregation in Paris' in Musterd, S. and Ostendorf, W. (eds) *Urban Segregation and the Welfare State*, London, Routledge.

Wirth, L. (1938) 'Urbanism as a way of life', *American Journal of Sociology*, vol.44, pp.1–24.

READING 2A
S. Branford and B. Kucinski: 'Profile: Benedita da Silva'

52-year old Benedita, Brazil's first black senator, remembers as a child delivering laundry to the house of President Juscelino Kubitschek in Leme in Rio de Janeiro. Her mother, a washer-woman, was a *mae-de-santo*, a priestess in the Afro-Brazilian candomblé religion. 'At that time, it wasn't respectable for public figures to be seen consulting a *mae-de-santo*', recalls Benedita, 'so they came secretly at night'. Benedita, one of 13 children, spent her childhood in a shanty town built on stilts in a flooded area of Rio de Janeiro. The family was poor and from the age of six Benedita worked, first in street markets and then as a maid. 'Then I got a job in a smart nursery. I cleaned the bottoms of several leading public figures, whom I now meet as an equal', she laughs.

Benedita married at 15, just after her mother's death. By the time she was 22, she had five children. Her first husband, a house painter, was a heavy drinker. Even so, she stayed with him until his death 22 years later. Life was hard. In 1968 she could no longer earn enough to support herself and her children. 'I belong to the poorest of the poor in Brazilian society', she says. 'I'm one of the da Silvas of this life'. [In Brazil, da Silva is the commonest working-class surname.] She felt suicidal until a friend took her to one of the evangelical churches, the Assembly of God. It got her over this difficult period and she has been a devoted follower ever since, even giving up Carnival which she used to enjoy enormously.

'After the bible, the PT [Workers' Party]' says Benedita. She was a founder member of the party and has been enormously active. Throughout her political career, she has turned the triple discrimination that she suffers into an electoral asset, using as her slogan, 'I am black, a woman and a shanty-town dweller'. She was a municipal councillor, then a federal deputy, becoming particularly active in the Constituent Assembly, which in 1987 drew up Brazil's new constitution. She presented 92 amendments, 25 of which were approved, including the controversial measure to make the job of maid a proper, regulated profession.

Benedita is a vehement defender of Brazil's black population. 'The Brazilian nation was forged through the rape of the black population', she says. 'Black families were destroyed. My grandparents and great-grandparents were slaves. They had children who were taken away from them and sold. We have no idea what happened to them.' Even today, Benedita suffers from discrimination. 'I go to the front entrance of apartment blocks and the porters still tell me to go round to the tradesmen's entrance.'

Benedita married for a second time in 1982. Her husband was a north-easterner with a long history of political involvement. In their heated political discussions, he cited Marx and Benedita replied with quotations from the Bible. He died in 1988 and Benedita married again, this time to the famous actor, Antônio Pitanga. Benedita waxes lyrical, 'I love my husband. I'm over the moon, passion 24 hours a day'.

Benedita has come up in the world. She is travelling widely and loves it. During a recent trip to Paris, she went to an official reception and was introduced to President François Mitterand. 'I spent so many years of my life working as a maid for rich madames in Brazil that I know how to sit at a banquet table and how to use the cutlery', she jokes.

Benedita still lives in the Chapeu Mangueira shanty town, where she brought up her children. Like most shanty towns in Rio, it is located on a hillside, forcing visitors to climb up 56 steep steps to get to Benedita's house. Though her three-roomed house is much more comfortable than most of the others, she still suffers from periodic police 'invasions', water shortages and electricity black-outs. A few years ago one of her nephews was killed in a shoot-out. Benedita justifies her decision to stay in the shanty town by saying, *Sou favelada, estou senadora*, using the two verbs in Portuguese for 'to be', to say that her condition of life is to be a shanty-town dweller, whereas she is only temporarily a senator.

Source: Branford and Kucinski, 1995, pp.32–3

READING 2B
Lea Jellinek: 'Displaced by modernity: the saga of a Jakarta street-trader's family from the 1940s to the 1990s'

A passion for hot lemon-juice and fried prawn-crackers brought me by chance to Sumira's stall. I loved sitting there deep into the night, watching the pedestrians go by. She parked her stall in the heart of the city, beside a water fountain opposite Merdeka (Freedom) Square. In the mid-1960s and early-1970s there was nothing unusual about street-stalls and traders in the centre of Jakarta. They were patronized by most of the city's population, rich and poor. The streets were alive with people passing to and fro. Most went on foot. Many travelled by bicycle or *becak* [a tricycle with seating for two passengers]. Only government employees and the very rich went by car.

Sumira's flourishing trade

Sumira started her day's work at 7 a.m. with a walk to a nearby market. She bought the foods she needed for her stall from the same few traders every day. Her survival as a trader depended on her ability to manage her modest finances very carefully. Her prices had to be competitive and profit margins were small. Most of what she earned from one day's trading had to be used to buy raw materials for the next day. The more she spent on herself, the less she could buy for her stall. There was little room for error. Sumira watched the market-traders keenly. She could accurately assess the price of her purchases before they were weighed. She began by offering a price she knew to be too low. Traders responded with prices they knew to be too high. On most occasions a compromise was soon reached and both would seem happy with the deal. Sometimes Sumira would hold out for a price which a trader refused to accept, or he would demand a price which she thought excessive. Sumira would move on to the next market-trader, who would have witnessed the unsuccessful negotiations, and start bargaining all over again. No one harboured ill feelings, though it was all in deadly earnest. Like Sumira, the market-traders also depended on the daily earnings for their next day's survival.

Sumira returned from the market in a *becak* laden down with fresh fruit, live chickens, and the other goods she had bought. The household was soon busily engaged in food preparation. They chopped and peeled and fried for the next few hours. Sumira's tiny house was taken over by delicious odours and baskets of banana fritters and fried chicken. The cramped and primitive kitchen did not seem to limit the output.

Cooking was completed by 3 or 4 o'clock in the afternoon. The food was carefully stacked into Sumira's cart which, loaded to the brim with the day's cooking, was then pushed to Sumira's trading site. Sumira operated her stall from 5 o'clock in the evening until the early hours of the morning, serving her prepared food or cooking afresh as required. After 1 a.m. Merdeka Square was reduced to the glow of kerosene lamps from the few night-traders, the bell-sounds of passing *becaks*, and an occasional pedestrian. Before dawn Sumira's tarpaulin was taken down and neatly folded and any left-over food carefully put aside for the next night's trade. With the help of her husband and another assistant the cart was packed up and pushed back home for storage near Sumira's house. By 4 a.m. she was asleep. Three hours later she would have to awaken to start the new day's work. Without her two-hour siesta in the early afternoon, while Tati the second wife did the cooking, Sumira would never have been able to endure these long hours of work …

Declining fortunes

As Sumira's aspirations and living expenses began to soar, the future of her stall, and thus of her income, became increasingly uncertain. Signs that her trade was doomed became more evident. The central city area was rapidly being transformed by new, wide, and ever-busier roads and the construction of numerous multi-storey offices, banks, and hotels. Sumira's community prospered from this building boom. Many labourers in the neighbourhood worked on construction sites, while other members of the community sold food or provided transport to workers engaged on the projects. But the community was gradually engulfed by the city it helped to create. In 1974 Sumira was directly affected, for the footbridge across the canal to Jalan Thamrin, the main highway running north south through the city, was dismantled and a high steel fence erected in its place. *Kampung* dwellers now had to make a long detour to get to the city centre. Sumira and her assistants had to find back routes away from the traffic to get to their trade location. On several occasions Sumira's cart was struck by speeding cars. The vacant plot of land beside Jalan Thamrin where Sumira and many other traders had stored their carts was fenced off for the construction of a multi-storey office block. As available land at the city centre became scarce, it became increasingly difficult and expensive for Sumira to find a place to store her cart.

Step by step, Sumira's stall was forced out of the central business district. She had started her trade in front of Sarinah department store less than five minutes from her home. After the first anti-trader campaigns in

1970, she had moved north along Jalan Thamrin to Merdeka Fountain, where I first met her. To avoid encountering the police clearance teams she became a night-trader. Her main clients were the soldiers and policemen who guarded the nearby offices, government buildings, and banks. Some of her customers were involved in the trader-clearance campaigns and they warned Sumira when it would be unsafe to set up her stall at Merdeka Fountain. As she provided a good service at a time and place where it was needed, her clients protected her against harassment. She in turn gave them generous helpings of food. While other traders went bankrupt, Sumira was able to survive and even thrive.

This happy state of affairs could not last. By 1975 the anti-trader campaign was being so firmly enforced that not even Sumira dared venture out to her normal location near the fountain. She had to be content with a more secluded spot along a back street between a number of banks and a military depot. As a result of her move she lost many of her valued customers. However, the move was not the only factor undermining her trade. Sumira began to find herself trapped in a cost–price squeeze from several directions. Her rising transport costs are a good illustration of this dilemma.

Sumira had always used a *becak* to ferry her from the market-place to her home when she was loaded down with her early morning shopping. Later she also used a *becak* to ferry her to her stall in style. Sumira used the same *becak*-driver each day. He was content to be paid with a meal from her stall. But just as the government was pursuing its anti-trader campaign, it was also implementing a ban on *becaks* in the streets of central Jakarta. As these policies came into effect, Sumira's driver could no longer operate in safety. Instead Sumira had to transport her goods from the market by *bajaj* (motorized *becak*) which were twice as expensive as *becaks*. Moreover, their drivers would not accept payment in kind.

Throughout Jakarta, the old markets of crowded muddy passages and makeshift stalls were being replaced by multi-storey concrete buildings. The stall-holders in the old markets had paid only nominal site rates, usually illegal taxes pocketed by local officials. They could hold their prices down because their overheads were minimal. As the markets were organized and rebuilt, additional fees were levied for the rent of kiosks, rubbish collection, the maintenance of law and order, and market administration.

These additional charges were passed on to buyers like Sumira, who in turn should have raised their prices. As the anti-trader campaigns progressively eliminated traders from large parts of Jakarta, those who were displaced re-clustered together in a few locations, near markets, major crossroads, and railway stations until they were dispersed again. Competition became intense. Despite rising costs, traders undercut one another in a desperate bid to attract customers.

By 1976, Sumira and her fellow street-traders were also competing with an increasing number of office canteens and small cafeterias which could buy in bulk and undercut their prices. Sumira's customers were increasingly reluctant to pay for a rickety seat at her roadside stall when, for a little more, they could dine elsewhere in greater comfort. Most of the offices which lined Jalan Thamrin set up their own canteens, so that employees no longer had to seek cheap food on the street. The government even launched a campaign in the newspapers and over television and radio to advise people against eating from street stalls which were said to be unhygienic. They did not conform with the city's 'modern' appearance and were accused of cluttering up the streets and hampering the flow of traffic …

Involuntary resettlement

… In 1981 the government announced that her neighbourhood was to be demolished. The government was proposing to replace the houses with apartments, which would be made available to the inhabitants of the *kampung*. Was this the same government that had banned their *becaks* and outlawed their petty trade? The *kampung* was full of households like Sumira's who had lost their livelihoods as a direct result of government policy. Was the government now going to help them? There were families in the *kampung* who had been the subject of earlier government demolition programmes. Sumira herself had once been forced to move as a result. Everyone knew that the government forcibly evicted *kampung* dwellers and either claimed that the *kampung* was illegally sited – and thus not entitled to compensation – or made such paltry payments that they scarcely covered the costs of transporting one's belongings to a new site, let alone buying another home. Fear and rumour swept the *kampung*.

For Sumira now isolated from her community the news was especially alarming. Once a week on her day off she made her way to her home wondering each time what she would find. Would her house still be standing? How would her infirm mother or her 9-year-old daughter face up to officials when they came? But the officials seldom came and the neighbourhood had little firm information to go on. Each time Sumira returned there was new speculation and new fears. The government held a public meeting at which it outlined its proposals. But the various officials contradicted each other. The brochures they distributed did not seem to agree with what they had said. The confusion and the apprehension increased.

The government proposals were in fact confused. The government announced its intentions before it had finalized its plans or properly defined its aims. The government planned to re-house Sumira's neighbourhood in subsidized apartments. The earliest announcements, however, mentioned prices for the proposed apartments that were beyond the financial capacity of all but a handful of the inhabitants. Moreover, many *kampung* dwellers used their houses to store and prepare materials for their livelihood, just as Sumira had prepared and cooked food in her home whilst she ran her stall. But the government was going to prohibit its apartments being used for commercial purposes. To its credit, the government did substantially modify its proposals to try to accommodate the objections of the *kampung* dwellers. It offered generous compensation and alternatives for those who did not want to accept an apartment. Unfortunately, it never overcame the suspicion of the community. People observed that the proposals continued to change no matter what was announced and believed that the junior officials with whom they had to negotiate were enriching themselves at the *kampung* dwellers' expense.

Source: Jellinek, 1997, pp.139–48

READING 2C
Ray Bromley: 'Working in the streets of Cali, Colombia: survival strategy, necessity, or unavoidable evil?'

This essay reviews some of the theoretical, moral, and policy issues associated with the low-income service occupations found in the streets, plazas, and other public places of most African, Asian, and Latin American cities. These 'street occupations' range from barrow-pushing to begging, from street-trading to night-watching, and from typing documents to theft. They are often grouped together in occupational classifications, and they are generally held in low esteem. The street occupations are frequently described by academics and civil servants as 'parasitic', 'disguised unemployment', and 'unproductive', and they are conventionally included within such categories as 'the traditional sector', 'the bazaar economy', 'the unorganized sector', 'the informal sector', 'the underemployed', and 'sub-proletarian occupations'. It seems as if everyone has an image and a classificatory term for the occupations in question, and yet their low status and apparent lack of developmental significance prevent them from attracting much research or government support.

This discussion of 'street occupations' is based mainly on 1976–8 research in Cali, then Colombia's third largest city with about 1.1 million inhabitants. The occupations studied in the streets and other public places of Cali are remarkably diverse, but they can be crudely described under nine major headings:

Retailing: the street-trading of foodstuffs and manufactured goods, including newspaper distribution.

Small-scale transport: moving cargo and a few passengers for payment, using *motocarros* (three-wheel motorcycles), horse-drawn carts, bicycles, tricycles, handcarts, or direct human effort as porters.

Personal services: shoe-shining, shoe repair, watch repair, the typing of documents, etc.

Security services: night-watchmen, car-parking attendants, etc.

Gambling services: the sale of tickets for lotteries and *chance*, a betting game based on guessing the last three digits of the number winning an official lottery.

Recuperation: door-to-door collection of old newspapers, bottles, etc., 'scavenging' for similar products in dustbins, rubbish heaps, and the municipal tip, and the bulking of recuperated products.

Prostitution: or, to be more precise, soliciting for clients.

Begging.

Property crimes: the illegal appropriation of movable objects with the intention of realizing at least part of their value through sale, barter, or direct use. This appropriation can be by the use of stealth (theft), by the threat or use of violence (robbery), or by deception ('conning').

Of these nine categories, 'retailing' was the largest in 1997, accounting for about 33 per cent of the work-force in the street occupations, followed by small-scale transport and gambling services, each accounting for about 16 per cent of the work-force. The six remaining categories each accounted for 2–10 per cent of the work-force in street occupations.

With the exception of small-scale public transport, all these occupations can be conducted in private locations as well as in public places, and private locations are generally considered more prestigious. Private premises give a business an aura of stability and security which is not available to most businesses conducted in streets, plazas, and other areas of public land. Those who work in public places may try to obtain a degree of stability by claiming a fixed pitch, and by building a structure there to give them some protection, but their tenure is almost always precarious, and their investment in 'premises' is likely to be very limited. Thus, the street occupations are classically viewed as 'marginal occupations', as examples of how the poor 'make out', or as the 'coping responses' of the urban poor to the shortage of alternative work opportunities and the lack of capital necessary to buy or rent suitable premises and to set up business on private property …

The streets of the city serve a wide variety of interrelated purposes: as axes for the movement of people, goods, and vehicles; as public areas separating enclosed private spaces and providing the essential spatial frame of reference for the city as a whole; as areas for recreation, social interaction, the diffusion of information, waiting, resting, and, occasionally, for 'down-and-outs and street urchins', sleeping; and as locations for economic activities, particularly the 'street occupations' (Anderson, 1978, pp.1–11 and 267–307). Within the functional complexity of the street environment, the street occupations are both strongly influenced by changes in other environmental factors, and also contributors to general environmental conditions. Thus, for example, street-traders and small-scale transporters depend upon the direction, density, velocity, and flexibility of potential customers' movements, and are immediately affected by changes in traffic flows and consumer behaviour. At the same time, they influence patterns of movement and overall levels of congestion …

Illegality is widespread in the street occupations, and serious problems and suffering may be associated with theft, robbery, conning, and prostitution. Most cases of illegality in the street occupations, however, are either trivial or common to many other groups across the social spectrum. The most frequent complaints against street-traders, for example, come from owners and managers of large shops and department stores: that they spoil the appearance of the street; that they cause congestion in downtown areas; and, that they pose unfair competition to indoor businesses because very few of them charge sales taxes to their customers and virtually all either drastically underestimate their earnings in income-tax declarations or do not make any declarations at all.

Many street enterprises provide useful services, but do not have some or all of the permits they are required to have, or fail to meet official specifications on receipting, taxation, equipment, uniforms, hygiene, etc. Just like big business corporations, wealthy taxpayers, and real-estate dealers, workers in the street occupations often cut corners and break a few rules here and there in order to make a living or a little extra profit. Complying with each and every official regulation can be costly and time-consuming. For minors and for adults who have lost or never obtained key identity papers, compliance may be impossible. In addition, many of the regulations are rarely enforced, and some officials will accept a bribe to stop enforcement …

Regulation and persecution of the street occupations

It is not surprising, and in total accordance with the legal system, that clearly illegal activities such as property crimes, trading in contraband goods, and the sale of prohibited gambling opportunities, are persecuted occupations in Cali. Indeed, many complain that these occupations are not persecuted enough. It is also hardly surprising that such occupations as prostitution and begging, viewed as 'disreputable' by most of the population, are officially regulated and suffer from periodic police harassment (see Bromley, 1982). In the case of prostitution, however, official attitudes are decidedly ambiguous, and there are many complaints that organizers of prostitution and upper-class prostitutes are free from harassment, while lower-class prostitutes are frequently persecuted.

The intervention of the authorities in occupations which are not clearly criminal, immoral, or anti-social according to conventional, élite-defined standards, is much more complex and diverse. In general, the concern is to regulate activities by introducing checks and controls on prices, standards, and locations, and by limiting entry to the occupations. Government personnel are appointed to enforce these regulations, and penalties are prescribed for offending workers. Thus, for many street occupations, e.g. the operation of a street stall, the commercial use of a *motocarro*, horse-drawn cart, or a handcart, and shoe-shining, registration procedures have been introduced, and regulations have been made as to when, where, and how these occupations should be practised. Hundreds of pages of official regulations … specify the municipal, departmental, and national government's regulations on street occupations, and substantial bodies of police and municipal officials are expected to administer these regulations. In reality, however, these regulations are excessively complex, little known, and ineffectively administered, resulting in widespread evasion, confusion, and corruption …

In general, those who administer law and order on the streets complain that there are so many people working in the street occupations, and that there is such widespread ignorance and disrespect for the official regulations, that controls must be very selective. The main objective is usually 'containment': to hold down the numbers of people working in the streets of such priority areas as the central business district, the upper-class shopping-centres and residential zones, and the main tourist zones.

Although there are occasional cases of assistance by the authorities to street occupations, as when help was given in improving street stalls and providing uniforms for street-vendors and shoe-shiners at the time of the Pan-American Games in Cali in 1971, official intervention in the street occupations is essentially negative and restrictive, rather than supportive. The basis of government policy is that 'off-street occupations' should be supported in the hope that they will absorb labour from the street occupations. However, this objective has not been achieved because insufficient investment funds have been mobilized, and because investment has concentrated in areas which generate relatively few work opportunities. In the mean time, the street occupations have often been persecuted, and opportunities to improve working conditions in these occupations have generally been neglected.

Street occupations conflict strongly with the prevailing approaches to urban planning. Although Cali has a warm, dry, and congenial climate for economic activities and social interaction in the open air, city planners have usually been concerned with reserving the streets for motorized transport and short-distance pedestrian movement, and concentrating economic activities into buildings. In general, no special provision has been made for street occupations, and restrictions on *motocarros*, non-motorized transport, and the sale of goods and services in the streets have been imposed to reduce traffic congestion, road accidents, and the incidence of these occupations. Cali is officially twinned with Miami and has strong links with other North American cities. Urbanistically, Cali is being planned along North American lines as a sprawling automobile-dependent metropolis, and the street occupations are, from the planners' point of view, an unfortunate embarrassment.

Women and children in street occupations

Among the urban poor, conventional official definitions of 'labour-force' and 'economically active population', based upon the idea that neither children nor housewives earn an income, are simply irrelevant. When personal incomes are low, when the membership of households is often unstable, and when many households with children are headed by a single, separated, divorced, or widowed woman, there is a strong pressure on all household members to seek work opportunities. Work is a form of personal security as well as a contribution to the household budget. Instability and insecurity of work and income opportunities are endemic among the urban poor (McGee, 1979; Rusque-Alcaino and Bromley, 1979), and reliance on only one breadwinner increases the risk of disaster. An adult male breadwinner may be the victim of theft, arbitrary arrest, or the eradication of job opportunities (Cohen, 1974), or he may choose to abandon family responsibilities and to spend heavily on tobacco, alcohol, drugs, or gambling to escape a depressing reality.

Despite all these pressures for female involvement in the labour-force, about 70 per cent of those working in the street occupations in Cali are males, at least three-quarters of these falling into the 18–55 age-range. Around 15 per cent of those working in the street occupations are aged under 18, about three-quarters of these being boys. Some occupations, for example virtually all work in transport and security services, shoe-shining, and most forms of street theft, are almost exclusively male preserves. In general, therefore, males and adults predominate in the street occupations, though females and children are numerically quite significant. Only prostitution is almost exclusively a female preserve, though women predominate in many forms of retailing, particularly the sale of fruit, vegetables, and cooked foodstuffs. Child workers are mainly concentrated in scavenging, newspaper-selling and other small-scale retailing, shoe-shining, and petty theft.

In general, women and children are especially concentrated in the least remunerative or lowest-status street occupations, and have less access to capital than men. Particular occupations are age- and sex-specific, and although this division of labour may at times be convenient, it mainly acts to reduce the range of work opportunities and the potential income available to women and children. Women and children working on the streets are much more liable than adult males to be subjected to physical threats and sexual harassment, and this factor sharply reduces the number of girls and young women in the street occupations. Children working in the streets have great difficulty in obtaining and keeping any significant capital equipment or merchandise for their occupation, and most young women without access to significant capital are aware that prostitution is potentially their most remunerative form of work.

Conclusions

This rapid summary of the characteristics of street occupations, based on the example of Cali, has emphasized the diversity of these occupations and the impossibility of applying a uniform set of policies to all street occupations. These occupations deserve a greater degree of attention and respect than they have received, and it is necessary to convert the predominance of negative policies to a predominance of more positive measures. Most potential improvements to the working conditions in street occupations are relatively inexpensive, and, as de Soto (1989) has argued for Peru, some would actually save government money by reducing the number of regulations and the costs of enforcing these regulations. There is no reason, therefore, why a 'humanization' of the street occupations should hold up vital investments in agriculture, manufacturing, and public services. An improvement of the working conditions in the street occupations will not greatly increase the number of workers in these occupations and lead to accelerated rural–urban migration if appropriate investments are made in all regions of the country, involving small cities and rural areas as well as the major metropolises.

For substantial numbers of the urban poor, working in the streets is a survival strategy. Their occupations fit de Soto's profile of 'informality' – performing useful functions and providing necessary services, but contravening numerous official regulations in the process. In many cases, such street occupations are the most viable alternatives to parasitic or anti-social occupations, or to destitution. Their negative features are reflections of much wider social *malaises* which cannot be resolved simply by regulating and persecuting the street occupations. Gross poverty and social inequality are institutionalized in Colombia, and it is unreasonable to blame the poor for their own situation and to fail to tackle the conditions which underlie their poverty.

References

Anderson, S. (1978) *On Streets*, Cambridge, Mass., MIT Press.

Bromley, R. (1982) 'Begging in Cali: image, reality and policy', *New Scholar*, 8, pp.349–70.

Cohen, D.J. (1974) 'The people who get in the way', *Politics*, 9, pp.1–9.

de Soto, H. (1989) *The Other Path: the Invisible Revolution in the Third World*, New York, Harper & Row.

McGee, T.G. (1979) 'The poverty syndrome: making out in the South-East Asian city' in Bromley, R. and Gerry, C. (eds) *Casual Work and Poverty in Third World Cities*, Chichester, Wiley.

Rusque-Alcaino, J. and Bromley, R. (1979) 'The bottle buyer: an occupational autobiography' in Bromley, R. and Gerry, C. (eds) *Casual Work and Poverty in Third World Cities*, Chichester, Wiley.

Source: Bromley, 1997, pp.124–38

READING 2D
Larissa Lomnitz: 'The social and economic organization of a Mexican shanty-town'

A common prejudice found in the sociological literature on poverty consists in portraying the urban poor as people bedevilled by a wide range of social pathologies, amounting to a supposed incapacity to respond adequately to social and economic incentives. More social scientists have directed their attention towards the material and cultural deprivation that meets the eye than towards the socio-cultural defence mechanisms which the urban poor have devised. My work in a Mexican shanty-town, as summarized in the present chapter, deals with a basic question: how do millions of Latin Americans manage to survive in shanty-towns, without savings or saleable skills, largely disowned by organized systems of social security?

The fact that such a large population can subsist and grow under conditions of extreme deprivation in Latin American cities has important theoretical implications. Obviously, the members of such a group can hardly be described as 'unfit' for urban life in any meaningful sense. On the contrary, the proliferation of shanty-towns throughout Latin America indicates that these forms of urban settlement are successful and respond to some sort of objective social need (Mangin, 1967; Turner and Mangin, 1968). My own work in Mexico City tends to support this view, by providing evidence that shanty-towns are actually breeding-grounds for a new form of social organization which is adaptive to the socio-economic requisites of survival in the city. In this essay, I show that the networks of reciprocal exchange among shanty-town dwellers constitute an effective stand-by mechanism, whose purpose is to provide a minimum of economic security under conditions of chronic underemployment …

When migrants reach the city, they normally move in with relatives. The presence of a relative in the city is perhaps the most consistent element within the migration process. The role of this relative determines the circumstances of the migrant family's new life in the city, including place of settlement within the metropolitan area, initial economic status, and type of work. There is no escaping the economic imperative of living near some set of relatives: the initial term of stay with a given kinship set may be variable, but subsequent moves tend to be made with reference to pre-existing groups of relatives elsewhere. Unattached nuclear families soon attract other relatives to the neighbourhood …

Social organization

The pattern of social organization which prevails in the shanty-town can be described as follows. Most nuclear families initially lodge with kin, either in the same residential unit (47 per cent), or in a compound arrangement (27 per cent). Compounds are groups of neighbouring residential units which share a common outdoor area for washing, cooking, playing of children, and so on. Each nuclear family in such a cluster forms a separate economic unit. Families in the compound are related through either consanguinity or marriage ties; each compound contains at least two nuclear families.

Extended families, e.g. two brothers with their wives and children, may share the same residential unit temporarily; in the case of newly married couples with the parents of either husband or wife, the arrangement may be more permanent. Any room or group of rooms having a single private entrance is defined as a residential unit: this excludes tenements of the '*vecindad*' type, consisting of a series of rooms opening on an alley with a public entrance gate, which may contain several independent family groups. Extended families contain at least two nuclear families; these share the rent expenses or own the property in common. Sometimes they also share living expenses ...

Thus, the social organization of the shanty-town may be described as a collection of family networks which assemble and disband through a dynamic process. There is no official community structure; there are no local authorities or mechanisms of internal control. Co-operation within the family networks is the basic pattern of social interaction ...

This pattern can be viewed as the outcome of a dynamic process, which depends on economic circumstances, the stage in the life cycle, the availability of housing vacancies, personal relationships with relatives, etc. The initial choice of moving in with the family of either spouse is usually an economic one. Since young husbands or wives often do not get along with their in-laws and conditions in an extended family may be very crowded, the couple tends to move out. However, new circumstances, such as the arrival of children, desertion of the husband, loss of employment, and so on frequently compel the family to return to the shelter of relatives ...

Local groups

The shanty-town is not organized around central institutions of any kind. Instead, there are several types of groupings, of unequal importance: (a) the family network; (b) football teams; (c) the medical centre; (d) temporary associations.

The family networks ... represent the effective community for the individual in the shanty-town. They are composed of members of an extended family, or a compound, but may include neighbours who are assimilated through fictive kinship. We shall see how these networks have developed into systems of reciprocal exchange of assistance, which are key to the survival of large numbers of people under the severe economic handicaps of shanty-town life.

Other forms of organization at the community level are relatively rudimentary. There are four football teams in Cerrada del Cóndor. Three of these teams belong in effect to a single large family network ... The fourth is a more recent team whose membership is recruited among young people of the shanty-town irrespective of family origins. Football teams represent one of the few vehicles of social contact between men of Cerrada del Cóndor and men who live in other parts of town. After a game, there are drinking sessions which reinforce the team spirit and friendship among members of the team.

The shanty-town's medical centre was organized and financed by a group of middle-class ladies from the neighbouring residential district, with some assistance from a nearby church. Later, the national Children's Hospital agreed to staff the centre, but this help has recently been withdrawn. In spite of the modest assistance offered, the centre has become an important part of shanty-town life. It is a place where children are welcome during most hours of the day, and when many girls and women receive guidance from an understanding social worker.

There is no local organization for solving the common problems of shanty-town life. Groups of neighbours may band together for specific issues; this has happened three or four times in the existence of Cerrada del Cóndor. The first time was to request the installation of a public water outlet. Another time, a group of women jointly requested an audience with the First Lady, in order to lodge a complaint about spillage of oil from a refinery that was causing brush fires in the ravine. These exceptional instances of co-operation merely serve to highlight the absence of any organized effort to solve community problems.

The residents of Cerrada del Cóndor have little contact with the city-wide or national organizations. Articulation with Mexican urban culture occurs mainly through work and through mass media such as radio and television. School is, of course, very important for the children. Adult reading is limited to sports sheets, comics, and photo-romance magazines. Only about one-tenth of the men belong to the social-security system. About 5 per cent are union members. In general, extremely few people belong to any organized group on a national level, such as political parties, religious organizations, and so on ...

Reinforcing mechanisms

The exchange of goods and services serves as the underpinning of a social structure: the network organization. When this exchange ceases to exist, the network disintegrates. The social structure which is erected on the basis of exchange depends on physical and social proximity of network members. Ideally, the networks are composed of neighbours related through kinship.

Actually, many networks contain non-kin members whose allegiance must be reinforced by means of fictive kinship (*compadrazgo*) and other means which will be analysed presently. Even among kin, relationships are far from secure: economic and personal differences arise frequently under conditions of extreme poverty and overcrowding. The reinforcing mechanisms to be discussed are therefore present in all networks.

Compadrazgo is widely used to reinforce existing or prospective network ties. In Cerrada del Cóndor, the *compadres* have few formal obligations towards one another. An informant says: 'When choosing a godfather for one's child one should look for a decent person and a good friend, if it's a couple, they should be properly married. They should be poor so no one can say that you picked them out of self-interest'. Among 426 *compadres* of baptism (the most important type of *compadrazgo* in the shanty-town), 150 were relatives who lived close by, and 200 were non-kin neighbours. Another 92 were relatives who lived elsewhere in the Federal District or in the countryside, i.e. prospective network affiliates. In most cases of *compadrazgo*, the dominant factors were physical proximity and kinship. This equalitarian pattern is at variance with the frequently observed rural pattern of selecting a *compadre* above one's station in life ...

The great importance of *compadrazgo* as a reinforcing mechanism of network structure is also reflected in the variety of types of *compadrazgo* that continue to be practised in the shanty-town. These types are, by order of decreasing importance: baptism (426 cases), confirmation (291), communion (79), wedding (31), burial (16), Saint's Day (13), fifteenth birthday (10), Divine Child (8), Gospels (8), grade-school graduation (4), habit (3), sacrament (2), scapulary (1), cross (1), and St Martin's (1). All these types of *compadrazgo* mark ritual or life-cycle occasions. The formal obligations among *compadres* can be described as follows: 'They must treat each other with respect at all times, and must exchange greetings whenever they meet.' Ideally some *compadres* should fulfil economic obligations, such as taking care of a godchild if the father dies; but these obligations are no longer taken very seriously in the shanty-town.

If *compadrazgo* formalizes and legitimizes a relationship between men and women, *cuatismo*, the Mexican form of male friendship, provides the emotional content of the relationship. *Cuates* (a Nahuatl term for 'twins') are close friends who pass time together, talking, drinking, playing cards or football, watching TV, treating each other in restaurants, and having fun together; above all, they are drinking companions. Women are totally excluded from the relationship. A wife 'would never dare' to approach a *cuate* of her husband's to request a favour.

Assistance among *cuates* is ruled by social distance. Among relatives, there will be more unconditional help than among neighbours. In general, the *cuates* borrow freely from one another, help one another in looking for work, give one another a hand in fixing their homes, and stand by one another in a fight. Like *compadrazgo*, *cuatismo* is practically universal: the man who has no *cuate* and no *compadre* is lost indeed ...

It is clear that these groups of *cuates* are based primarily on the male sector of the networks described above, even though neighbours, work companions, or friends not affiliated with one's special network may be included. The existence of *cuatismo* to reinforce network affiliation is evidence that the networks are not simply built around the wives and mothers, as might be supposed from a superficial analysis. On the contrary, many networks appear to be male-dominated. If networks were based exclusively on the more visible forms of daily exchange of goods and services practised by women, the substantial overlap between networks and groups of *cuates* would be rather puzzling. Networks are constituted of nuclear families as entities; all members of each family participate actively in the relationship.

Drinking relationships among *cuates* are exceedingly important and usually take precedence over marital relationships. From a psychological point of view, drinking together is a token of absolute mutual trust which involves a baring of souls to one another (... Butterworth, 1972). From the economic point of view, *cuatismo* implies a mechanism of redistribution through drink which ensures that all *cuates* remain economically equal. And from a social point of view, it reinforces existing social networks and extends the influence of networks in many directions, since a drinking-circle may contain members of several networks.

The *ideology of assistance* is another important factor in network reinforcement. When questioned, most informants are reluctant to describe their own requests for assistance; yet they are unanimous in claiming to be always ready to help out their own relatives and neighbours in every possible way.

The duty of assistance is endowed with every positive moral quality; it is the ethical justification for network relations. Any direct or indirect refusal of help within a network is judged in the harshest possible terms and gives rise to disparaging gossip. People are constantly watching for signs of change in the economic status of all members of the network. Envy and gossip are the twin mechanisms used for keeping the others in line. Any show of selfishness or excessive desire for privacy will set the grapevine buzzing. There will be righteous comments, and eventually someone will find a way to set the errant person straight …

Social networks in the context of marginality

Residents of shanty-towns such as Cerrada del Cóndor are often counted among the 'marginal' sector of the urban population in Latin America. The emergence of urban marginal populations is not, of course, exclusive to underdeveloped societies. In advanced industrial nations, such populations result from the displacement of certain social strata from the labour market through mechanization and automation of the means of production. These growing population sectors have no expectation of absorption into productive occupations, and become increasingly dependent on welfare. They represent *surplus* population (rather than a labour reserve) and are, therefore, an unwanted by-product of the system.

… this situation is considerably aggravated in underdeveloped countries, because the rate and pattern of industrial development are imposed from abroad. Economic dependence introduces a factor of instability, because of the hypertrophic growth of large industrial cities at the expense of the countryside. Easy access to sources of raw materials and cheap labour attract an overflow of hegemonic capital into formerly pre-industrial societies. As a result, (a) there is an increasing gap between 'modern' cities and 'traditional' rural areas on the verge of starvation; (b) new skills required by industrial growth are monopolized by a relatively small labour élite, while the great mass of unskilled peasants and artisans is displaced from their traditional sources of livelihood; and (c) superficial modernization has caused a sudden population explosion, which increases the rate of rural-to-urban migration, thus offsetting any efforts at promoting the gradual absorption of surplus populations into the industrial labour force. Thus, the process of marginalization is not transitional, but, rather, intrinsic to the system.

References

Butterworth, D. (1972) 'Two small groups: a comparison of migrants and non-migrants in Mexico City', *Urban Anthropology*, vol.1, no.1, pp.29–50.

Mangin, W. (1967) 'Latin American squatter settlements: a problem and a solution', *Latin American Research Review*, vol.2, no.3, pp.65–98.

Turner, J.C. and Mangin, W. (1968) 'The Barriada Movement', *Progressive Architecture*, May, pp.154–62.

Source: Lomnitz, 1997, pp.204–17

CHAPTER 3

Walled cities: surveillance, regulation and segregation

by Eugene McLaughlin and John Muncie

1 Introduction

Even the most poetic representations of 'the open city' acknowledge that the diversity and intensity of the central inner city are capable of also producing disorientating, disconcerting and painful lived realities. As we have seen in the last chapter, there are fearful 'shadow' representations of the city, in different locations and times, as a breeding ground for crime, decay and disorder. Contemporary celluloid representations of the city as a storm-darkened, post-industrial carceral, or as an 'urban jungle' in the grip of dangerous strangers, often seem anything but products of the paranoid imagination of science fiction authors. Such totalizing dystopian visions of the city as a place of fear and insecurity need, however, to be continually read against the attractions of the city as a place of pleasure, entertainment and intensified stimulation.

This chapter explores the nature of this paradox by focusing on how order and disorder are inscribed in and produced through various spatial reorderings of the city. Demands for orderly and stable spatial formations and social processes are frequently generated by perceptions of the city as an unruly 'crime-scape'. Yet the production of ordered space, which at core represents an attempt to *de-intensify* the points and sources of urban friction, gives rise to a host of related questions. For example, do particular forms of spatial ordering pose a serious threat to the ideals of open access and freedom of movement? Does 'putting people in their place' exacerbate fear and anxiety by imposing rigid distinctions of orderly/disorderly on otherwise diverse and plural social groups? More broadly, is the distinction between order and disorder clear-cut?

The central arguments of this chapter are that:

- all forms of order, and indeed disorder, are relational and in the process of being constantly renegotiated, challenged and reconstituted;
- the boundaries of order and disorder are historically, socially and spatially contingent;
- order and disorder are ambiguously interconnected so that order creates disorder and new disorders create new reorderings.

The contemporary histories of many cities are littered with examples of how attempts to produce order have paradoxically intensified the processes of disorder. For example, the differentiation of urban space, represented by the building of (real and metaphoric) walls to contain and control particular forms of disorder, entrench particular relations of power which enhance specific perceptions of safety and marginalize or discipline certain groups, recasting them as the undesirable 'other'. Initiatives promising to banish disorder create new versions of dangerous people and dangerous places.

The uncomfortable reality is that the complexities of urban life themselves generate multiple relations and practices of spatial domination and resistance. This chapter explores these themes by:

- highlighting what it is about the city that both historically and currently brings forth the double movement of intensification and de-intensification of social relations (section 2);
- considering the pivotal role played by images of 'eyes on the street' in cementing particular notions of 'order' and 'disorder' (sections 2 and 5.2);
- examining the alternative 'flows of order' produced by the rebuilding of city walls, whether in the form of the entire remapping of cities such as Paris and Rio de Janeiro in the nineteenth and early twentieth centuries (section 3), the building of residential and commercial fortified enclaves in such mega-cities as Los Angeles and São Paulo in the late twentieth century (section 4), or enhancing means of spatial surveillance through new forms of 'beat' policing in New York and the gaze of surveillance cameras in UK cities (section 5).

Whilst it may be tempting to interpret contemporary forms of ordering through surveillance, enclosure, segregation, regulation and so on as somehow novel and unique, this chapter attempts to contextualize such developments historically. In particular, important precursors to the present can be found in the spatial disorganization and reordering that engulfed cities in the nineteenth and early twentieth centuries. These 'swarming' modernist discourses and techniques of spatial reordering have left deeply etched marks in cities around the globe and, as we shall see, continue to seep into our present.

2 Safety, insecurity and street life: the company of strangers and 'eyes on the street'

Let us begin by reminding ourselves about why people feel safe or insecure in cities. As Steve Pile noted in Chapter 1, coexisting in close quarters with large numbers of people from completely different walks of life whom one does not know is perhaps the defining 'felt intensity' of city life. The sheer volume of potential and actual face-to-face encounters generated by the flow of strangers in a city's public spaces – that is, 'those areas of the city to which, in the main, all persons have legal access … the city streets, its parks, its places of public accommodation … its public buildings or the "public sectors" of its private buildings' (Lofland, 1973, p.19) – pose particular problems because there appears no choice but to 'press ourselves upon others without a hierarchy of established recognition' (Sennett, 1990, p.17). Risk, uncertainty, exposure, self-consciousness and vulnerability are inevitable and unresolvable aspects of our being in city spaces which are perpetually in motion. We can never completely suspend doubt about the motives, intentions and desires of others. As a result, there is a *paradox* implicit in the tangled web of predominantly transient social encounters and bodily contacts characteristic of city life: we must hold the 'double tension' of 'within' (self-trust and self-doubt) and 'without' (trusting and doubting others).

City dwellers are therefore involved in a delicate balancing act of assessing and negotiating the fears, dangers, risks and pleasures associated with the presence of strangers in public places. Very obvious negative risks range from loss of time, having to help out, generating obligations that one does not want, through to possible damage to one's self-esteem and assault on one's identity and person. There are many moments, of course, when we are attracted to and want to share crowded public places precisely because we are seeking the excitement and intensity associated with the possibility of risk-laden, unpredictable encounters. As Charles Jencks (1993, p.18) notes, the urban self is partly constructed and defined 'through interaction with others, by trying out new languages and attitudes; in short by entering into new social situations'.

Lyn Lofland (1973, p.22) has pointed out that city dwellers must, of necessity, transform 'a potentially chaotic and meaningless world of strangers into a knowable and predictable world of strangers'. We must, in effect, *de-intensify* our relations with others. This is accomplished by acquiring and routinely deploying a range of highly refined scanning skills, defined by Goffman (1971, p.238) as 'smells, sounds, sights, touches, pressures' which allow us to read, interpret and negotiate the multiple encounters and situations that the city affords. Our cognitive understandings of public spaces are continually being remade in order to:

- create defensible private spaces in the midst of crowded streets;
- 'read' the character and intentions of others; and
- maintain the social and psychic 'self' in the context of falling under the regulatory, and potentially disciplining, gaze of anonymous others and being 'read' and commented upon.

The city dweller has to trust in the kindness of strangers and believe that strangers will behave appropriately or with civility, respecting each other's 'right to be' in a public place and trust that if a public encounter goes wrong, passers-by will not just passively observe but act and intervene to protect that 'right to be'. Our sense of safety and insecurity in the city, we would argue, is not a given but a set of spatially situated relational understandings and reflective judgements that have to be constantly renewed in daily encounters. In order to avoid or manage difficult situations and encounters – to keep a safe distance – we have to take each other into account. Hence we enter into an intricate social contract, or what Lofland defines as 'a whispered enjoinder', that enables us to safeguard each other so that we can carry on the business of living. In public places we have to learn, therefore, to hold a delicate balance between intimate anonymity, public privacy, involved indifference and distant interdependence (Karp *et al.*, 1991). Hence the ultimate goal for many people in public places is minimum involvement but maximum reassurance. In sum, intricate 'streetwise' scanning skills and mapping strategies and a 'protective' wall-like skin need to be acquired and nurtured if we are to order and represent ourselves in public spaces, relatively free of unwarranted fear.

It is important here to note Elspeth Probyn's (1995) point that space presses upon bodies differently. Distinct social categories of people, depending on gender, sexuality, age, 'race', ethnicity, caste, religion or physical ability, have different perceptions and different ways of representing themselves and managing the risks and social controls associated with being in cities (see Chapter 5). For example, in many cities attachment to street life is greatest among the groups of young men who make disproportionate use of public spaces. By hanging out at the shopping arcades, congregating on street corners, using public squares as cruising grounds, playing soccer and so

FIGURE 3.1 *People in public places desire minimum involvement but maximum reassurance*

on, they physically mark certain areas as 'their' territory. As a consequence of such practices associated with heterosexual street masculinities, women, particularly unescorted women, are acutely aware of the attention they are likely to attract when passing through these streets and areas at particular times of the day and night. Indeed, researchers have found that the sense of insecurity and vulnerability is so strong that these 'dangerous' streets and areas are avoided whenever possible. Those unaccompanied women who assert their right to spend time in public spaces at certain times risk unwanted male attention and adverse comment. Thus women's ordering and embodiment of self is constructed to seek out the safest and most secure public spaces in many cities (Fincher, 1990; Valentine, 1992).

ACTIVITY 3.1 Turn now to Reading 3A, 'The uses of sidewalks: safety' by Jane Jacobs. Jacobs' observations of the movements in street life in New York in the 1950s exemplify the continual interplay of such themes as order/disorder, friends/strangers and safety/insecurity. As you read the extract, focus on the following questions:

- Why is the ordering of street life complex?
- How does she conceptualize surveillance?
- What roles do strangers play on the streets? ◆

The key to Jacobs' argument is her belief that the 'bedrock attribute' of a humane and compassionate urban civilization is one of providing both safety for and from strangers. Safe public spaces manage to produce a network of relationships which balance people's right to privacy and their coexistent desire for degrees of contact, excitement or assistance. Public spaces must have the 'magical' capacity to bring strangers together and transform their presence into 'a safety asset'. Safety on the streets and in urban neighbourhoods can only be ensured (according to Jacobs) through 'natural' surveillance – the 'eyes on the street' – mutual inspection, and the 'soft' ordering of our own and each other's behaviour. The safety of the street works best when people are using the city streets voluntarily and are least conscious that they are policing or being policed (Jacobs, 1961, p.46). (You might, at this point, want to look back at Chapter 1 to remind yourself of the discussion there of this quote from Jacobs.)

For Jacobs it is how public spaces are conceived, used and watched, not the location, that matters. Streets with a good mix of stores, offices, bars, clubs, cafés and so on which are in fairly constant use, work in mutually supportive, interconnected ways to enhance public safety and trust by attracting the kind of public attention, interaction, circulation and watching and listening that keeps them naturally safe. Physically diverse areas ensure that:

- a variety of people have reason to be on and use the streets;
- the street is 'travelled and peopled' in an interconnected and routine manner;

- the owners of commercial enterprises – the 'sidewalk guardians' – have a vested interest in ensuring the 'good order' and respectability of the streets;
- a lively, convivial street-life acts as an attraction for others who may either want to connect with the action or just watch. Hence its very success as an active collectivizing space widens the circle of users.

One of the important points made by Jacobs is that closely grained diversity, density and a degree of benign disorder associated with the presence of strangers are crucial for the vitality and safety of public places: 'City areas flourishing in diversity sprout strange and unpredictable uses and peculiar scenes. But this is not a drawback of diversity. This is the point or part of it. Cities have the capacity of providing something for everybody, only because, and only when, they are created by everybody' (Jacobs, 1961, pp.250–1).

Jacobs is also only too aware that cities and areas within cities differ considerably in their capacity to facilitate and manage safe interactions among strangers. Indeed, she stresses that the breakdown of trust between strangers and the heightening of a sense of personal insecurity and vulnerability can be disastrous for a city. Richard Sennett (1990, p.xiii) argues that the knee-jerk response to insecurity in many cities is an attempt to 'wall off differences between people, assuming that these differences are more likely to be mutually threatening than mutually stimulating'. In the rest of this chapter (and in Chapter 4) we examine some of the different forms of walling that take place when relations of trust between strangers break down and when the gaze of 'eyes on the street' is perceived as indifferent or dangerous.

3 Building walls I: the modernist project

Lewis Mumford (1961) has detailed the crucial role that walls and gates have played in the development of urban settlements. The establishment of physical walls was a way of securing, defending and *projecting* space, while strong, effective fortifications generated a feeling of safety, unity and sense of territorial identity. In the medieval city, for example, walls came to embody not just spatial enclosure but the concurrent processes of inclusion and exclusion. In many instances, citizens residing within the defensive walls enjoyed governance, culture, commerce and protection long before nation states were imagined. Those residing outside the city walls were 'outlaws' – excluded from the obligations and privileges of urban life.

However, the industrial revolution effectively ended the symbolic potency of city walls and gates in many parts of the world. Technological advances in military weaponry meant that no matter how well fortified city walls were, they could be breached. It is also clear that the unravelling of the tightly knit urban fabric generated by the coming of a new socio-economic order meant that the threat to the existing urban order was more likely to originate inside rather than outside the city walls. It is not surprising, therefore, that from the early nineteenth century a multitude of new walls and gates – physical and metaphorical – were built, not around, but *inside* the internally divided modernist city.

3.1 SEPARATION AND DISTANCE: PARIS AND RIO DE JANEIRO

The multiple contradictions associated with the coming of the nineteenth-century city called into question the preconditions of free circulation and predictable economic exchange, and generated new forms of spatial differentiation in an attempt to sort out who and what belonged where. This section begins with a brief examination of how nineteenth-century Paris and Rio de Janeiro were radically reordered in an attempt to de-intensify the urban crises associated with their most dramatic moments of growth.

Paris constitutes an appropriate point of departure for several reasons. In the course of the nineteenth century the city was transformed into the first modern European metropolis, and became the 'paradigmatic city' to be emulated by urban planners in capital cities in various parts of the world right up to the present day. Between the 1850s and 1870s, thanks to Baron Haussmann, Emperor Napoleon III's master planner, Paris acquired a modern physical infrastructure in the form of new residential and public buildings, effective systems of sewage disposal and water supply, public lighting, public parks,

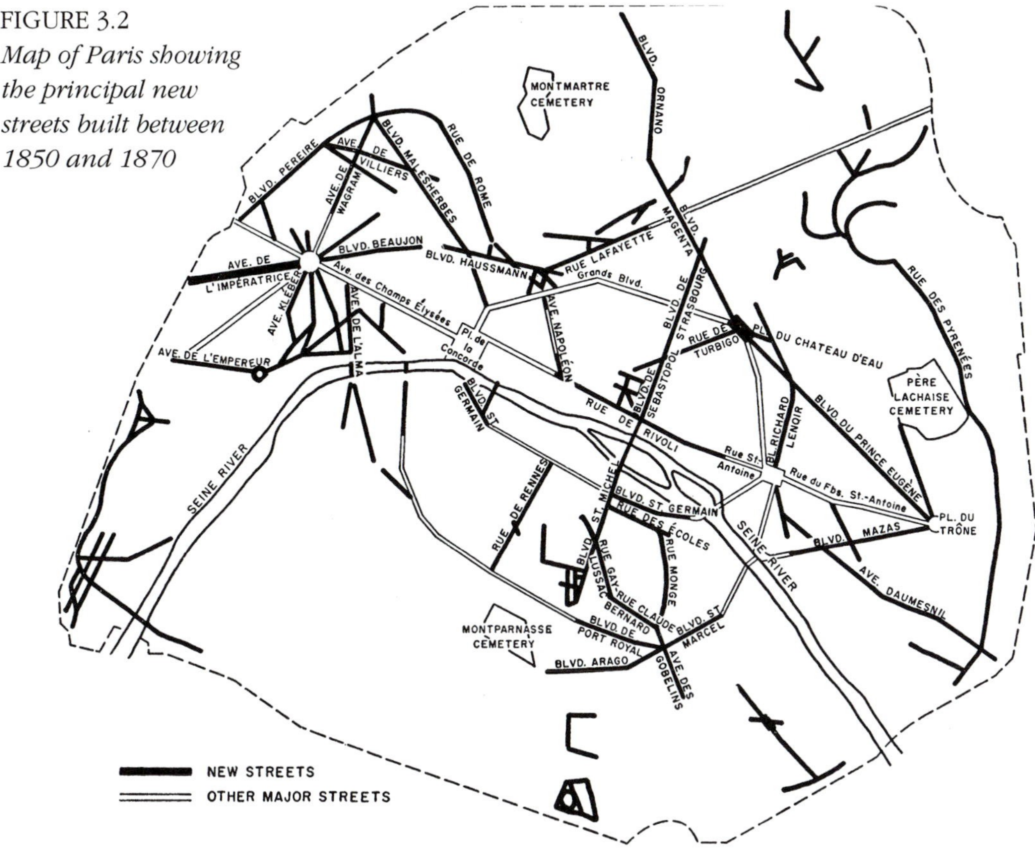

FIGURE 3.2
Map of Paris showing the principal new streets built between 1850 and 1870

market-places and new railway-stations. Haussmann forcefully completed the transformation of Paris that had begun during the Restoration by driving a network of wide, straight boulevards with expansive pavements through the economically and socially integrated historic core of the old medieval city.

The stated intention behind the wholesale reordering of the urban status quo (which effectively created the Paris we see today) was to construct a spatially integrated city that would allow its inhabitants and traffic to circulate rapidly both within and through neighbourhoods. The expanded urban scale and the liberation, cleansing and reordering of key cultural spaces were the key reasons why foreign visitors gazed on *Nouveau Paris* as a unique work of art:

> In the years following Haussmann's renovations, the boulevard became the symbol of modern Paris. City life was essentially a public life, and the street was the stage on which the urban drama was played. All the variety and vitality of Paris – its social range, its material abundance, its sense of fashion – seemed to be visible in the streets. To artists the boulevard was an abundantly stimulating subject; probably no other city sat for its portrait as often as the French capital.
>
> (Evenson, 1979, p.21)

FIGURE 3.3 *The new boulevards:* Paris Street; Rainy Day, *a painting by Gustave Caillebotte (1876/7, oil on canvas, 212.2 by 276.2 cm)*

Other commentators have been more ambivalent in their reading of Haussmann's reordering, insisting that he transformed Paris into an 'introverted' city by:

- delineating concentrations of commercial, business, manufacturing, administrative and entertainment areas;
- socially segregating residential neighbourhoods; and
- replacing integrated, semi-public spaces with privatized ones.

The differentiation of urban spaces (which had previously carried parallel public and private, formal and informal activities) and the de-intensification of social relations was central to the bourgeoisie's attempts to develop its own economic power and social hegemony. The spatial reordering in conjunction with a highly rigid regulatory framework enabled the bourgeoisie to erect new walls of cultural distance – in dialect, dress, manners, eating habits, leisure and consumption – which marked it off as different from, and superior to, the plebeian lower orders.

Haussmannization or 'boulevardization' facilitated the emergence of enclosed and semi-enclosed 'interior streets', such as new glass-covered galleries, arcades and department stores (pre-figuring the shopping malls we shall be discussing in section 4.2), which allowed the bourgeoisie to avoid unwanted contact with the lower classes. Paris was civilized and purified in those illuminated 'cultural corridors' where the bourgeoisie 'presented' themselves to enjoy refined perception, taste and performance. This, by definition, meant that the poor – the transgressive non-citizens who would offend the sensibilities of urban modernity – had to be *forcibly disconnected* from these public spaces and quarantined elsewhere. The reordering of the city was also crucial to the construction and normalization of particular gendered zones of access and visibility. As Hooper has noted,

> Female factory workers and department store clerks, housewives and matrons, female utopianists, socialists and revolutionaries, bourgeois women whose first 'respectable' unescorted excursions into public space were into department stores – all these women, in addition to prostitutes, had their movements throughout the city restricted, controlled and monitored by what masculinist power produced. Wherever they moved, inside or outside the borders of bourgeois respectability, women performed, concretized in the flesh, their very being, the abjected, excluded other of hegemonic masculinist order.
>
> (Hooper, 1998, p.244)

Walter Benjamin (1968) goes so far as to insist that it was the incessant clash of alternative conceptions of social order that lay at the core of Haussmann's grand plans. Barricades had been thrown up by the Parisian working class on eight occasions between 1827 and 1848, and the authorities were only too aware that the tangle of narrow streets and tightly packed tenements of central Paris were the epicentres of a revolutionary threat to the dominant social order. By breaking up working-class *quartiers*, eliminating the volatile street life that had

FIGURE 3.4 *Making way for* Nouveau Paris*: slum clearance in Rue Champlain*

occupied the heart of the old city and by displacing thousands of people to the northern and eastern margins of the new city, the administrative heart of Napoleon III's Second Empire could be secured against serious civil disorder. The cutting of the boulevards:

- made the erection of 'traditional' barricades more difficult;
- constructed more effective sightlines for the military;
- created direct routes to facilitate the rapid movement of troops between the newly established barracks and the now clearly defined working-class areas.

Nouveau Paris was to be defined not by the insurrectionary actions of a subordinate working class on the city streets, but by a dominant bourgeois conception of social order as represented by the boulevards (Rabinow, 1989).

Variants of boulevardization swarmed across Europe, North America, the cities of the French Empire and Latin America. Intense spatial struggles also accompanied the deliberate attempt to reorder the fundamental character of those cities subjected to Haussmannization. Rio de Janeiro's race-conscious élites were well aware of the contradictory strains pulling and pushing the capital in different directions (Needell, 1989; Meade, 1997). Threatening the élite's white European future were the congested and unregulated tenements or *corticos*, which stood in intimate proximity to respectable and commercial areas. They were spatially stigmatized as places of disorder and disorganization. A wide range of social pathologies – vice, crime, infectious disease and visible poverty – supposedly associated with their inside world were defined as a danger to the outside world of commerce, trade and high culture.

For Rio's élites, the *corticos* also symbolized an alternative ordering of Brazil's capital city because they, along with the unregulated shanty towns, or *suburbios*, on the edges of the city were home to an ever-increasing multitude of poor European immigrants and former plantation slaves. The collapse of the old forms of social order and racial boundaries associated with slavery resulted in racial intermingling, cross-class socializing, permissive sexual unions and an open-ended African-Brazilian street culture consisting of raucous entertainment, all-night cabarets, and the disturbing proto-political meanings embedded in the *entrudos* (wild pranks) and *cordoes* (rowdy processions) associated with *Carnaval*. For the europhile élites, these intensifications signalled that Rio de Janeiro stood at the edge of darkness and foretold its future either as a 'black city' or a hybrid city of many hues.

With a sense of urgency and purpose, from 1902 Rio de Janeiro was subjected to an ambitious, highly racialized programme of urban modernization and public health reform that was intended to erase the sights, sounds and smells associated with the popular customs and practices of 'Old Rio' and silence any enunciation of a non-European urban future. By 1906, new wide and straight thoroughfares, consciously based on Haussmann's design principles, were cut through the *cidade velha* and, in conjunction with the extension of the electric trolley lines, connected the central district with the outlying docks and the wealthy areas of *zona norte*. In order to free up the downtown area, the 'unsafe'

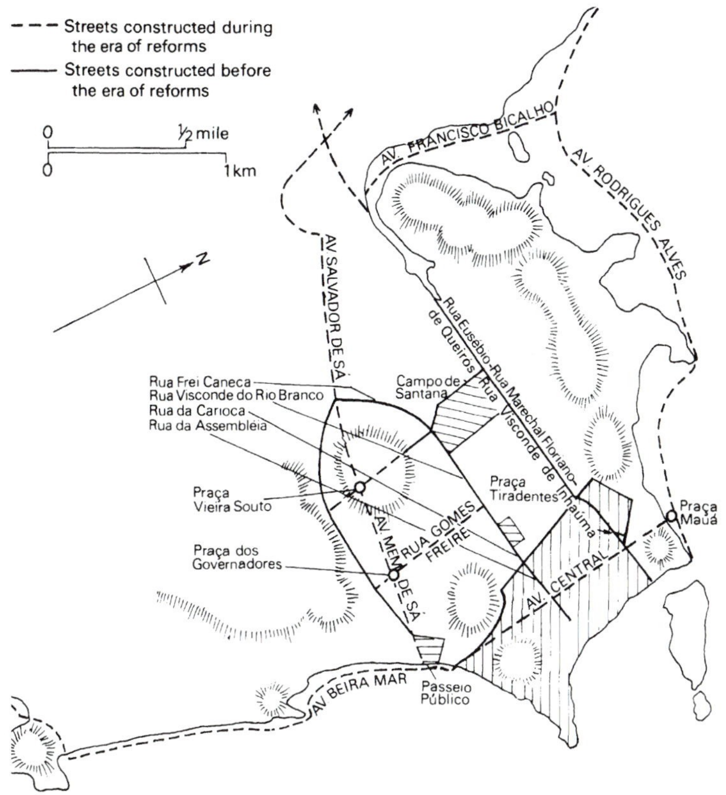

FIGURE 3.5 *Plan of Rio de Janeiro, showing the impact of the 1902–6 reforms*

FIGURE 3.6 *The wide boulevard of Avenida Centrale in Rio de Janeiro (later renamed Avenida Rio Branca)*

and 'unsanitary' *corticos* (like the working-class areas of central Paris) were demolished, and the majority of their inhabitants were displaced to the overcrowded peripheral northern suburbs. The downtown area of *belle époque* Rio was to be reserved for the europhile local élite, foreign merchants and visitors. New by-laws were also enacted to outlaw, for example, hawking food on the streets, spitting in public, the presence of certain animals, certain forms of architectural styles and street festivals. The opulent, safe Avenida Centrale boulevard was symbolic of the newly frenchified Latin American city.

The following important points follow from our discussion of Paris and Rio de Janeiro:

- The *intensities* associated with urban modernization in the nineteenth century produced particular class, racial and gender conceptions of order and disorder which were expressed spatially in the built environment.
- The differentiation of urban space produced by boulevardization attempted to *de-intensify* social relations in a variety of urban settings and produced 'flows of order' premised on the processes of separation, exclusion and regulation.
- Formal and informal segregation did not eradicate the subordinate 'other', their links to the city centre, or their distinctive cultural value systems. Indeed, their displacement to the periphery only heightened bourgeois fears because ghettoization inevitably generated new opportunities for collective unity and group identification. As Chapter 4 makes clear, the greater the freedom from official surveillance, the more that novel strategies of resistance can be developed.

Building walls II: postmodern cantonments

Since the early 1980s, radically new and complex logics of segregation and displacement have emerged in mega-cities through representations of dramatic increases in violent crime, immorality, chronic insecurity and extreme conflicts and struggles associated with the intensification of material inequalities and/or racial, ethnic or cultural divisions. The intensified fears associated with unwanted or unpredictable juxtapositions of different people have produced a crisis of confidence in public spaces and resulted in the reordering of the heart and environs of many cities. In an ever-increasing number of global contexts, the middle and upper classes in cities are opting, like their nineteenth-century counterparts, to live, shop and work in privately guarded, security-conscious, fortified enclaves. The latest attempts to de-intensify urban space are producing the emergence of postmodern 'fortress cities' riddled with sharply demarcated privatized walled and gated enclaves (see the discussion of these elsewhere in the series: **Allen, 1999**, and **Amin and Graham, 1999**).

4.1 GATED COMMUNITIES AND THE PRODUCTION OF DEFENSIBLE SPACE

The term 'no-go area' was once used to characterize only poor inner-city ghettos or shanty towns which were beyond the writ of the central state and where the police would not enter except in force. However, at the beginning of the twenty-first century, it can perhaps be better applied to an ever-increasing number of urban neighbourhoods in an ever-increasing number of cities which are opting for *voluntary ghettoization* and self-segregation.

The following discussion draws upon and seeks to develop Blakely and Snyder's (1997) analysis of three different forms of gated and walled residential communities: 'lifestyle', 'prestige' and inner-city/suburb 'security zone' gated communities. These residential formations are spreading rapidly across certain parts of the USA because the daily catalogue of crime and disorder is perceived to have crept through the invisible but supposedly impenetrable walls and unspoken boundaries that separated respectable neighbourhoods from the 'mean streets' of inner-city ghettos. It is worth noting that in South Africa too, after the racial restrictions on residential housing were repealed, black professionals began to leave the townships for the formerly white suburbs. Their arrival triggered a flight by whites to US-style gated communities (see also Chapter 4).

'Lifestyle' gated communities, first built in the 1970s, deployed walls and gates to provide not only secure residential accommodation but exclusive access to the all-weather leisure activities and amenities within. They attracted those who wanted to distance themselves completely from the city, desired a socially

homogeneous, ordered and manicured environment, and were willing to pay for separate private services and amenities. Some 'lifestyle' communities grew so popular that they were transformed into self-contained neo-medieval cities, with their own extensive shopping facilities and commercial office developments.

A second generation of *'prestige' gated communities* developed from the late 1980s. These employ increasingly sophisticated high-security walls, electrified fences, remote-controlled gates and audio-visual screening devices to project an image of high-status exclusivity and signify a protected luxury lifestyle and secure property values. They can feature elaborate and ornate gatehouses, monumental entrances and sweeping driveways, and are designed to take advantage of a 'natural' attraction, such as the 'original' purpose of the development, renovated communal courtyards, landscaped waterfront locations and/or stunning views. Unlike 'lifestyle' gated communities, these luxury developments are located in close proximity to cities, or perhaps more significantly can be positioned on a 'stand alone' basis within gentrified or even run-down and derelict but highly valued parts of the inner city. Property developers have gone to great lengths to stress that the security-conscious architectural design will allow residents of these prime-site 'urban villages' and 'city squares' to enjoy all the excitement and buzz of an urban lifestyle and a high degree of spatial insulation without the aggravation of living in the city. Indeed, some of the gated apartment blocks sell the idea of being able to enjoy their space as if in other places and times (past and future). Thus, we are left with the strong impression that these residential developments are international enclaves bearing a remarkable resemblance to the imperial cantonments of the nineteenth-century colonial city. Indeed, the residents of enclaves in London, Los Angeles, São Paulo, Cape Town, Guangzhou and Istanbul, because they are networked into global economic flows and corresponding lifestyles, are in

FIGURE 3.7 *A heavily fortified house in a suburb of Johannesburg*

certain respects culturally closer to each other than to their own cities. Teresa Caldeira (1996), for example, notes how the glossy advertisements for fortified condominiums in São Paulo in Brazil present them as an opportunity for increasing numbers of the middle classes to construct and defend 'First World' lifestyles in 'Third World' cities. (See also Chapter 2, section 4.1 on similar developments in Istanbul.)

Inner-city and inner-suburb *'security zone' gated communities* are the third form of gated community identified by Blakely and Snider, who suggest that the origins of these particular 'walls and gates' lie in the notion of 'defensible space' developed by the American architect Oscar Newman. Newman's (1972) research in Chicago and New York suggested that the differential risks of crime and disorder that exist within and between different urban neighbourhoods were the direct result of planning and design decisions which built high-density housing developments at the lowest cost and with little consideration of the social consequences. He was particularly scathing of the high-rise modernist tower block, which he argued was a prime example of a residential development that had no 'defensible space': that is, the mix of 'real and symbolic barriers, strongly defined areas of influence, and improved opportunities for surveillance' that combine to bring an area under the control of its residents (Newman, 1972, p.3). Newman's solution to design out crime and create secure urban residential environments was fourfold:

1 To enhance territoriality by subdividing neighbourhoods into semi-private 'zones of influence' to discourage outsiders and to encourage residents to identify with and defend their areas.
2 To increase natural surveillance and the number of eyes on public spaces by positioning windows so that residents could survey the exterior and interior public areas of their environment.
3 To improve the image of the immediate environment by redesigning residential buildings to avoid the stigma of low-cost or public housing.
4 To enhance residents' safety by placing new public housing projects within 'safe' parts of the city.

Newman's 'common-sense' emphasis on the importance of the physical elements of residential space and the territorial values and interactions produced by these particular elements was influential in a number of ways. Certain architectural design features, such as introducing access control systems, improving external lighting and lines of visibility, reducing anonymous walkways, enclosing open and green spaces, increasing pedestrian access and the flow of residents, and the resetting of windows to allow for greater surveillance, have been implemented in numerous urban housing projects across Europe and North America. Newman's estimate that much crime and disorder can be curtailed by reducing the spatial opportunities for its committal has also been widely influential in a battery of 'target-hardening' measures to prevent crime through environmental design by employing locks, bolts, safes, guard dogs, security screens and alarms in a multitude of public and private places (Clarke, 1993).

Because the complete closure of all entrances to poorer neighbourhoods by 'walls and gates' is often architecturally impossible, residents of cities in the USA have recombined space and security through the erection of barricades and by pressurizing the municipal authorities to close off, 'cul de sac' or 'calm' as many streets as possible to restrict the movement of 'outsiders' – drug dealers, burglars, gangs, prostitutes, joy riders and so on. In Belfast, residents of republican and loyalist working-class neighbourhoods barricaded off the entry streets to their territory in an attempt to de-intensify inter-communal violence through the establishment of not just 'clean' sectarian borders but ethnically pure enclaves. Eventually the authorities were forced to acknowledge this particular form of spatial apartheid and agreed to build permanent 'peace walls' at symbolic and strategic interfaces to disconnect the two communities.

The strength of the logic of defensible space is such that once a neighbourhood's boundaries have been 'cleaned up' and streets taken out of public use, entry controls can be maintained through a variety of surveillance strategies ranging from increased public policing, through employing private security guards, to volunteer 'Street Watch', vigilante or para-military patrols. Such neighbourhoods have also been enthusiastic supporters of dusk-to-dawn curfews, of identity checks on young people to deter delinquency, gang formation, loitering, truancy, under-age drinking and joy riding, and of the introduction of legislation to evict neighbours considered to be 'anti-social'.

ACTIVITY 3.2 From what you have read so far in this chapter, for whom do you think 'walls and gates' work? Are they a good thing? Take a little time to note down the reasons for your answer. ◆

Although commentators agree that the proliferation of 'walls and gates' is generating a hyper reordering of certain cities, they disagree on the implications of this shift for urban citizenship. Communitarian-type arguments, stressing the positive 'greater good' aspects, maintain that:

- Walls and gates can bind residents together, producing a spatial consciousness and new forms of highly localized 'small town' politics of place and common purpose.
- They provide residents with a very real sense of self-determination – they control and manage their neighbourhoods and localities.
- In heterogeneous cities, walls and gates may be the only realistic way of protecting, negotiating and living with the conflicting rights associated with incompatible social, cultural and ethnic differences. Hence, walls can act as a form of settlement and stability in an unsettled and unstable urban environment.
- They are inclusive in structure, with residents enjoying legally established rights and responsibilities.
- Behind the walls of these safe havens residents can develop a heightened sense of belonging and new networks of sociability because they are more

likely to interact with those whom they can now assume to be like-minded in safe 'open' spaces.

- The walls ensure that residents can enjoy the ultimate urban liberty – that is, being able to feel secure in their homes and feel confident about the safeness of their streets.

Those critical of this enclosure and privatization of public space argue that it is no longer a response, but contributes to and deepens class/racial/ethnic segregation and polarization. Consider, for example, Caldeira's (1996) argument that:

> In cities where fortified enclaves produce spatial segregation, social inequalities become quite explicit ... in these cities, residents' everyday interactions with people from other social groups diminish substantially, and public encounters primarily occur inside of protected and relatively homogeneous groups. In the materiality of segregated spaces, in people's everyday trajectories, in their uses of public transportation, in their appropriations of streets and parks, and in their constructions of walls and defensive façades, social boundaries are rigidly constructed. Their crossing is under surveillance. When boundaries are crossed in this type of city, there is aggression, fear and a feeling of unprotectedness; in a word, there is suspicion and danger. Residents from all social groups have a sense of exclusion and restriction. For some, the feeling of exclusion is obvious as they are denied access to various areas and are restricted to others. Affluent people who inhabit exclusive enclaves also feel restricted; their feelings of fear keep them away from regions and people that their mental maps of the city identify as dangerous.
>
> Contemporary urban segregation is complementary to the issue of urban violence. On the one hand, the fear of crime is used to legitimate increasing measures of security and surveillance. On the other, the proliferation of everyday talk about crime becomes the context in which residents generate stereotypes as they label different social groups as dangerous and therefore as people to be feared and avoided. Everyday discussions about crime create rigid symbolic differences between social groups as they tend to align them either with good or with evil. In this sense, they contribute to a construction of inflexible separations in a way analogous to city walls. Both enforce unforgiving boundaries. In sum, one of the consequences of living in cities segregated by enclaves is that while heterogeneous contacts diminish, social differences are more rigidly perceived and proximity with people from different groups considered as dangerous, thus emphasizing inequality and distance.
>
> (Caldeira, 1996, p.352)

Critics also argue that the aggressive territorialization embedded in such fortified enclaves

- represents a narrowing or breaking of traditional notions of urban citizenship and governance;
- produces extreme forms of insular subjectivity;

- generates a less balanced view of trust and risk and fewer opportunities to cross boundaries;
- allows property developers and transnational private security companies to play on people's fears in order to sell them a false sense of security;
- normalizes a paranoid attitude towards strangers where cloistered communities define themselves by what they are against, and sanction racist discourses of 'outsiders' and the 'criminal other'.

There is also a very real possibility that 'walling and gating' can destabilize and disorder neighbouring localities through spatial *displacement*: that is, the movement of criminal and anti-social activity from protected to unprotected parts of the city. For example, in cities around the world, resident campaigns to drive prostitutes, pimps, drug dealers and customers off their streets and to stop their neighbourhoods becoming unofficial 'red light' areas or 'tolerance zones' appear to be offset by increases in prostitution in adjacent areas (see **McDowell, 1999**).

Finally, there is a distinct possibility that urban problems and discarded populations will be concentrated and contained in those parts of the city or its hinterland with the least political and economic power. Researchers have recorded instances where the fencing off of 'white' working-class neighbourhoods in various US cities has meant that adjacent African-American and Hispanic neighbourhoods are being turned into 'disconnected' hyper-ghettos (see Wacquant, 1997, and Chapter 2, section 5). In these 'carceral ghettos', '*zones à risque*', 'symbolic locations', 'reservations' and 'bantustans' there is the possibility that the mantle of ordering will be taken over by criminal gangs (see also Chapter 4). For others, living in these marginalized areas, outside of crucial socio-economic networks and without the privileges of legal order, any semblance of communal commitment will be lost.

ACTIVITY 3.3 Turn now to read Reading 3B by Mike Davis on life in the carceral ghettos of Los Angeles in the 1980s. Compare his dystopian vision of street life to that offered some forty years earlier by Jacobs (Reading 3A). Which do you think provides the most accurate portrayal of the future? Are things necessarily and systematically getting worse? ◆

4.2 THE FORTIFIED TOWERS OF THE SHOPPING MALL

During the 1980s and 1990s, diversified shopping malls became a dominant force in retailing, replacing many traditional shopping streets and thoroughfares and drawing business offices, leisure facilities and retail outlets into privately regulated and controlled environments. The first regional shopping centre was built in Seattle in 1950, while the first shopping mall, the Southdale Centre, appeared in the Minneapolis suburb of Edina in 1956 as a response to the volatile weather conditions of the American Mid West. Its designer, Victor Gruen, specifically modelled it on nineteenth-century European covered arcades, such as the Galleria Vittorio Emanuel in Milan. In the 1980s, the inner faces of shopping malls were dramatically remerchandised and repackaged in Japanese cities and the city states of Hong Kong and Singapore as the ultimate spatially condensed pleasure domes. They subsequently reappeared in hyper-form in North America, Europe, Australia and Latin America, with certain mega-malls looking like medieval walled cities.

Commentators such as Kowinski (1985), in seeking to account for the runaway commercial success of the new wave of shopping malls, have concluded that it resides in the coming together in space and time of a number of cultural, demographic and social factors. Architectural designers imploded traditional conceptions of urban form by managing to gather all the social amenities and shopping experiences of the 'traditional' downtown city street in the suburbs by playing with space, light, time, representation and perceptions of safety. City streets, squares, plazas and markets, in many respects the defining sites of urban circulation and exchange, could be carefully miniaturized within the walls of the shopping centres and reimagined as idealized public spaces free from the inconvenience of the weather, traffic pollution and the unwelcome juxtapositions constituted through the presence of poor people or 'threatening' ethnicities.

Crawford (1992) and Davis (1992) argue that urban planners found that they could copy the architectural epitome of US suburban values and lifestyles in virtually any part of the city and construct them to meet a variety of local conditions. Certain property developers in the USA also realized that they could utilize 'target-hardening' and 'brutalist' architectural techniques to re-enter abandoned or high-risk inner-city retail markets (Davis, 1992).

ACTIVITY 3.4 Turn now to Reading 3C, 'Take a walk on the safe side' by Andy Beckett, which is an extract from an article in *The Independent on Sunday*. As you read the article, consider the following issues:

- How is social order imagined and maintained in CityWalk?
- What is the relationship of CityWalk to the rest of Los Angeles?
- What do you think Jane Jacobs would make of CityWalk? ◆

For the thousands of shoppers who use CityWalk, the integral architectural advantage is that it promises a safe, privatized, highly controlled version of the crowded street, free from 'contamination' and the benign disorder desired by Jane Jacobs. Those who can afford such a lifestyle can drive from their protected neighbourhoods in their private cars and park within the guarded, enclosed shopping mall (see **Hamilton and Hoyle, 1999**). Commentators note that as more and more of the city has *relocated* inside the stockade walls of the shopping mall, the question of the status of these differently constituted types of space has become intense. The ways in which commercial corporations govern and regulate significant expanses of urban space merits closer examination because they raise important questions about who and what has the right to be in such spaces (see the discussion in **Hinchliffe, 1999**, where there is an interesting case study of a plan to establish a 'Universal City' on Rainham Marshes to the east of London). At heart, shopping malls are schizophrenic sites of contradictory qualities. Whilst they may be imagined as civilized, downtown public spaces, they also represent at core a privatized, commercial form of governance, and those enjoying the many attractions of the shopping malls are endowed only with *customer* rights and responsibilities rather than those associated with full *citizenship*.

The construction of downtown shopping malls has in many cities served to intensify the segregated urban shopping experience. Those whose market position disqualifies them from participation in the postmodern consumer citadels are economically and spatially excluded. White shoppers from LA's inner suburbs flocked to CityWalk, effectively abandoning core parts of the 'original' LA to mostly black and Hispanic patrons. Such outcomes have led Goss (1993, pp.26–7) to conclude that the shopping mall is 'a strongly bounded or purified social space that excludes a significant minority of the population and so protects patrons from the moral confusion that a confrontation with social difference might provoke and reassures preferred customers that the unseemly and seamy side of the real public world will be excluded'.

However, it is important to note that the malls have also produced new forms of cultural resistance. Corrigan (1994) provides us with an example of the very active struggle over space that takes place in the 'mega-mall of America' in the Minneapolis suburb of Bloomington. Whilst white, middle-class nuclear families

FIGURE 3.8 *A shopping mall in the USA: 'a safe, privatized, highly controlled version of the crowded street'*

shop and passively consume, private security guards periodically monitor closed-circuit television screens for 'undesirables', mostly groups of teenagers who do not belong:

> After school hours and at the weekends, the shopping atmosphere is less tranquil. That's when the omnipresent 'mall rats' – teenagers who loiter near the mall's food courts – arrive, looking for the usual adolescent thrills of pizza, sex and trouble. Mall rats wage an almost gleeful war with mall security, winding up the rent-a-cops that constitute the faux arm of the law by committing minor infractions against the rules. Bloomington's police force conducts truancy sweeps to make sure that mall rats confine their activities to non-school hours. Too young to drive – and too short of funds, perhaps, to be ideal customers – the mall is one of the only places American teenagers can go to flaunt their youth.
>
> (Corrigan, 1994, pp.43–4)

So far in this chapter we have focused on the ways in which order is imposed on cities through the building of real or metaphoric walls which regulate, segregate and displace their populations. However, as Peter Marcuse (1994) points out, we are left with a series of paradoxes concerning the purpose of walls:

- Walls protect and offer shelter *and* confine and imprison.
- Walls increase security *and* exacerbate a loss of trust and a resurgence of fear.
- Walls close certain parts of the city *and* open others within.

These paradoxes are amplified when we extend our vision of 'walls' to include a much wider range of means of spatial surveillance afforded by the rapid dissemination of intensive policing practices and information and communication technologies at the turn of the twentieth century. But again we need to ask: who is being watched, who is doing the watching, and for what purpose? In the following section we look in some detail at the paradoxes involved in how policing practices and technologies of surveillance support, extend and network with pre-existing configurations of walled confinement but are simultaneously the means through which city spaces are being freed of their physical barriers, thereby inaugurating a new openness.

5 *Building walls III: intensified circuits of policing and surveillance*

In the 1980s and 1990s, report after report painted a picture of city centres in the USA and UK as ungoverned cultural, social and economic wastelands, especially in the evening and after dark (see Mulgan, 1989; Lovatt, 1994; Warpole, 1997). Many visitors noted that, unlike cities in many other parts of the world, fears for public safety and personal security and the fundamental distrust of strangers were widespread; these fears and distrust were intensified by poor lighting, boarded-up shop fronts, squalid streets, offensive graffiti, vandalized public utilities, the lack of child-friendly spaces, inadequate public transport and the presence of the homeless, the mentally ill, drug dealers, prostitutes and, in the case of the UK, large groups of young males moving from pub to pub. The reports suggested that many city centre streets had become fear-filled 'no-go areas' for many women, older people and visible minorities. The prevailing atmosphere was also drawing many social groups away from city centres and downtown areas to the protected 'interior' street life of 'out of town' shopping malls and metrocentres.

In an attempt to reverse the vicious downward spiral of fear and distrust of empty, uncared-for public spaces, these reports recommended that metropolitan authorities and city-centre traders should radically rethink urban planning policies to revalue and reorder city centres as places of consumption. Politicians and planners were urged to reproduce the humane spaces representative of a continental European urban culture by introducing 'bridging activities' which would keep office workers in and also attract a broader base of local people, tourists and investors to the city. The means adopted have included late-night opening of shops, museums, cafés, restaurants, art galleries, 'themed' heritage areas, libraries, educational institutions, markets and so on (see Kearns and Philo, 1993). These reports implicitly echoed Jane Jacobs' belief that crime, anti-social disorder and the fear of strangers could be naturally and softly 'crowded out' by re-establishing public confidence in the city-centre streets as neutral, convivial 'safe corridors', free of harassment and fear. But such democratic mingling, and market-led urban regeneration and a corresponding *intensification* of socio-spatial relations could not take place without some means of de-intensification. In this section we examine two different approaches to the officially defined problem. We take the notion of 'panopticism' as our starting point.

In the late eighteenth century the utilitarian philosopher Jeremy Bentham coined the phrase 'panopticon' to describe his redesign of prison architecture to allow for the uninterrupted inspection, observation and surveillance of prisoners. Bentham, according to Foucault (1977), was first and foremost championing a new conception of surveillance. The panopticon reversed the principles of penality: the old dark and dank prison would be replaced by a light, visible architectural form in which the inmates would be separated off from each other and placed permanently in view. Its purpose was:

> To induce in the inmate a conscious and permanent visibility that assures the automatic functioning of power. So to arrange things that the surveillance is permanent in its effects, even if it is discontinuous in its action; that the perfection of power should tend to render its actual exercise unnecessary; that this architectural apparatus should be a machine for creating and sustaining a power relation independent of the person who exercises it; in short, that the inmates should be caught up in a power situation of which they themselves are the bearers.
>
> (Foucault, 1977, p.201)

In pure form the panopticon was never built, being abandoned in the early nineteenth century on technical grounds, expense and the belief that prisoners were best dealt with 'elsewhere' (notably via transportation to Australia). However, the core principles of panopticism identified in the quote from Foucault had widespread influence in the emergent 'disciplinary sites' of the early nineteenth-century city, such as factories, schools, barracks and hospitals, as well as prisons. Foucault refers to this as an emergent carceral society in which every irregularity or departure from the norm was capable of being monitored and regulated. The aim was to produce a new kind of individual subjected to habits and rules that were exercised continually around him and upon him, and which he must allow to function automatically within him' (Foucault, 1977, p.129). In this, the surveillance afforded by panopticism was central because it induces a state of 'conscious and permanent visibility that assures the automatic functioning of power' (Foucault, 1977, p.201). What Foucault misses from his analysis of the 'swarming' of micro disciplinary mechanisms is any detailed discussion of the new proactive urban policing that emerged in the course of the nineteenth century. We would argue that, along with Haussmannization, this was one of the truly radical reforms of nineteenth-century urban governance. The new police produced what turned out to be an irreversible *spatialization* of Bentham's concept of surveillance.

5.1 ASSERTIVE BEAT POLICING: FROM LONDON TO NEW YORK AND BACK AGAIN

At 6 o'clock on the evening of Tuesday 29 September 1829, approximately 3,000 uniformed officers of the Metropolitan Police marched for the first time out of strategically located stations in London. To grasp the novelty of this initiative, it is important to keep in mind that the new policing, which had first been tested in Dublin, necessitated a very specific remapping of London and the coining of a new spatial metaphor – 'the beat'. Police districts were mapped out into numbered and clearly defined divisions, sub-divisions, sections and beats. Every constable was provided with a beat card and 'pounded' his beat in a fixed time according to an appointed route night and day.

By connecting urban space with law and order, the 'beat' resulted in the production of new 'policed' spaces. A report in the *Edinburgh Review* of 1852 explained that because the new police could not make their presence felt evenly throughout the metropolis, the new beat mapping targeted resources in particular locations where the contradictions and conflicts associated with the new urban order were most intense:

> The beats vary considerably in size; in those parts of the town which are open and inhabited by the wealthier classes, an occasional visit from a policeman is sufficient, and he traverses a wide district. But the limits of the beat are diminished, and of course the frequency of the visits increased, in proportion to the character and density of the population, the throng and pressure of traffic, the concentration of property, and the intricacy of the streets. Within a circle of six miles from St. Paul's, the beats are ordinarily traversed in periods varying from 7 to 25 minutes, and there are points which, in fact, are never free from inspection. Nor must it be supposed that this system places the wealthier localities at a disadvantage, for *it is an axiom in police that you guard St. James by watching St. Giles.*
>
> (*Edinburgh Review*, July 1852, p.5, emphasis added)

From the outset the new police were a bureaucratic instrument for the ordering of the promiscuity of the urban crowd by de-intensifying the disorderly and disreputable edges of main streets and central thoroughfares of the capital. Police beat practices formalized the idea that streets were rule-governed places and that their principal purpose was to facilitate the free circulation and movement both of goods and individuals (Cohen, 1981, p.126).

If a police officer pounding his beat deemed an individual to be 'out of place', he or she would be stopped, questioned and searched. If people did not 'move on' when they were told to, they ran the risk of being arrested for being suspicious persons, loitering with intent, causing an obstruction, committing a nuisance, using abusive or insulting language, importuning or soliciting, being drunk and disorderly, or engaging in behaviour likely to cause a breach of the peace. The movement of street traders, prostitutes, beggars, entertainers and pamphleteers received a disproportionate amount of police attention. They quickly became the visible 'subjects' of the power relations embedded in police work because the police were legally empowered to discipline and criminalize the presence of certain groups on the streets through defining 'what were the wrong people, wrong age, wrong sex, in the wrong place and at the wrong time' (Cohen, 1981, p.127). The new police were in effect able to impose a system of unofficial curfew, on the main streets through regular 'street-sweeps' or 'culls'.

The available evidence suggests that rather than erasing 'disorderly' activities, the new police practices displaced them to other places and in certain instances refashioned their form. Storch (1981) argues that it was the intense pressure of the new forms of proactive police surveillance that ultimately impacted on the public thoroughfares of British cities:

> The technique was well-chosen. As far as the policed were concerned, the impression of being watched or hounded was not directly dependent on the presence of a constable on every street corner and at all times. What produced this effect was the knowledge that the police were always near and likely to appear at any time. This it seems was – and still is – the main function of the pressure of surveillance.
>
> (Storch, 1981, p.292)

The *principles* of the new urban policing were such a success that in the course of the nineteenth century they were exported to various cities. However, in the 1990s, British cities looked to New York for a new twenty-first century model of high-profile policing that would improve the quality of life on the streets. Aggressive 'zero tolerance' policing strategies were introduced in New York in 1994 after complaints that the police and the city authorities had not just lost control of the streets, but had lost interest in preventing the city from collapsing under the weight of its lawlessness:

> New York is staggering. The streets already resemble a New Calcutta, bristling with beggars and sad schizophrenics tuned into inner voices. Crime, the fear of it as much as the fact, adds overtones of a New Beirut. Many New Yorkers now think twice about where they can safely walk; in a civilised place that should be as automatic as breathing.
>
> (*New York Times*, 30 December 1990, section 4, p.10)

The *New York Times* argued that it was the duty of the city authorities to make the city's streets safe again: 'Safe streets are fundamental; going out on them is the simplest expression of the social contract; a city that cannot maintain that side of the contract will choke'. Cops on the beat were told by Mayor Rudolph Giuliani and Commissioner William Bratton to wrest back the public spaces dominated by 'dangerous strangers', as represented by aggressive beggars, squeegee merchants, hustlers, abusive drunks, litter louts, prostitutes, petty drug dealers and junkies. Police precincts were told to turn the streets into 'no-go areas' for criminals and deviants by deploying task forces to 'cool off' any 'hot spots' identified by up-to-date crime-mapping of their districts. As part of his high-profile campaign to attract respectable people back to central New York, Mayor Giuliani also launched a campaign to transform Times Square from a sleazy neighbourhood full of X-rated shops into a respectable, up-market commercial district. New zoning restrictions were used to drive the porn shops and brothels into the peripheral industrial areas. Police forces from many global cities visited New York to see 'zero tolerance' policing in action. While the results were impressive both in terms of the falling crime rate and the greatly improved 'look' of the streets, concerns were expressed about its appropriateness for other cities. Critics warned that if the police clampdown on the streets was too authoritarian, it would:

- result in allegations of police brutality;
- provoke disorder in areas where relations between the police and certain groups were fraught;

- displace disorder to other times and places; and
- destroy the very essence and vitality of inner-city life.

If we think back to Jane Jacobs' conceptualization of the 'soft' ordering of street life, we can see how the Robocop version of beat policing could quite easily destroy the 'ballet of the street' and 'benign disorder' that she believes are so crucial to a vital street life.

5.2 ELECTRONIC 'EYES ON THE STREET': HYPER- OR POST-PANOPTICISM?

Perhaps more significant for the reordering of city spaces in the twenty-first century was the appearance in the UK in the 1990s of banks of overt and covert closed-circuit television (CCTV) cameras watching over streets, shops, offices, clubs, car parks, residential streets and 'inside' urban housing estates on a twenty-four-hour basis. The demand for such electronic surveillance was instigated by heightened public anxiety over the safety of city centres, evidence that the public was reassured by shopping malls with private security guards and surveillance cameras, and calls from insurance companies, store owners, city centre managers and property developers for effective protection from risks ranging from shoplifting to terrorist bombs. By the end of the decade, UK cities had the most intensive concentration of electronic 'eyes on the street' in the world. The location of surveillance cameras in an expanding range of urban sites fuelled the demand for the development of ever more sophisticated information and communication technologies, including biometric scanning devices (smartcards, fingerprints, eyescans, hand geometry scans, voice recognition, DNA testing and digitized facial recognition) and multiplex pictorial databases to enhance the *monitoring, tracking and identification* of entire populations.

ACTIVITY 3.5 What implications do you think the technological and conceptual advances in surveillance may hold for the reordering of city spaces in the twenty-first century? On balance, do you think the new surveillance infrastructure represents a 'soft' or 'hard' flow of ordering? Refer back to our previous discussions of 'eyes on the street' in section 2 when thinking about these questions. ◆

There are, very broadly, two competing perspectives that we need to consider. The most widely adopted perspective – 'hyper-panopticism' – argues that in terms of intensity and pervasiveness, we are seeing the realization of the principles of population observation and control underpinning Bentham's panopticon. Corbett and Marx (1991) believe that we are witnessing the emergence of surveillance capabilities which mean that our fundamental understandings of barriers and boundaries – distance, darkness, time, walls and even skin, no longer have any meaning. In a similar manner, Gandy (1993) argues that the integration of new surveillance technologies into everyday life has become relentless and irresistible because each application generates new uses and justifies others (see Figure 3.9). Throughout the UK, rural towns and villages, driven by the fear that urban crime and criminals would be displaced from 'watched' to 'unwatched' places, have 'hard-wired' themselves to the 'new surveillance' infrastructure. Such a development may indeed herald the penetration of the city into the suburbs and countryside and an unprecedented blurring of the boundaries between traditional images of urban, suburban and rural order. There is also the distinct possibility of a wide array of private and public surveillance networks swarming transnationally in a variety of key capital and regional cities in order to monitor and record the movement of displaced 'suspect populations'. Such developments in the technologies of surveillance would seem to lend support to Bogard's contention

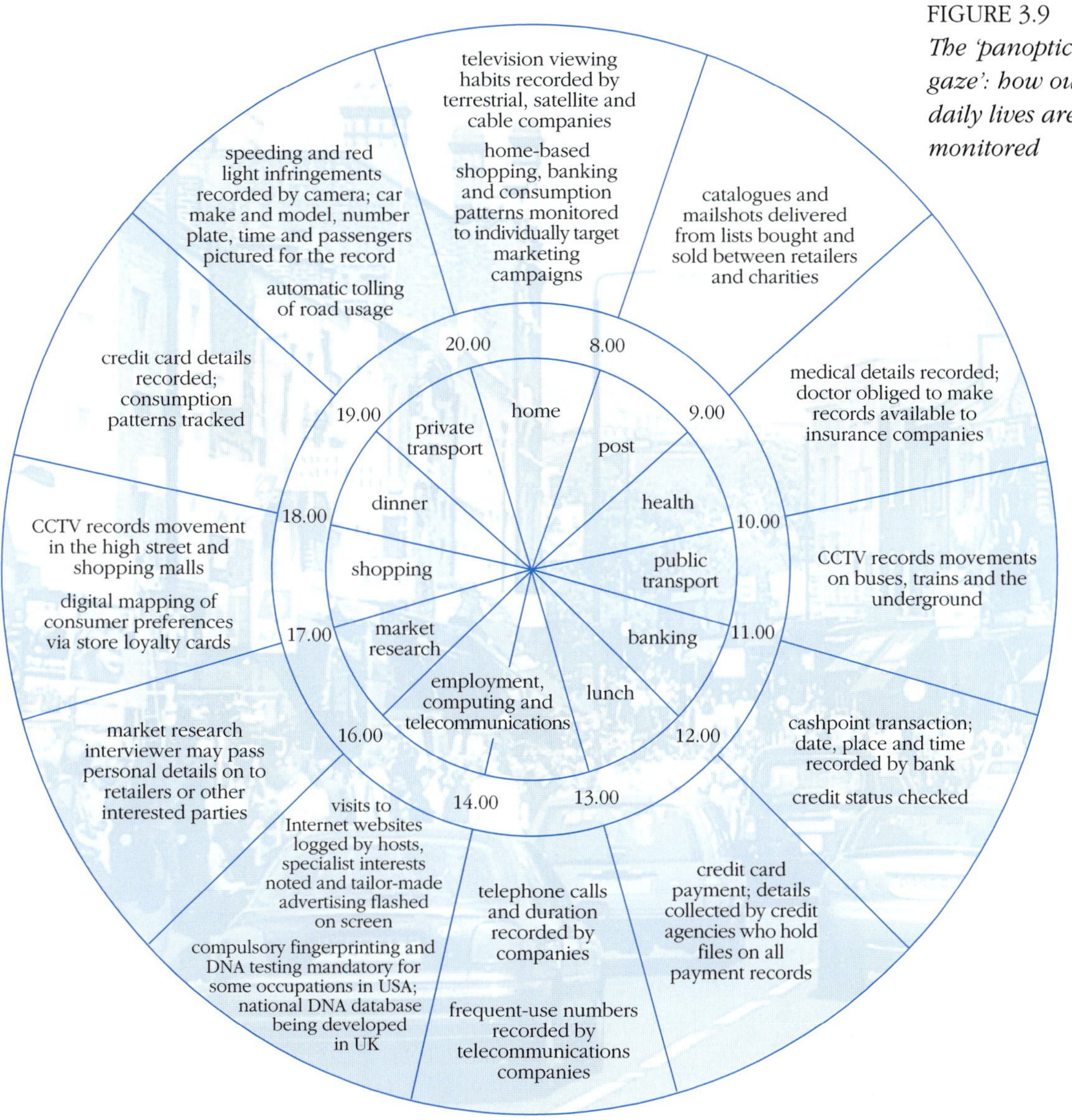

FIGURE 3.9 *The 'panoptic gaze': how our daily lives are monitored*

that in the twenty-first century we will move towards *non-geographical* 'telematic societies' characterized by

> forms of *hyper-surveillant* control, where the prefix 'hyper' implies not simply an intensification of surveillance but the effort to push surveillance technologies to the absolute limit. That limit is an imaginary line beyond which control operates, so to speak in 'advance' of itself and where surveillance – a technology of exposure and recording – evolves into a technology of *pre*-exposure and *pre*-recording ... a fantastic dream of seeing everything capable of being seen, recording every fact capable of being recorded, and accomplishing these things, whenever and wherever possible prior to the event itself.
>
> (Bogard, 1996, pp.4–5)

FIGURES 3.10 and 3.11 *Surveillance cameras are a common feature in UK cities, whether in residential areas (top) or city centres*

However, other commentators working within this general framework insist that the new surveillance infrastructure is not universal or free-floating in its gaze but is impacting differently on different populations at different times and in different parts of the city. They pose key questions about who exactly is being monitored, by whom, and for what purpose. For example, available research on the deployment of CCTV in the UK suggests that in the absence of any objective information as to whom they should monitor, surveillance camera operators in UK city centres, like their counterparts in the shopping malls, use an electronic ordering of people that corresponds to their 'common-sense' understandings of those social groups they believe to be or look 'deviant', 'peculiar', 'undesirable', 'out of place' or 'interesting' (Groombridge and Murji, 1994; Coleman and Sim, 1996; Fyfe and Bannister, 1996; Graham, 1997; Norris and Armstrong, 1997). The 'usual suspects' who are 'screened in' are groups of young men, the poorly attired and visible ethnic minorities. Ironically, given the concern accorded to gender relations in many city centre rejuvenation plans and public safety strategies, young women are observed quite intensely, but voyeuristically, by male camera operators.

Thus, while the 'soft' ordering of the electronic surveillance cameras may be imperceptible and relatively harmless for certain social groups, a kaleidoscope of stereotypical 'online' images is being reproduced by the electronic lenses. This 'augmented reality' is, in both spatial and temporal terms, reinforcing boundaries and existing social relations, effectively transforming city centres and business/ tourist districts into 'no-go areas' for disadvantaged groups and exacerbating aggressive policing practices.

If we shift our angle of perspective to include what we might call the 'post-panoptic' approach of social theorists such as Lyon (1994), we find that the 'Big Brother' significance of the new surveillance infrastructure is played down. Drawing upon a complex reading of Foucault's work, they insist that a re-theorization of the concept of surveillance allows us to recognize that, for the following reasons, there is no automatic connection between technological developments and authoritarian forms of urban control:

- There is no exercise of power without potential contestation and resistance.
- Strategies of power-knowledge are never complete or determining but always in a process of becoming – partial, contingent, unevenly developed, enabling as well as constraining – and full of specificities, hybridities, subtleties and 'open-ended cosmologies' (Pile and Thrift, 1995; Hannah, 1997).
- The logic of panopticism also assures that there is a 'double gaze' – observers themselves will also be open to observation and thus, in principle, subject to forms of democratic control.
- There will always be fractures and 'blindspots' because technology cannot generate a multi-dimensional mode of representation that is equal to the complexity of human subjectivity.
- Those who advocate technological solutions to certain problems are invariably disappointed because of their unintended consequences.

Lyon (1994) emphasizes that modern surveillance is paradoxical and ambiguous and is therefore capable of bringing real benefits as well as real harms. By linking the expansion of surveillance to a simultaneous growth in civic citizenship, he argues that it is the very improvements in civil rights and political entitlement that have generated demands for a greater documentary identification, which in turn depend on more sophisticated methods of regulation, recording and surveillance. Thus, there is no single omnipotent Big Brother, but a variety of dispersed 'Little Brothers' (Lyon, 1994, p.53).

From this perspective it is feasible to argue that what Mike Davis calls 'electronic guardian angels' are a vital part of the long-term 'civilizing mission' to rejuvenate beleaguered city centres in order to preserve lawful freedoms and increase freedom of movement by making it safe to walk the streets. This is why the cameras enjoy the overwhelming support of members of the public. The intensive, totalizing sweep of the ever-vigilant electronic eye over its 'beat' can warn off would-be predators, zoom in on anti-social elements, and direct police officers and private security guards to places where intrusions are taking place. Hence, we can expect many forms of street crime to 'virtually disappear' from city centres.

Certain theorists take this analysis one step further, arguing that a complex 'post-panoptic' culture is emerging within image-saturated societies. Obsessed with the visual representation of the self, chameleon- like individuals in these societies accept electronic surveillance as a normal part of everyday life. Poster points out that the 'reciprocal control of the population by itself' is already in place. Populations have

> been disciplined to surveillance and to participating in the process. Social security guards, drivers licenses, credit cards, library cards and the like – the individual must apply for them, have them ready at all times, use them continuously. Each transaction is recorded, encoded and added to the databases. Individuals in many cases fill out the forms; they are at once the source of information and the recorder of information. Home networking constitutes the streamlined culmination of this phenomenon: the consumer, by ordering products through a modem directly linked to the producer's database, enters data about himself or herself directly into the producer's database in the very act of purchase.
>
> (Poster, 1990, p.93)

However, conformity and control may not be the only future scenario. We can imagine, for example, how certain individuals, including anti-social 'others', will actually get a buzz out of being active participants 'appearing' in the digital crime-scape that is city life or producing 'hazy' spaces beyond the power of vision. For example, post-panopticon subjectivities constituted through electronic disobedience (such as 'hacking') continually threaten to disrupt the ordering represented by the new surveillance technologies.

6 Conclusion: the paradoxes of reordering

This chapter has explored various attempts to de-intensify and regulate the tensions and disorders that inevitably play themselves out in urban spaces. However, as we have seen, we are confronted with a series of uneasy paradoxes that might be schematically listed as follows:

- Constructions of 'disorder' and 'order' are inextricably linked, and in any given urban context they frequently appear as 'idealized imaginaries'. However, it is only in the company of strangers in city spaces that they are symbiotically realized.
- The gaze of the 'eyes on the street' can be benign and reassuring, indifferent or predatory. It depends on the nature of those power relationships constituted through the gaze.
- Strategies of 'defensible space' exacerbate, rather than resolve, processes of spatial segregation and social polarization. There is a clear redistribution of 'risks' involved in the entrenchment of 'suspect populations' in outcast under-protected, disordered 'ghettos', and of 'innocent populations' in over-protected, ordered consumerist 'citadels' and residential 'enclaves'.
- Attempts to purify public spaces through privatization, zoning or repressive policing may reduce the fear of the street for some whilst generating a cycle of spatial displacement that is never-ending for certain social groups and individuals.
- The resort to 'walls and gates' and the proliferation of electronic 'eyes on the street' represents a 'solution' to the 'fear of strangers' but is also symptomatic of a decline in communal trust and mutual support.

As a consequence, 'building walls' can both enhance public confidence and also generate chronic levels of anxiety, lower levels of trust and indifferent or defensive social encounters. The hard task is to imagine relatively more inclusive forms of urban order within which the diverse exchanges of communal life can take place freely and safely, whilst limiting the practices of spatial domination and exclusion ushered in by 'walling' and hyper surveillance. Unless there is a dramatic reversal in urban governance in a variety of global contexts, it seems that an intensification of self-declared zones of exclusion and officially imposed safety zones will remain a defining feature of city life. Some of these may rely on crude means of enforcement, whilst in others a high-tech infrastructure of surveillance and regulation will continue to be grafted onto existing grids of separation, partition and exclusion. This analysis of the production of differentiated urban spaces is taken a step further in the next chapter.

References

Allen, J. (1999) 'Worlds within cities', in Massey, D., Allen, J. and Pile, S. (eds) *City Worlds*, Routledge/The Open University (Book 1 in this series).

Allen, J., Massey, D. and Pryke, M. (eds) (1999) *Unsettling Cities*, Routledge/The Open University (Book 2 in this series).

Amin, A. and Graham, S. (1999) 'Cities of connection and disconnection', in Allen, J. *et al.* (eds).

Beckett, A. (1994) 'Take a walk on the safe side', *Independent on Sunday*, 27 February.

Benjamin, W. (1968) 'Paris – capital of the 19th century', *New Left Review*, no.48.

Blakely, E.J. and Snyder, M. (1997) *Fortress America*, Washington, Brookings Institute.

Bogard, W. (1996) *The Simulation of Surveillance: Hypercontrol in Telematic Societies*, Cambridge, Cambridge University Press.

Caldeira, T. (1996) 'Fortified enclaves: the new urban segregation', *Public Culture*, vol.8, no.2, pp.329–54.

Clarke, R. (ed.) (1993) *Situational Crime Prevention: Successful Case Studies*, Albany, Harrow and Heston.

Cohen, P. (1981) 'Policing the working-class city', in Fitzgerald, M. *et al.* (eds).

Coleman, R. and Sim, J. (1996) 'From the docklands to the Disney store: surveillance risk and security in Liverpool city centre', paper presented at Law and Society Association Conference, University of Strathclyde, July 1996.

Corbett, R. and Marx, G.T. (1991) 'Critique: no soul in the new machine; techno fallacies in the electronic monitoring movement', *Justice Quarterly*, vol.8.

Corrigan, S. (1994) 'Mall', *Life*, 27 November.

Crawford, M. (1992) 'The world in a shopping mall', in Sorokin, M. (ed.) *Variations on a Theme Park*, New York, Hill and Wanq.

Davis, M. (1992) *City of Quartz*, London, Verso.

Davis, M. (1994) 'Beyond Blade Runner: urban control: the ecology of fear', Open Magazine Pamphlet Series, Westfield, NJ.

Evenson, N. (1979) *Paris: A Century of Change, 1879–1978*, New Haven, Yale University Press.

Fincher, R. (1990) 'Women in the city', *Australian Geographical Studies*, vol.28, pp.29–37.

Fitzgerald, M., McLennan, G. and Pawson, J. (eds) (1981) *Crime and Society*, London, Routledge in association with The Open University.

Foucault, M. (1977) *Discipline and Punish*, London, Allen Lane.

Fyfe, N. and Bannister, J. (1996) 'City watching: closed circuit television surveillance in public spaces', *Area*, vol.28, no.1, pp.37–46.

Gandy, O. (1993) *The Panoptic Sort*, Boulder, Westview Press.

Goffman, E. (1971) *Relations in Public: Microstudies of Public Order*, New York, Doubleday.

Goss, J. (1993) 'The magic of the mall: an analysis of form, function and meaning in the contemporary retail environment', *Annals of the Association of American Geographers*, vol.83, pp.18–47.

Graham, S. (1997) 'Spaces of surveillant-simulation', unpublished paper, Centre for Urban Technology, University of Newcastle.

Groombridge, N. and Murji, K. (1994) 'As easy as A, B and CCTV', *Policing*, vol.10, no.4, pp.283–90.

Hamilton, K. and Hoyle, S. (1999) 'Moving cities: transport connections', in Allen, J. *et al.* (eds).

Hannah, M. (1997) 'Imperfect panopticism: envisioning the construction of normal lives', in Benko, G. and Stohmayer, U. (eds) *Space and Social Theory*, Oxford, Blackwell.

Hinchliffe, S. (1999) 'Cities and natures: intimate strangers', in Allen, J. *et al.* (eds).

Hooper, B. (1998) 'The poem of male desires: female bodies, modernity and "Paris, capital of the nineteenth century"', in Sandercock, L. (ed.) *Making the Invisible Visible*, Berkeley, University of California Press.

Jacobs, J. (1961) *The Death and Life of Great American Cities*, Harmondsworth, Penguin.

Jencks, C. (1993) *Heteropolis*, London, Academy Books.

Karp, D.A., Stone, L. and Yoels, W.C. (1991) *Being Urban: A Sociology of City Life*, New York, Praeger.

Kearns, G. and Philo, C. (eds) (1993) *Selling Places: The City as Cultural Capital, Past and Present*, Oxford, Pergamon Press.

Kowinski, W.S. (1985) *The Malling of America*, New York, William Morrow.

Lofland, L. (1973) *A World of Strangers*, New York, Basic Books.

Lovatt, A. (1994) *More Hours in the Day*, Manchester, Manchester Institute for Popular Culture.

Lyon, F. (1994) *The Electronic Eye: The Rise of Surveillance Society*, Cambridge, Polity Press.

McDowell, L. (1999) 'City life and difference: negotiating diversity', in Allen, J. *et al*. (eds).

Marcuse, P. (1994) 'Walls as metaphor and reality', in Dunn, S. (ed.) *Managing Divided Cities*, Keele, Keele University Press.

Meade, T.A. (1997) *'Civilizing Rio': Reform and Resistance in a Brazilian City*, University Park, Pennsylvania University Press.

Mulgan, G. (1989) 'The changing shape of the city', in Hall, S. and Jacques, M. (eds) *New Times*, London, Lawrence and Wishart.

Mumford, L. (1961) *The City in History*, Harmondsworth, Penguin.

Needell, J.D. (1989) *A Tropical Belle Epoque: Elite Culture and Society in Turn of the Century Rio de Janeiro*, Cambridge, Cambridge University Press.

Newman, O. (1972) *Defensible Space: People and Design in the Violent City*, London, Architectural Press.

Norris, L. and Armstrong, G. (1997) 'The unforgiving eye: the social construction of suspicion and intervention in CCTV systems', paper presented at British Criminology Conference, Belfast.

Pile, S. and Thrift, N. (1996) 'Mapping the subject', in *Mapping the Subject: Geographies of Cultural Transformation*, London, Routledge.

Poster, M. (1990) *The Mode of Information: Poststructuralism and Social Context*, Cambridge, Polity Press.

Probyn, E. (1995) 'Queer belongings: the politics of departure', in Grosz, E. and Probyn, E. (eds) *Sexy Bodies: The Strange Carnalities of Feminism*, London, Routledge.

Rabinow, P. (1989) *French Modern: Norms and Forms of the Social Environment*, Cambridge, Mass., MIT Press.

Sennett, R. (1990) *The Conscience of the Eye*, New York, Alfred A. Knopf.

Storch, R. (1981) 'The plague of blue locusts: police reform and popular resistance in Northern England, 1840–57', in Fitzgerald, M. *et al.* (eds).

Valentine, G. (1992) 'Images of danger: women's sources of information about the spatial distribution of male violence', *Area*, vol.24, no.1, pp.22–9.

Wacquant, L.J.D. (1997) 'The new urban color line: the state and the fate of the ghetto in post Fordist America', in Agnew, J. (ed.) *Political Geography*, London, Arnold.

Warpole, K. (1997) *Nothing to Fear? Trust and Respect in Urban Communities*, London, Comedia/Demos.

READING 3A
Jane Jacobs: 'The uses of sidewalks: safety'

Under the seeming disorder of the old city, wherever the old city is working successfully, is a marvellous order for maintaining the safety of the streets and the freedom of the city. It is a complex order. Its essence is intricacy of sidewalk use, bringing with it a constant succession of eyes. This order is all composed of movement and change, and although it is life, not art, we may fancifully call it the art form of the city and liken it to the dance – not to a simple-minded precision dance with everyone kicking up at the same time, twirling in unison and bowing off *en masse*, but to an intricate ballet in which the individual dancers and ensembles all have distinctive parts which miraculously reinforce each other and compose an orderly whole. The ballet of the city sidewalk never repeats itself from place to place, and in any one place is always replete with new improvisations.

The stretch of Hudson Street where I live is each day the scene of an intricate sidewalk ballet. I make my own first entrance into it a little after eight when I put out the garbage can, surely a prosaic occupation, but I enjoy my part, my little clang, as the droves of junior high school students walk by the centre of the stage dropping candy wrappers. (How do they eat so much candy so early in the morning?)

While I sweep up the wrappers I watch the other rituals of morning: Mr Halpert unlocking the laundry's handcart from its mooring to a cellar door, Joe Cornacchia's son-in-law stacking out the empty crates from the delicatessen, the barber bringing out his sidewalk folding chair, Mr Goldstein arranging the coils of wire which proclaim the hardware store is open, the wife of the tenement's superintendent depositing her chunky three-year-old with a toy mandolin on the stoop, the vantage point from which he is learning the English his mother cannot speak. Now the primary children, heading for St Luke's, dribble through to the south; the children for St Veronica's cross, heading to the west, and the children for PS 41, heading towards the east. Two new entrances are being made from the wings: well-dressed and even elegant women and men with brief-cases emerge from doorways and side streets. Most of these are heading for the bus and subways, but some hover on the curbs, stopping taxis which have miraculously appeared at the right moment, for the taxis are part of a wider morning ritual: having dropped passengers from midtown in the downtown financial district, they are now bringing downtowners up to midtown. Simultaneously, numbers of women in house dresses have emerged and as they criss-cross with one another they pause for quick conversations that sound with either laughter or joint indignation, never, it seems, anything between. It is time for me to hurry to work too, and I exchange my ritual farewell with Mr Lofaro, the short, thick-bodied, white-aproned fruit man who stands outside his doorway a little up the street, his arms folded, his feet planted, looking solid as earth itself. We nod; we each glance quickly up and down the street, then look back to each other and smile. We have done this many a morning for more than ten years, and we both know what it means: all is well.

The heart-of-the-day ballet I seldom see, because part of the nature of it is that working people who live there, like me, are mostly gone, filling the roles of strangers on other sidewalks. But from days off, I know enough of it to know that it becomes more and more intricate. Longshoremen who are not working that day gather at the White Horse or the Ideal or the International for beer and conversation. The executives and business lunchers from the industries just to the west throng the Dorgene restaurant and the Lion's Head coffee-house; meat-market works and communications scientists fill the bakery lunch-room. Character dancers come on, a strange old man with strings of old shoes over his shoulders, motor-scooter riders with big beards and girl friends who bounce on the back of the scooters and wear their hair long in front of their faces as well as behind, drunks who follow the advice of the Hat Council and are always turned out in hats, but not hats the Council would approve of. Mr Lacey, the locksmith, shuts up his shop for a while and goes to exchange the time of day with Mr Slube at the cigar store. Mr Koochagian, the tailor, waters the luxuriant jungle of plants in his window, gives them a critical look from the outside, accepts a compliment on them from two passers-by, fingers the leaves on the plane tree in front of our house with a thoughtful gardener's appraisal, and crosses the street for a bite at the Ideal where he can keep an eye on customers and wigwag across the message that he is coming. The baby carriages come out, and clusters of everyone from toddlers with dolls to teenagers with homework gather at the stoops.

When I get home after work, the ballet is reaching its crescendo. This is the time of roller skates and stilts and tricycles, and games in the lea of the stoop with bottle-tops and plastic cowboys; this is the time of bundles and packages, zigzagging from the drugstore to the fruit stand and back over to the butcher's; this is the time when teenagers, all dressed up, are pausing to ask if their slips show or their collars look right; this is the time when beautiful girls get out of MGs; this is the time when the fire engines go through; this is the time when anybody you know around Hudson Street will go by.

As darkness thickens and Mr Halpert moors the laundry cart to the cellar door again, the ballet goes on under lights, eddying back and forth but intensifying at the bright spotlight pools of Joe's sidewalk pizza dispensary, the bars, the delicatessen, the restaurant, and the drugstore. The night workers stop now at the delicatessen to pick up salami and a container of milk. Things have settled down for the evening but the street and its ballet have not come to a stop.

I know the deep night ballet and its seasons best from waking long after midnight to tend a baby and, sitting in the dark, seeing the shadows and hearing the sounds of the sidewalk. Mostly it is a sound like infinitely pattering snatches of party conversation and, about three in the morning, singing, very good singing. Sometimes there is sharpness and anger or sad, sad weeping, or a flurry of search for a string of beads broken. One night a young man came roaring along, bellowing terrible language at two girls whom he had apparently picked up and who were disappointing him. Doors opened, a wary semi-circle formed around him, not too close, until the police came. Out came the heads, too, along Hudson Street, offering opinion, 'Drunk ... Crazy ... A wild kid from the suburbs.'

Deep in the night, I am almost unaware how many people are on the street unless something calls them together, like the bagpipe. Who the piper was and why he favoured our street I have no idea. The bagpipe just skirled out in the February night and, as if it were a signal, the random, dwindled movements of the sidewalk took on direction. Swiftly, quietly, almost magically a little crowd was there, a crowd that evolved into a circle with a Highland fling inside it. The crowd could be seen on the shadowy sidewalk, the dancers could be seen, but the bagpiper himself was almost invisible because his bravura was all in his music. He was a very little man in a plain brown overcoat. When he finished and vanished, the dancers and watchers applauded, and applause came from the galleries too, half a dozen of the hundred windows on Hudson Street. Then the windows closed, and the little crowd dissolved into the random movements of the night street.

The strangers on Hudson Street, the allies whose eyes help us natives keep the peace of the street, are so many that they always seem to be different people from one day to the next. That does not matter. Whether they are so many always-different people as they seem to be, I do not know. Likely they are. When Jimmy Rogan fell through a plate-glass window (he was separating some scuffling friends) and almost lost his arm, a stranger in an old T-shirt emerged from the Ideal bar, swiftly applied an expert tourniquet and, according to the hospital's emergency staff, saved Jimmy's life. Nobody remembered seeing the man before and no one has seen him since. The hospital was called in this way: a woman sitting on the steps next to the accident ran over to the bus stop, wordlessly snatched the dime from the hand of a stranger who was waiting with his fifteen-cent fare ready, and raced into the Ideal's phone booth. The stranger raced after her to offer the nickel too. Nobody remembered seeing him before, and no one has seen him since. When you see the same stranger three or four times on Hudson Street, you begin to nod. This is almost getting to be an acquaintance, a public acquaintance, of course.

I have made the daily ballet of Hudson Street sound more frenetic than it is, because writing it telescopes it. In real life, it is not that way. In real life, to be sure, something is always going on, the ballet is never at a halt, but the general effect is peaceful and the general tenor even leisurely. People who know well such animated city streets will know how it is. I am afraid people who do not will always have it a little wrong in their heads – like the old prints of rhinoceroses made from travellers' descriptions of rhinoceroses.

On Hudson Street, the same as in the North End of Boston or in any other animated neighbourhoods of great cities, we are not innately more competent at keeping the sidewalks safe than are the people who try to live off the hostile truce of Turf in a blind-eyed city. We are the lucky possessors of a city order that makes it relatively simple to keep the peace because there are plenty of eyes on the street. But there is nothing simple about that order itself, or the bewildering number of components that go into it. Most of those components are specialized in one way or another. They unite in their joint effect upon the sidewalk, which is not specialized in the least. That is its strength.

Source: Jacobs, 1961, pp.60–5

READING 3B
Mike Davis: 'Urban control: the ecology of fear'

Is there any need to explain *why* fear eats the soul of Los Angeles?

The current obsession with personal safety and social insulation is only exceeded by the middle-class dread of progressive taxation. In the face of unemployment and homelessness on scales not seen since 1938, a bipartisan consensus insists that the budget must be balanced and entitlements reduced. Refusing to make any further public investment in the remediation of underlying social conditions, we are forced instead to make increasing private investments in physical security. The rhetoric of urban reform persists, but the substance is extinct. 'Rebuilding LA' simply means padding the bunker.

As city life, in consequence, grows more feral, the different social milieux adopt security strategies and technologies according to their means. Like Burgess' original dart board, the resulting pattern condenses into concentric zones. The bull's eye is Downtown.

In another essay I have recounted in detail how a secretive, emergency committee of Downtown's leading corporate landowners (the so-called Committee of 25) responded to the perceived threat of the 1965 Watts Rebellion. Warned by law-enforcement authorities that a black 'inundation' of the central city was imminent, the Committee of 25 abandoned redevelopment efforts in the old office and retail core. They then used the city's power of eminent domain to raze neighborhoods and create a new financial core a few blocks further west. The city's redevelopment agency, acting virtually as their private planner, bailed out the Committee of 25's sunk investments in the old business district by offering huge discounts, far below market value, on parcels in the new core.

Key to the success of the entire strategy (celebrated as Downtown LA's 'renaissance') was the physical segregation of the new core and its land values behind a rampart of regraded palisades, concrete pillars and freeway walls. Traditional pedestrian connections between Bunker Hill and the old core were removed, and foot traffic in the new financial district was elevated above the street on pedways whose access was controlled by the security systems of individual skyscrapers. This radical privatization of Downtown public space – with its ominous racial undertones – occurred without significant public debate or protest.

Last year's riots, moreover, have only seemed to vindicate the foresight of Fortress Downtown's designers. While windows were being smashed throughout the old business district along Broadway and Spring streets, Bunker Hill lived up to its name. By flicking a few switches on their command consoles, the security staffs of the great bank towers were able to cut off all access to their expensive real estate. Bullet-proof steel doors rolled down over street-level entrances, escalators instantly stopped and electronic locks sealed off pedestrian passageways. As the *Los Angeles Business Journal* recently pointed out in a special report, the riot-tested success of corporate Downtown's defenses has only stimulated demand for new and higher levels of physical security.

In the first place, the boundary between architecture and law enforcement is further eroded. The LAPD have become central players in the Downtown design process. No major project now breaks ground without their participation, and in some cases, like the recent debate over the provision of public toilets in parks and subway stations (which they opposed), they openly exercise veto power.

Secondly, video monitoring of Downtown's redeveloped zones has been extended to parking structures, private sidewalks, plazas, and so on. This comprehensive surveillance constitutes a virtual *scanscape* – a space of protective visibility that increasingly defines where white-collar office workers and middle-class tourists feel safe Downtown. Inevitably the workplace or shopping mall video camera will become linked with home security systems, personal 'panic buttons', car alarms, cellular phones, and the like, in a seamless continuity of surveillance over daily routine. Indeed, yuppies' lifestyles soon may be defined by the ability to afford *electronic guardian angels* to watch over them. (In the meantime, these hard times are boom years for the makers of video surveillance technology. The leading manufacturer, a Swedish conglomerate, is now the official sponsor of the huge London marathon.)

Thirdly, tall buildings are becoming increasingly sentient and packed with deadly firepower. The skyscraper with a computer brain in *Die Hard I* (actually F. Scott Johnson's Fox-Pereira Tower) anticipates a possible genre of architectural anti-heroes as *intelligent buildings* alternately battle evil or become its pawns. The sensory system of the average office tower already includes panoptic vision, smell, sensitivity to temperature and humidity, motion detection, and, in some cases, hearing. Some architects now predict the day when the building's own AI security computer will be able to automatically screen and identify its human population, and, even perhaps, respond to their emotional states (fear, panic, etc.). Without dispatching security personnel, the building itself will manage crises both minor (like ordering street people out of the building or preventing them from using toilets) and major (like trapping burglars in an elevator).

When all else fails, the smart building will become a combination of bunker and fire-base. When the federal Resolution Trust Corp. recently seized the assets of Columbia Savings and Loan Association they discovered that the CEO, Thomas Spiegel, had converted its Beverly Hills headquarters into a secret, 'terrorist-proof' fortress. In addition to elaborate electronic security sensors, a sophisticated computer system that tracked terrorist incidents over the globe, and an arms cache in its parking structure, the 8900 Wilshire building also has Los Angeles' most unusual executive washroom:

> Tom Spiegel's office, in addition to the bullet-proof glass, was designed to have an adjoining bathroom with a bullet-proof shower. In the event an alarm was sounded, secret panels in the shower walls would open, behind which high-powered assault rifles would be stored.

Free fire zone

Beyond the scanscape of the fortified core is the halo of barrios and ghettos that surround Downtown Los Angeles. In Burgess' original Chicago-inspired schema this was the 'zone in transition': the boarding house and tenement streets, intermixed with old industry and transportation infrastructure, that sheltered new immigrant families and single male laborers. Los Angeles' inner ring of freeway-sliced Latino neighborhoods still recapitulate these classical functions. Here in Boyle and Lincoln Heights, Central-Vernon and MacArthur Park are the ports of entry for the region's poorest immigrants, as well as the low-wage labor reservoir for Downtown's hotels and garment sweatshops. Residential densities, just as in the Burgess diagram, are the highest in the city. (According to the 1990 Census, one district of MacArthur Park is nearly 30% denser than Midtown Manhattan!)

Finally, just as in Chicago in 1927, this tenement zone ('where an inordinately large number of children are crowded into a small area') remains the classic breeding ground of teenage street gangs (over one-hundred according to LA school district intelligence). But while 'Gangland' in 1920s Chicago was theorized as essentially *interstitial* to the social organization of the city – 'as better residential districts recede before the encroachments of business and industry, the gang develops as one manifestation of the economic, moral, and cultural frontier which marks the interstice' – a gang map of Los Angeles today is coextensive with the geography of social class. Tribalized teenage violence now spills out of the inner ring into the older suburban zones; the Boyz are now in the 'Hood where Ozzie and Harriet used to live.

For all that, however, the inner ring remains the most dangerous sector of the city. Ramparts Division of the LAPD, which patrols the salient just west of Downtown, regularly investigates more homicides than any other neighborhood police jurisdiction in the nation. Nearby MacArthur Park, once the jewel in the crown of LA's park system, is now a free-fire zone where crack dealers and street gangs settle their scores with shotguns and Uzis. Thirty people were murdered there in 1990.

By their own admission the overwhelmed inner-city detachments of the LAPD are unable to keep track of all the bodies on the street, much less deal with common burglaries, car thefts or gang-organized protection rackets. Lacking the resources or political clout of more affluent neighborhoods, the desperate population of the inner ring is left to its own devices. As a last resort they have turned to Messieurs Smith and Wesson, whose name follows 'protected by …' on many a porch.

Slumlords, meanwhile, are mounting their own private reign of terror against drug-dealers and petty criminals. Faced with new laws authorizing the seizure of drug-infested properties, they are hiring goon squads and armed mercenaries to 'exterminate' crime in their tenements. The *LA Times* recently described the swashbuckling adventures of one such crew in the Pico-Union, Venice and Panorama City (San Fernando Valley) areas.

Led by a six-foot-three 280-pound 'soldier of fortune' named David Roybal, this security squad is renown amongst landlords for its efficient brutality. Suspected drug-dealers and their customers, as well as mere deadbeats and other landlord irritants, are physically driven from buildings at gunpoint. Those who resist or even complain are beaten without mercy. In a Panorama City raid a few years ago, the *Times* notes, 'Roybal and his crew collared so many residents and squatters for drugs that they converted a recreation room into a holding tank and handcuffed arrestees to a blood-spattered wall.' The LAPD knew about this *private jail* but dismissed residents' complaints 'because it serves the greater good.'

Roybal and his gang closely resemble the so-called *matadors*, or hired gunslingers, who patrol Brazilian urban neighborhoods and frequently, while the police deliberately turn their backs, execute persistent criminals, even street urchins. Their common coda is that 'they get the job done [after] all else has failed.' As one of Roybal's most aggressive competitors explains:

> Somebody's got to rule and when we're there, we rule. When somebody says something smart, we body slam him, right on the floor with all of his friends looking. We handcuff them and kick them and when the paramedics come and they're on the stretcher, we say: 'Hey, sue me.'

Apart from these rent-a-thugs, the Inner City also spawns a vast cottage industry that manufactures bars and grates for home protection. Indeed most of the bungalows in the inner ring now tend to resemble cages in a zoo. As in a George Romero movie, working-class families must now lock themselves in every night from the zombified city outside. One inadvertent consequence has been the terrifying frequency with which fires immolate entire families trapped helpless in their barred homes.

The *prison cell house* has many resonances in the landscape of the inner city. Before the Spring uprising most liquor stores, borrowing from the precedent of pawnshops, had completely caged in the area behind the counter, with firearms discretely hidden at strategic locations. Even local greasy spoons were beginning to exchange hamburgers for money through bullet-proof acrylic turnstiles. Windowless concrete-block buildings, with rough surfaces exposed to deter graffiti, have spread across the streetscape like acne during the last decade. Now insurance companies may make such *riot-proof bunkers* virtually obligatory in the rebuilding of many districts.

Local intermediate and secondary schools, meanwhile, have become even more indistinguishable from jails. As per capita education spending has plummeted in Los Angeles, scarce resources have been absorbed in fortifying school grounds and hiring armed security police. Teenagers complain bitterly about overcrowded classrooms and demoralized teachers on decaying campuses that have become little more than daytime detention centers for an abandoned generation. The schoolyard, meanwhile, has become a killing field. Just as their parents once learned to cower under desks in the case of an atomic bomb attack, so students today are 'taught to drop at a teacher's signal in case of … a driveby shooting – and stay there until they receive an all-clear signal.'

Federally subsidized and public housing projects, for their part, are coming to resemble the infamous 'strategic hamlets' that were used to incarcerate the rural population of Vietnam. Although no LA housing project is yet as technologically sophisticated as Chicago's Cabrini-Green, where retinal scans (c.f. the opening sequence of *Blade Runner*) are used to check id's, police exercise increasing control over freedom of movement. Like peasants in a rebel countryside, public housing residents of every age are stopped and searched at will, and their homes broken into without court warrants. In one particularly galling incident, just a few weeks before the Spring 1992 riots, the LAPD arrested more than fifty people in the course of a surprise raid upon Watts' Imperial Courts project.

In a city with the nation's worst housing shortage, project residents, fearful of eviction, are increasingly reluctant to claim any of their constitutional protections against unlawful search or seizure. Meanwhile national guidelines approved by Housing Secretary Jack Kemp (and almost certain to be continued in the Clinton administration) allow housing authorities to evict *families* of alleged drug-dealers or felons. This opens the door to a policy of *collective punishment* as practiced, for example, by the Israelis against Palestinian communities on the West Bank.

The half-moons of repression

In the original Burgess diagram, the 'half-moons' of ethnic enclaves ('Deutschland', 'Little Sicily', 'the Black Belt', etc.) and specialized architectural ecologies ('residential hotels', 'the two flat area', etc.) cut across the 'dart board' of the city's fundamental socio-economic patterning. In contemporary metropolitan Los Angeles, a new species of special enclave is emerging in sympathetic synchronization to the militarization of the landscape. For want of a better generic appellation, we might call them 'social control districts' (SCDs). They merge the sanctions of the criminal or civil code with land-use planning to create what Michel Foucault would undoubtedly have recognized as further instances of the evolution of the 'disciplinary order' of the twentieth-century city.

As Christian Boyer paraphrases Foucault:

> Disciplinary control proceed[s] by distributing bodies in space, allocating each individual to a cellular partition, creating a functional space out of this analytic spatial arrangement. In the end this spatial matrix became both real and ideal: a hierarchical organization of cellular space and a purely ideal order that was imposed upon its forms.

Currently existing SCDs (simultaneously 'real and ideal') can be distinguished according to their juridical mode of spatial 'discipline.' *Abatement* districts, currently enforced against graffiti and prostitution in sign-posted areas of Los Angeles and West Hollywood, extend the traditional police power over nuisance (the legal fount of all zoning) from noxious industry to noxious behavior. Because they are self-financed by the fines collected or special sales taxes levied (on spray paints, for example), abatement districts allow homeowner or merchant groups to target intensified law enforcement against specific local social problems.

Enhancement districts, represented all over Southern California by the 'drug-free zones' surrounding public schools, add extra federal/state penalties or 'enhancements' to crimes committed within a specified radius of public institutions. *Containment* districts are designed to quarantine potentially epidemic social problems, ranging from that insect illegal immigrant, the Mediterranean fruit fly, to the ever increasing masses of homeless Angelenos. Although Downtown LA's 'homeless containment zone' lacks the precise, if surreal, sign-posting of the state Department of Agriculture's 'Medfly Quarantine Zone', it is nonetheless one of the most dramatic examples of a SCD. By city policy, the spillover of homeless encampments into surrounding council districts, or into the tonier precincts of the Downtown scanscape, is prevented by their 'containment' (official term) within the over-crowded Skid row area known as Central City East (or the 'Nickle' to its inhabitants). Although the recession-driven explosion in the homeless population has inexorably leaked street people into the alleys and vacant lots of nearby inner-ring neighborhoods, the LAPD maintains its pitiless policy of driving them back into the squalor of the Nickle.

The obverse strategy, of course, is the formal *exclusion* of the homeless and other pariah groups from public spaces. A spate of Southland cities, from Orange County to Santa Barbara, and even including the 'People's Republic of Santa Monica', recently have passed 'anti-camping' ordinances to banish the homeless from their sight. Meanwhile Los Angeles and Pomona are emulating the small city of San Fernando (Richie Valens' hometown) in banning gang members from parks. These 'Gang Free Parks' reinforce non-spatialized sanctions against gang membership (especially the recent Street Terrorism Enforcement and Prevention Act or STEP) as examples of 'status criminalization' where group membership, even in the absence of a specific criminal act, has been outlawed.

Status crime, by its very nature, involves projections of middle-class or conservative fantasies

THE ECOLOGY OF FEAR

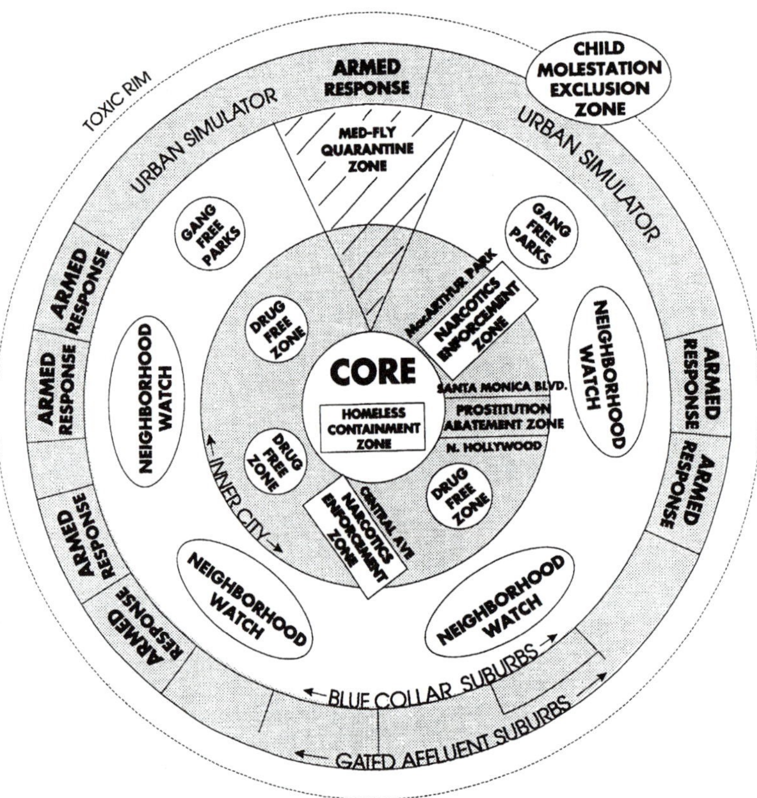

about the nature of the 'dangerous classes.' Thus in the 19th century the bourgeoisie crusaded against a largely phantasmagorical 'tramp menace', and, in the 20th century, against a hallucinatory domestic 'red menace.' In the middle 1980s, however, the ghost of Cotton Mather suddenly reappeared in suburban Southern California. Allegations that local daycare centers were actually covens of satanic perversion wrenched us back to the seventeenth century and the Salem witch trials. In the course of the McMartin Preschool molestation case – ultimately the longest and most expensive such ordeal in American history – children testified about molester-teachers who flew around on broomsticks and other manifestations of the Evil One.

One legacy of the accompanying collective hysteria, which undoubtedly mined huge veins of displaced parental guilt, was the little city of San Dimas' creation of the nation's first 'child molestation exclusion zone.' This Twin-Peaks-like suburb in the eastern San Gabriel Valley was sign-posted from stem to stern with the warning: 'Hands Off! Our children are photographed and fingerprinted for their own protection.' I don't know if the armies of lurking pedophiles in the mountains above San Dimas were actually deterred by these warnings, but any mapping of contemporary urban space must acknowledge the existence of such dark, Lynchian zones where the *social imaginary* discharges its fantasies.

Meanwhile, post-riot Southern California seems on the verge of creating yet more SCDs. On the one hand, the arrival of the federal 'Weed and Seed' program, linking community development funds to anti-gang repression, provides a new set of incentives for neighborhoods to adopt exclusion and/or enhancement strategies. As many activists have warned, 'Weed and Seed' is like a police-state caricature of the 1960s War on Poverty, with the Justice Dept. transformed into the manager of urban redevelopment. The poor will be forced to cooperate with their own criminalization as a precondition for urban aid.

On the other hand, emerging technologies may give conservatives, and probably neo-liberals as well, a real opportunity to test cost-saving proposals for *community imprisonment* as an alternative to expensive programs of prison construction. Led by Heritage Institute ideologue Charles Murray – whose polemic against social spending for the poor, *Losing Ground* (1984), was the most potent manifesto of the Reagan era – conservative theorists are exploring the practicalities of the *carceral city* depicted in sci-fi fantasies like *Escape from New York* (which, however, got the relationship of land-values all wrong).

Murray's concept, as first adumbrated in the *New Republic* in 1990, is that 'drug-free zones for the majority' may require social-refuse heaps for the criminalized minority. 'If the result of implementing these policies [landlords' and employers' unrestricted right to discriminate in the selection of tenants and workers] is to concentrate the bad apples into a few hyper-violent, antisocial neighborhoods, so be it.' But how will the underclass be effectively confined to its own 'hyper-violent' super-SCDs and kept out of the drug-free shangri-las of the overclass?

One possibility is the systematic establishment of discrete *security gateways* that will use some biometric criterion, universally registered, to screen crowds and bypassers. The 'most elegant solution', according to a recent article in the *Economist,* 'is a biometric that can be measured without the subject having to do anything at all.' The individually unique cart-wheel pattern of the iris, for example, can be scanned by hidden cameras 'without the subject being any the wiser'. 'That could be useful in places like airports – to check for the eye of a Tamil Tiger, or anybody else whose presence might make security guards' pupils dilate.'

Another emerging technology is the police utilization of LANDSAT satellites linked to Geographical Information Systems (GIS). Almost certainly by the end of the decade the largest US metropolitan areas, including Los Angeles, will be using geosynchronous LANDSAT systems to manage traffic congestion and oversee physical planning. The same LANDSAT-GIS capability can be cost-shared and time-shared with police departments to surveil the movements of tens of thousands of electronically tagged individuals and their automobiles.

Although such monitoring is immediately intended to safeguard expensive sports cars and other toys of the rich, it will be entirely possible to use the same technology to put the equivalent of an electronic handcuff on the activities of entire urban social strata. Drug offenders and gang members can be 'bar-coded' and paroled to the omniscient scrutiny of a satellite that will track their 24-hour itineraries and automatically sound an alarm if they stray outside the borders of their *surveillance district.* With such powerful Orwellian technologies for social control, community confinement and the confinement of communities may ultimately mean the same thing.

Source: Davis, 1994, pp.3–12

READING 3C
Andy Beckett: 'Take a walk on the safe side'

If you drive up to Universal Studios' over-watered hilltop above Los Angeles, and turn right past their busy turnstiles, you come to Universal CityWalk. From across its great sweep of sun-baked car park CityWalk looks like a fortified funfair. Pink and green neon-clad towers poke up from behind the thick curtain of its blank, high walls. Mounted police circle its perimeter.

CityWalk is a shopping citadel: 38 boutiques, restaurants and amusements shelter from the rest of the city behind its grey walls. Neatly dressed families walk busily (no cars are allowed inside) in the shade of numbered palm trees, past smiling buskers and tidy flowerbeds. Californian rock and Californian coffee waft in the warm winter air. Kids play in a fountain against a backdrop of dry hills that could be Tuscany. There's no graffiti – only a long children's mural asking 'What do you hope for in the future?' 'To live away from Big Cities' is the answer.

Built on rock with seismic joints and maximum safety margins, CityWalk's consumer idyll was shielded from the earthquake that twisted and cracked the San Fernando Valley just below it. A few tiles fell into the cosy canyon between the shops, a couple of them had to close for a day or two, some staff couldn't get into work because of freeway damage – and custom was back to normal in a fortnight.

Now CityWalk is expecting to benefit from the earthquake. 'People will need something to take their minds off what they've been through', says Jack Kyser, a Los Angeles County economist. But it isn't open for postquake partying. The first thing you see as you approach one of its narrow entrances is a list of rules, on a board strapped to a palm tree. 'Boisterous activities' are forbidden on the premises; so are 'rummaging in the trash receptacles', wearing clothes 'likely to create a disturbance', 'sitting on the ground for more than five minutes', and any 'expressive activity without the prior written permission of the management'. There are lots of security guards about too: sauntering in pairs, cruising the car park in regular circles. CityWalk has three separate forces, and an on-site police station. 'The security presence keeps troublemakers away in the first place', says the CityWalk public relations office. There are security floodlights fixed above the shop fronts.

Entrance to CityWalk is free, but parking is expensive. If you don't have a car, public transport stops a strenuous 15-minute climb below CityWalk's hilltop, at a little-used bus stop overgrown with ivy. It's on Cahuenga Avenue, the nearest public street and a generic strip of poor North Hollywood: offices, petrol stations, fast food restaurants, and old men fussing in a liquor store as the traffic speeds past relentlessly outside. North Hollywood is an area in slow economic decay, the small businesses of its new Latino and Asian immigrants not making up for all the investment and customers going up the hill to CityWalk.

CityWalk cost Universal's parent company MCA \$100m to build. Since it opened last spring, a year after the Los Angeles riots, CityWalk's promise of 'urban thrills without the ills' has drawn up to 25,000 people a day, jamming approach roads and restaurant waiting lists. The Universal Studios tour next door has also benefited, with several all-time record attendances. It was already drawing five million visitors a year.

'CityWalk's a home run', says Kyser. 'It's exceeded MCA's expectations.' The powerful *Los Angeles Times* reports favourably ('It's going to change the recreational center of gravity of … maybe even the whole of LA') about CityWalk almost every day.

The sheer cost of CityWalk, twice that of *Jurassic Park*, has dictated how it operates. Rents are twice the local average, so its tenants need to attract what management calls 'an upscale demographic'. This isn't easy in a Los Angeles economy traumatised by earthquake, fire, riots, recession and defence cuts. People with money do come to CityWalk – even on a winter weekday afternoon, it's busy with lunching film executives, families from the leafy Westside of the city, and comfortably dressed San Fernando Valley retirees, spending steadily – but they have expensive needs.

First of all, CityWalk has to draw its customers with skilfully targeted 'specialty shopping'. So it has a sports memorabilia store selling basketballs autographed by Magic Johnson for \$399.95; a giant novelties store with dozens of two-foot slices of rubber toast for \$69.95; somewhere to have yourself inserted into stills from famous films; a book shop offering free gourmet coffee with every purchase over \$19.95; and a University of California annex offering courses in – surprise – media and entertainment, right next door to Crabtree & Evelyn. CityWalk has lots of shops for its customers' children too: one with its own play area and personalising option on clothes and books, another with an artificial rainforest, complete with a halo of damp mist. And when the kids get bored, they can go and mess around with video cameras in the Steven Spielberg-designed interactive pavilion down the road.

Eating at CityWalk is 'upscale' too. The cheapest place is Jody Maroni's Sausage Kingdom, which sells duck and sun-dried tomato sausages for the price (\$4) of a whole breakfast in the rest of Los Angeles, from a polished and tidied version of a popular Venice Beach shack. In fact, there are cleaned-up, miniature pieces of Los Angeles all over CityWalk, arranged as a westward journey from its Hollywood (an 18-screen cinema) to its Pacific Ocean (a wave pool outside a surf shop). The promenade, the outdoor seating, the entire *faux*-Mediterranean concept are all openly borrowed from other popular local shopping areas like Venice, Melrose Avenue and Third Street, Santa Monica. But 'CityWalk will be safer than those places', says its architect John Jerde, 'because this is private property.' CityWalk's publicity emphasises this repeatedly.

Universal's hilltop has been a private, self-contained fiefdom since the film company bought a 130-acre chicken ranch there in 1914, installed its own power and water, and won semi-independent status from the rest of Los Angeles as Universal City. Today, Universal City still gets special tax and planning concessions, and has grown – like Disneyland to the south – into a substantial company town, with CityWalk its newest suburb.

Jerde says he designed CityWalk for locals, not tourists, as a *replacement* for Los Angeles – a compact, clean alternative to the scattered, dirty city, 'the city that Los Angeles never had'. He says he wants CityWalk to 'address the needs of everybody … for a centre … an emotional place to relate to'. But there's a hard edge beneath his community rhetoric, behind CityWalk's much-publicised charity events and planning consultations with local residents. Jerde emphasises that CityWalk is 'an enclave'. The barely veiled promise of social insulation from what its PR calls 'troublemakers' is a key to CityWalk's appeal.

Half an hour away at Third Street, Santa Monica, there are original branches of many of CityWalk's shops and restaurants, set in a wide, open promenade lined with wooden benches. This is a public street in a liberal part of Los Angeles often ridiculed by the rest of the city for being soft on crime. Third Street has lots of homeless people: they push shopping trolleys full of returnable bottles between the cafés and ask browsers at the news stands for change. Punters seem to be just hanging around, not necessarily shopping. They're younger and less wealthy than those at CityWalk. There aren't many families here.

'The homeless cast a pall', says economist Jack Kyser, who's a big CityWalk enthusiast. Mike Davis, a big critic, and professor of urban theory at the Southern California Institute of Architecture, agrees. In Los Angeles today, he says, 'Wealthy people think they own an insulation from the poor as a kind of right', and CityWalk is in the business of 'filtering people by category'.

Social control was in the CityWalk plan from the start. MCA first thought of the project as a means of directing and slowing the flow of people from Universal's huge car parks to its Studios. Once commissioned, CityWalk's architect John Jerde saw his work 'as designing the experiences rather than the buildings, designing all the things that happen to you … One of the things we do best is create environments'. Manager Tom Gilmore calls the result 'safely chaotic'. That perhaps gives a clue to CityWalk's backless stone benches, and its hard paving instead of grass that can be sprawled on. No amount of lush subtropical planting can disguise the fact that CityWalk doesn't encourage loitering, and people generally don't – they criss-cross its streets from shop to shop without quite looking at home.

CityWalk's buildings look forbidding underneath their veneer of kitsch (a giant pizza slice as a facade, a giant yellow surfboard as a roof), like concrete-and-steel bunkers. Shops are ornately fronted but surprisingly shallow inside, with great voids of corridors and store rooms behind them. Jerde designed CityWalk using computer-compiled and altered traces of local Los Angeles architecture; it reproduces both the freeway and beach symbols of the mythical Los Angeles – a lost world today, with curfews stopping cruising and the beaches shut at night by the police – and also the paranoia increasingly infecting the real city. The extravagant CityWalk branch of Gladstone's seafood restaurant may have its own artificial beach, but it also has its own security guards.

Fortification isn't new to Los Angeles: the original 1900s *Los Angeles Times* building was a citadel with sentry boxes and an armoury. Shopping malls in Southcentral Los Angeles had their own police stations years before CityWalk. The sheer size of the city, and the fact that so many of its white residents came here to escape the growing integration of eastern cities, has always made it economically and racially segregated. Egalitarian 19th-century ideals about socially 'promiscuous' cities, developed by academics and planners in Chicago and New York, never really rooted here.

But the process really took off after the Watts riots of 1965, the predecessors in scale and impact of those in 1992. As nervous investors left poor parts of the city and gangs moved in, so a demand was created for 'gated' communities, private housing estates separated from the outside world by fences, walls, and guardhouses. Today, some of them have become virtual city states, with their own lists of rules and ruling committees …

CityWalk's cultural rhetoric can be as sophisticated as its cultural politics are brazen. John Jerde outflanks critics who point to CityWalk's Disney-like unreality by agreeing with them. He calls it 'a simulacrum', the French postmodern thinker Jean Baudrillard's term for a copy with no original. And to people like Mike Davis who attack CityWalk's élitism, Jerde says segregated private spaces like CityWalk with 'a public feel' – and their own public institutions like libraries and schools – are the only, regrettable solution to America's rising crime and fear of crime, which make real public spaces unworkable. In New Jersey, a public park was recently fitted with undersoil sensors to detect after-hours trespassers.

Of course, Jerde's is a very middle class notion of social workability. Everyone else still throngs Los Angeles's remaining public spaces at weekends, without constant fear for their lives. But, as CityWalk's attendances show, Jerde has picked up on a growing desire to hide from America's apparently insoluble social problems in leisure enclaves – what Canadian social critic Douglas Coupland calls 'bunkering' – at least among those who can afford to.

Jerde wants CityWalk to be just the start, the hilltop hub of a new 'Urbanopolis'. MCA has won permission to double the site, and wants to build residential towerblocks too. Excavators are already clanking and grinding on the slopes below, and a half-built concrete skeleton partly obscures the distant hills, the last piece of the outside world still visible from inside CityWalk. Houston and San Diego already have their own CityWalk-style developments, and Las Vegas – America's fastest-growing city, and site for Jerde's new projects – is moving away from cheap-and-cheerful gambling to a series of huge fantasy-themed resorts.

But the logic of this escapism is self-fulfilling, says Davis. Developments like CityWalk siphon jobs and investment from the rest of Los Angeles, which deteriorates further, thus creating a demand for more segregation. This certainly seems to have happened since the 1992 riots: rather than trying to revitalise poor, dangerous areas like Southcentral Los Angeles, wealthier residents have bought guns (665,000 more across the state last year) and poured into places like CityWalk. And Davis predicts that as its economic problems worsen, the territorial carve-up of LA will lead to a Brazilian or Filipino-style 'final solution': private death squads protecting the enclaves of the super-rich from everyone else.

There's a chance that the earthquake may bring belated recognition, and federal aid, for the city's general deterioration. But this didn't happen after the riots, or after last years' fires. Earthquake relief may well be the latest priority to jump the queue ahead of relieving Los Angeles's social tensions – a much more politically difficult task, with the state short of money and its leaders generally unsympathetic to the poor, few of whom vote.

The earthquake struck hardest at Los Angeles' most vulnerable citizens: the Asians and Latinos living in unreinforced stucco houses and tenements, like the one that collapsed and killed 16 people in Northridge, about 15 miles from CityWalk. This means more homeless people – only a quarter of those who lost their homes could afford earthquake insurance. Moreover, the physical and psychological damage done – an acceleration in the decay in the city's infrastructure, and of its sense of crisis – may hasten what Davis calls a 'collective nervous breakdown' in Los Angeles. Meanwhile, CityWalk's fantasy looks more attractive with every news broadcast.

Source: Beckett, 1994, pp.10–12

CHAPTER 4

Divisive cities: power and segregation in cities

by Jenny Robinson

1 *Introduction: inside cities*

At the same time as cities bring together people, resources and ideas, intensifying social relations and creating intense experiences, they generate and enable different responses to these sometimes overwhelming phenomena. The apparent disorderliness and complexity of cities has contributed to the emergence of all sorts of strategies which help people cope with city life or which try to order and regularize social interactions and urban space. From the studied indifference of the passer-by to the drawing of hard spatial boundaries in gated communities (see **Allen, 1999a**; Chapter 3), from an enthusiastic embracing of difference to a disdainful rejection of different ways of living as disorderly (see **McDowell, 1999**; Chapter 2), from the stimulation of creative enclaves to the disconnection of some areas from city economies (Chapter 1), city spaces bear witness to the different ways in which people have responded to the intensity and diversity of city life.

As we saw in Chapter 1, heterogeneous histories and geographies drawn together in cities produce *differentiated urban spaces* – a series of differentiated, relatively homogeneous areas, with particular kinds of buildings and streets, often lived in by people with different kinds of backgrounds, sometimes located next to one another, but sometimes cut off from one another in more or less definite ways. These differentiated spaces could be the inner city or the suburb, the commercial area or the industrial district, the gated community or the ghetto. This chapter will explore further some of the ways in which cities come to be patterned internally, divided into differentiated neighbourhoods, zones or areas which are more or less separated from one another by boundaries, borders or distance. Separation, borders and boundaries, then – as much as coming together, mixing and flow – are characteristic of city space.

This chapter will delve more deeply into understanding why it is that alongside the cosmopolitanism and heterogeneity associated with cities and the diverse encounters which cities stimulate, there are all sorts of *differentiations, separations and divisions* between citizens who apparently share the same city.

ACTIVITY 4.1 Examine the maps of ethnic segregation in London (Figures 4.1(a)–(h), produced from the 1991 census by the London Research Centre. Take a look, too, at the map of Johannesburg (Figure 4.2). Think about the extent to which these maps show evidence of differentiation of urban space. ◆

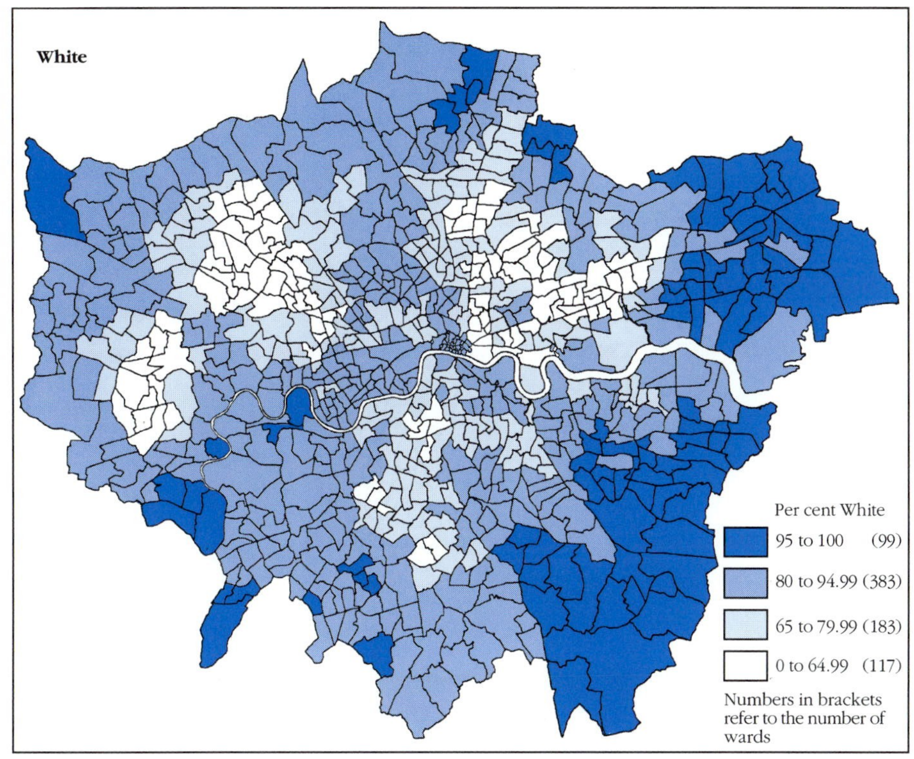

FIGURE 4.1(a) *Percentage of population in ethnic groups in London, by ward: White*

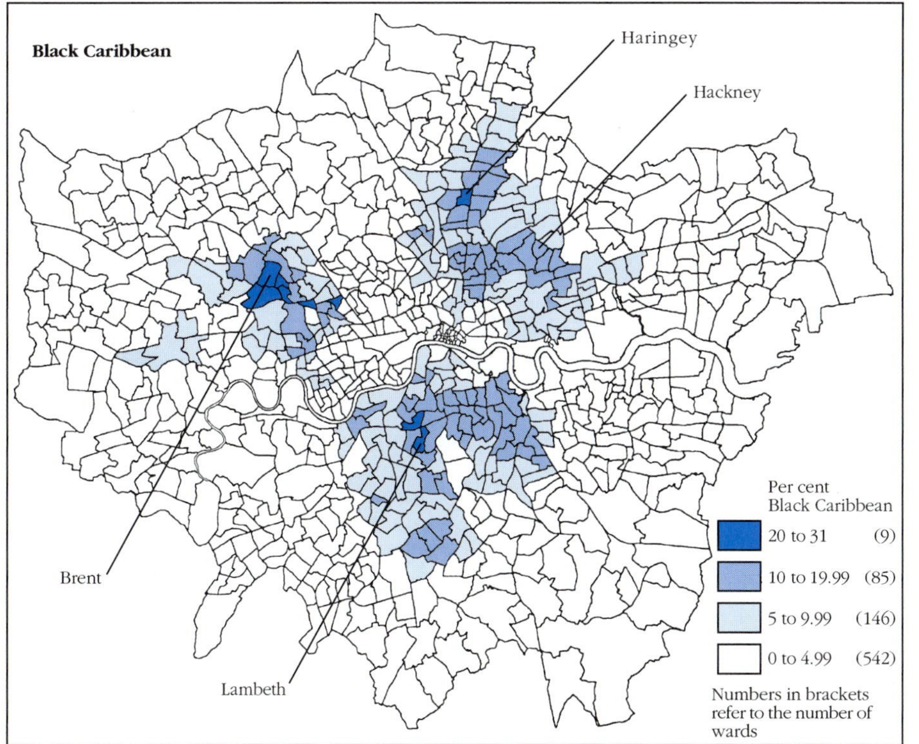

FIGURE 4.1(b) *Percentage of population in ethnic groups in London, by ward: Black Caribbean*

FIGURE 4.1(c)
Percentage of population in ethnic groups in London, by ward: Black African

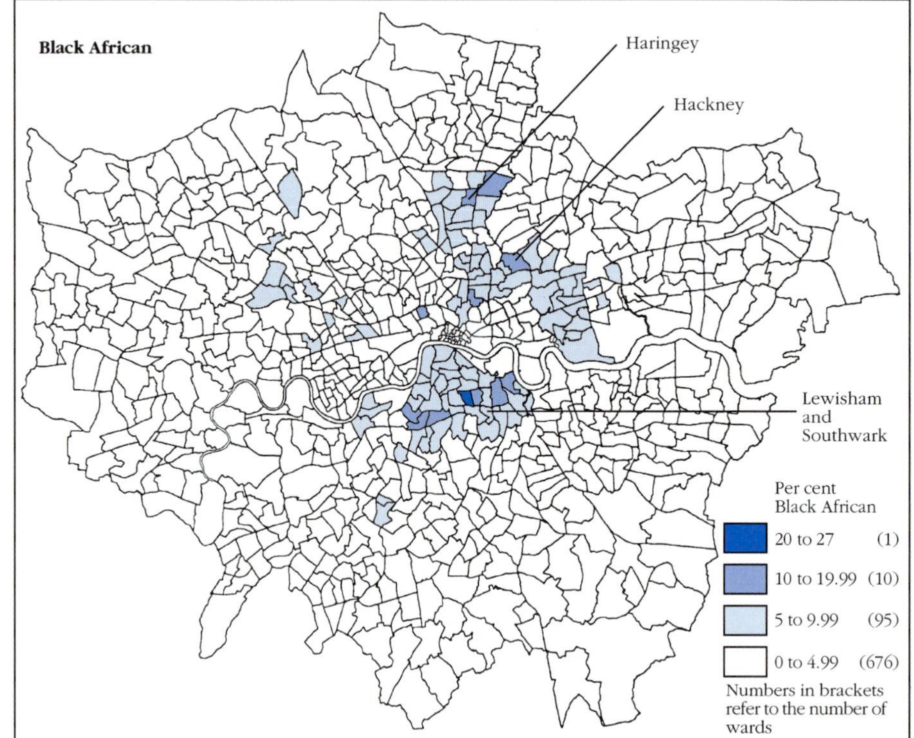

FIGURE 4.1(d)
Percentage of population in ethnic groups in London, by ward: Black other

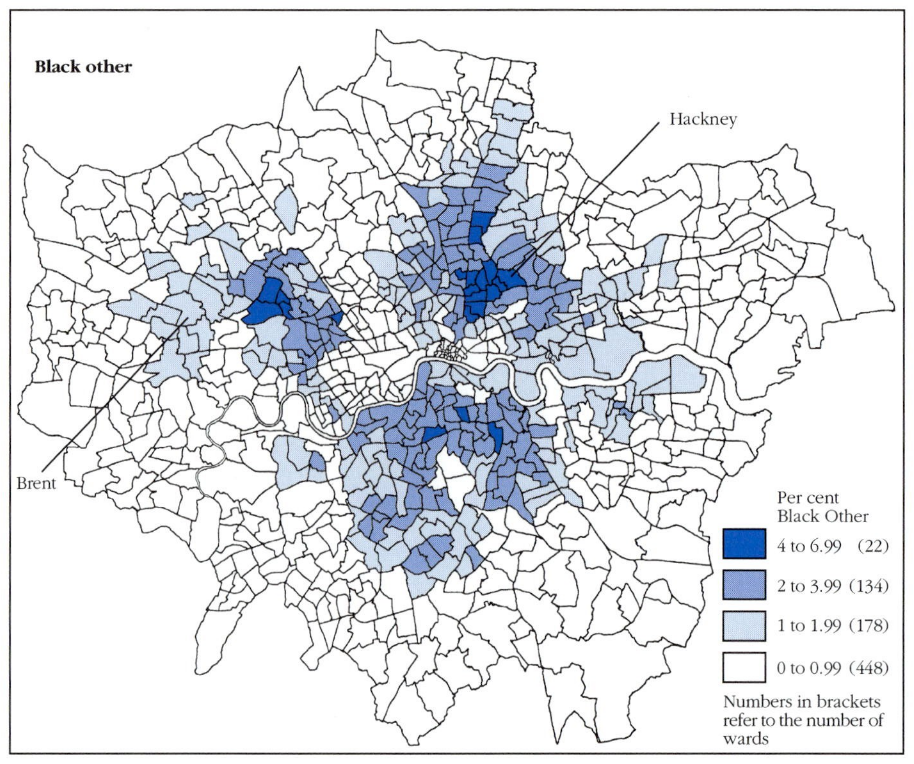

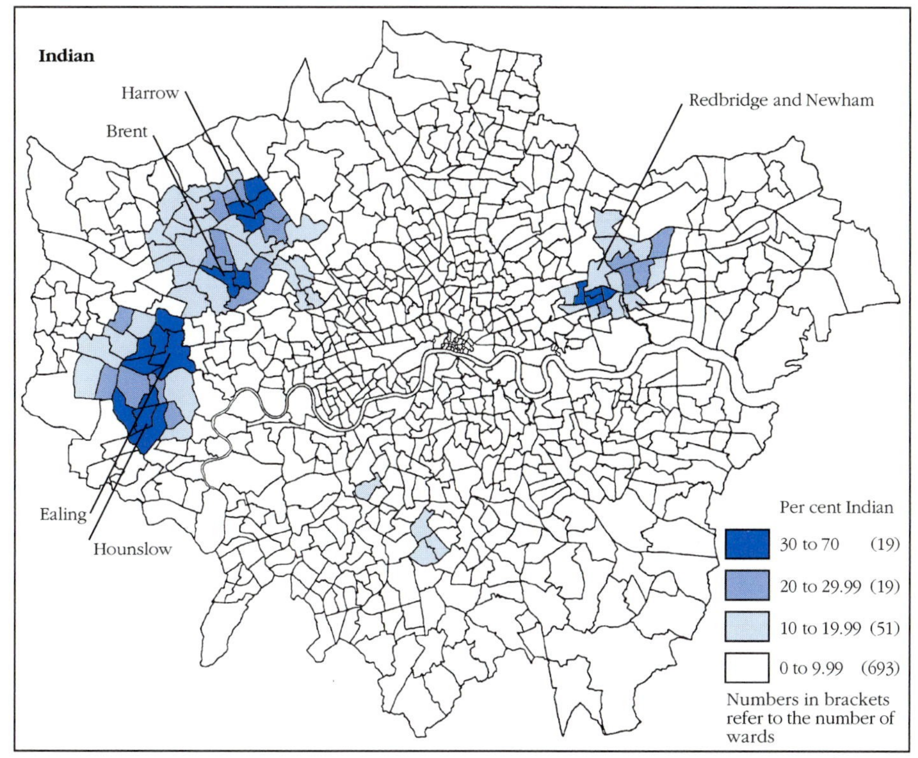

FIGURE 4.1(e) *Percentage of population in ethnic groups in London, by ward: Indian*

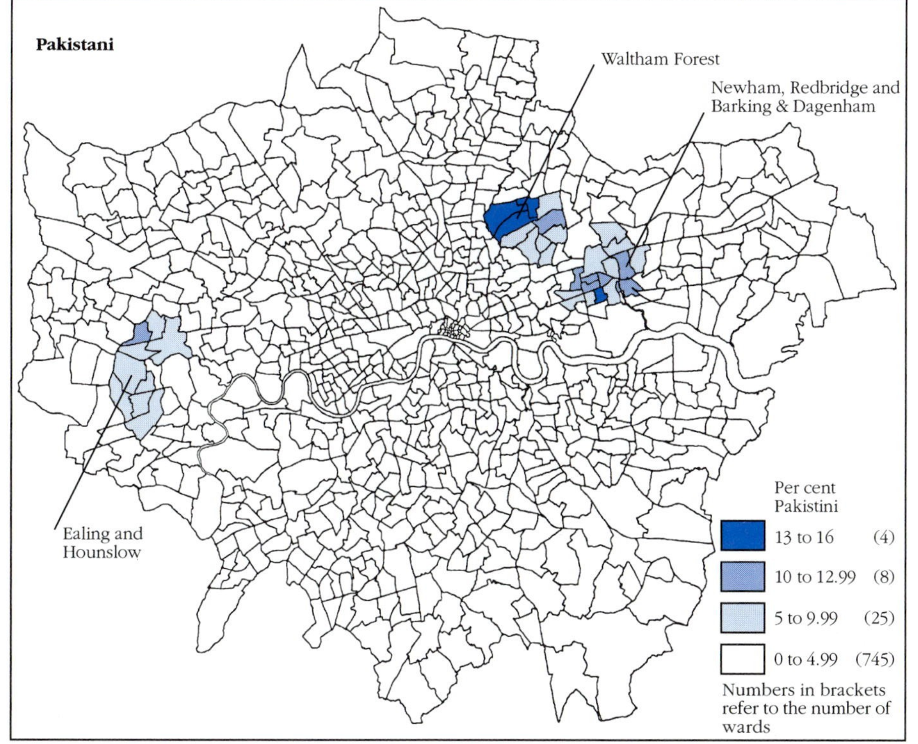

FIGURE 4.1(f) *Percentage of population in ethnic groups in London, by ward: Pakistani*

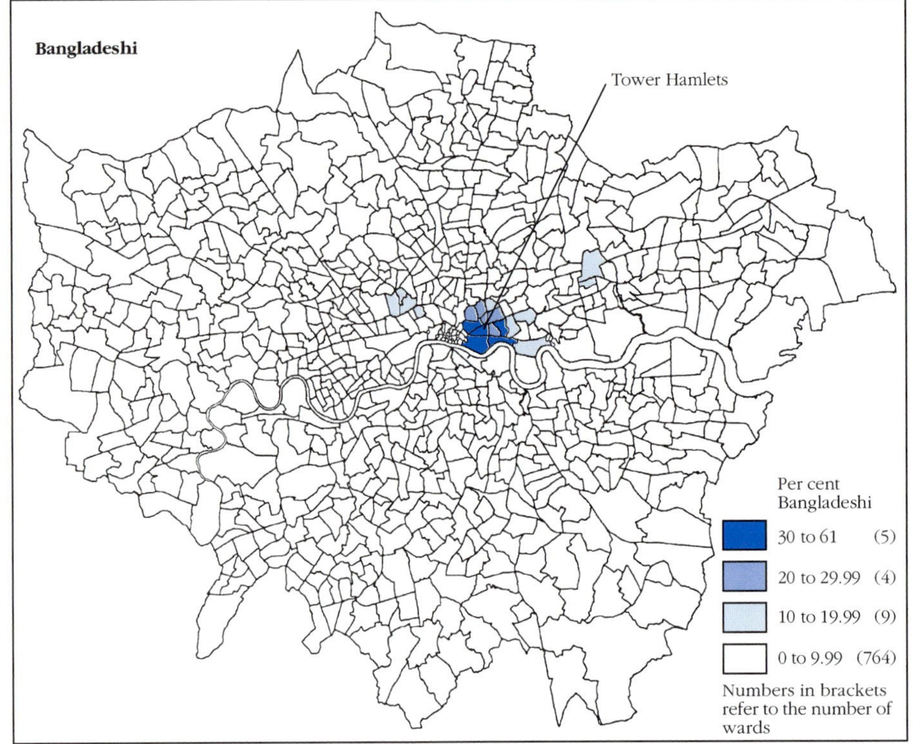

FIGURE 4.1(g) *Percentage of population in ethnic groups in London, by ward: Bangladeshi*

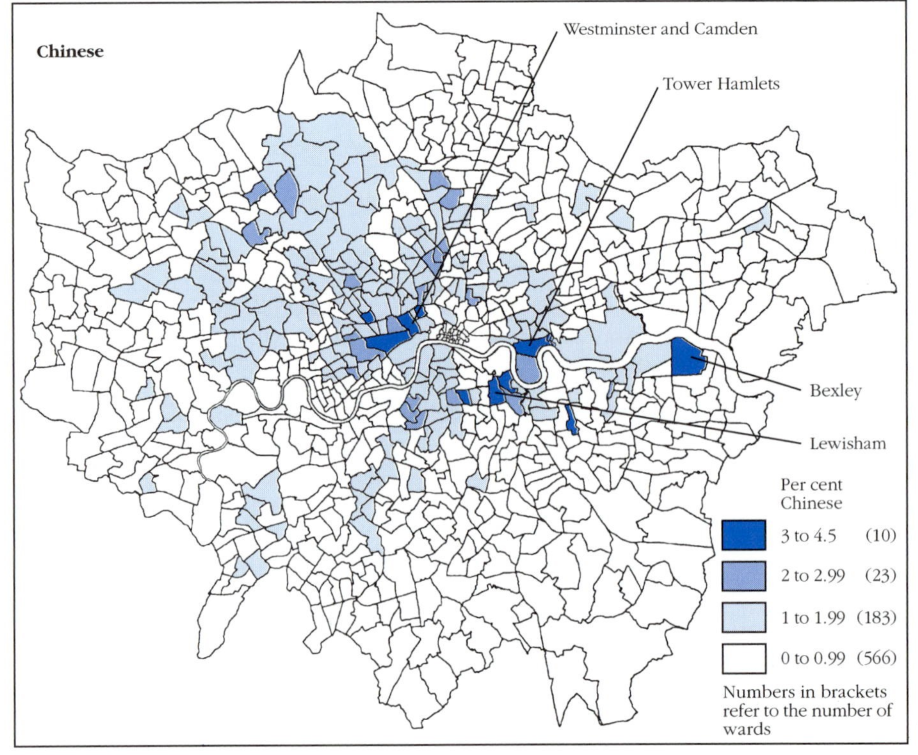

FIGURE 4.1(h) *Percentage of population in ethnic groups in London, by ward: Chinese*

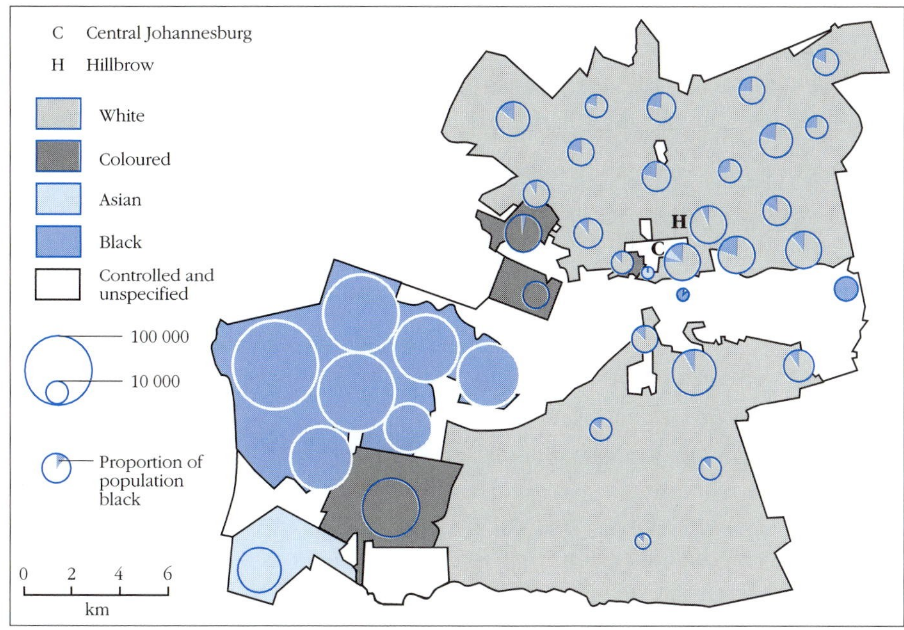

FIGURE 4.2
Johannesburg: population distribution by group area, 1990

As I was looking at the maps of ethnic groups in London one of the things that struck me about them was that they created something of an illusion! At first sight, it looks as if there are really clear divisions of the city associated with different ethnic groups. However, from the keys to the maps, it is apparent that very different levels of segregation are involved in each case and that some ethnic communities are only a very small minority of the population in the area (as Starkey, 1994, points out). Bangladeshi people living in Tower Hamlets, for example, made up 30–61 per cent of the population in that area, whereas Pakistani people in Waltham constituted only 13–16 per cent – yet both groups were indicated by the same shading on their respective maps. Compare this to the map of Johannesburg in 1990, where the levels of racial concentration are generally over 90 per cent. Although the maps may have given the impression that we were looking at quite similar spatial phenomena, the levels of differentiation involved are very varied and the reasons why these social concentrations emerged are also quite diverse. In South Africa laws and police action ensured the very high levels of segregation recorded, whereas in Britain we will see that it was generally failures of the state to act that enabled the levels of differentiation by ethnic origin shown in Figures 4.1(a)–(h).

In addition, the maps of social differentiation do not make clear what the spatial concentrations mean for people living in the different areas. In some neighbourhoods people living among others from a similar ethnic background may find that this offers a sense of belonging and access to services and goods which they enjoy. For example, in one London neighbourhood with a large number of Turkish residents, the local supermarket has a special section with traditional Turkish foods and there are Turkish grocery stores, restaurants and bakeries nearby. However, in other cases residential concentrations of people from a specific ethnic background may give rise to racist attacks from antagonistic neighbours, or may be the result of a situation in which people have very little choice about where to live and in which they are faced with poorer services and facilities than those of other communities.

With this in mind, we should be aware of

- the *different causes* of different kinds of spatial differentiation in the city, and be alert to
- the *different meanings* that separations might have for the people concerned.

As we try to understand this patterning of the city, we should also bear in mind that:

- As well as being divided up and parcelled out among different social groups or activities, different city spaces are also *connected* to one another through numerous interactions, paths, routes, journeys or networks. The various rhythms of urban life often work to draw people from different areas together. Even the heavily divided spaces of South Africa's cities have links and connections running between and across them.
- These interconnections frequently mean that even quite different, and quite separate, urban spaces are in fact *heterogeneous*, as we saw in relation to the suburb in Chapter 1, for example.

The patterning or differentiation of cities is not always as clear or as settled as is implied by the image of the map. But first we need to consider how and why some of these differentiations and separations come to be put in place at all.

2 Creating differentiated spaces

2.1 IMPOSING RACIAL DOMINATION IN THE COLONIAL CITY

In explaining experiences of racialized exclusion, imposition of borders, boundaries and spatial differentiation, or segregation in cities, it is the 'power' of some groups to impose these on others which often attracts attention. In relation to this, let us look at Extract 4.1 from Kay Anderson, writing about the beginnings of Vancouver's Chinatown.

EXTRACT 4.1
Kay Anderson: 'Cultural hegemony and the race-definition process in Chinatown, Vancouver: 1880–1980'

Amid the growth and change wrought by the completion of the trans-Canada railway at its western terminus [Vancouver] in 1885, the once peaceful relations between the small settlements of whites, Chinese, and Indians on Burrard Inlet grew competitive and violent. The first Mayor of Vancouver, Mr A Maclean, lost little time when a 'small horde of unemployed Chinamen' tried to locate in the emerging business section on Water Street. It was the 'thin end of the wedge', a news report stated (*Vancouver News* 2 June 1886, p.2), and, in the same month, he and Alderman Hamilton presided over three street meetings convened to find means of preventing Chinese settlement. ... By November, with Chinese scattered in various locations, Hamilton pressed in Council for further 'action in trying as far as possible to prevent Chinese from locating within the city limits' (City of Vancouver Archives (CVA), *Council Minutes* volume 1, 8 November 1886, p.164). To aid the cause, Council prohibited the employment of 'any person of Chinese race' on municipal contracts or City-assisted projects (see CVA *Incoming Correspondence* volume 1, 8 August 1887, p.82).

The City was helpless, however, to constrain people like Mr J McDougall who hired 'batches of Mongolians' in 1886 for his clearing contract in Vancouver's West End (*Vancouver News* 9 January 1887, p.4). Tension built among the city's white population, and by late February, after a crammed meeting at City Hall, some 300 of them tried their own hand at ousting the Chinese. While the authorities looked on, they attacked the West End camp, after which they burned the homes of a settlement of Chinese on Dupont Street (*Vancouver News* 25 February 1887, p.1). Ironically, it was the Provincial Government in Victoria, not known for its sympathy to the

Chinese, that intervened on behalf of the victims. The Attorney-General looked unfavourably on Vancouver's decline into mob rule, and, within days of the riot, he suspended the Vancouver City Charter and sent constables to take charge of law and order (British Columbia Statutes (BCS), 1887).

With this protection, the dispersed Chinese returned to Vancouver and established a concentrated pattern of settlement in the swampland around Dupont Street ... Mr Justice Crease explained their 'tendency to congregate' as 'directly owing to the fact that as foreigners, held in dangerous disesteem by an active portion of whites, they naturally cling together for protection and support' (Report of the Royal Commission on Chinese Immigration (RCCI), 1885, p.143). Chinese spatial concentration was also a product of the host of *secondary* effects that their racial definition supported. The shanties around Dupont (later Pender) Street were home to a pool of politically vulnerable, single, and poorly paid men who were often dependent on Chinese bosses for employment, lodgings, and in some other cases, their head-tax payment (Chan, 1983). Merchants lived more comfortably, often in elegantly decorated homes (see *Province* 5 February 1910, p.4), but they too were constrained in their residential choices and forms of livelihood by the prevailing culture of race.

References

BCS (1887) *British Columbia Statutes,* 50, Victoria, chapter 33.

Chan, A. (1983) 'Chinese bachelor workers in nineteenth century Canada' *Ethnic and Racial Studies,* 5, 513–534.

CVA *Council Minutes* (miscellaneous years) Office of the City Clerk.

CVA *Incoming correspondence* (miscellaneous years) Office of the City Clerk.

Province 1906–67 (miscellaneous issues) Main Library, University of British Columbia, Vancouver.

RCCI (1885) 'Sessional Paper', 54a, report of the Royal Commission on Chinese Immigration, printed by order of the Commission, Ottawa.

Vancouver News 1886–87 (miscellaneous issues) Main Library, University of British Columbia, Vancouver.

Source: Anderson, 1988, pp.134–5

The relative power of the white residents of Vancouver to impose their will on the Chinese, either by 'mob rule' or through laws and practices of the white council based on their racist attitudes towards the Chinese, was directly responsible for the emergence of the segregated area of Chinatown in Vancouver. These practices were influenced by prevailing opinions in Europe about racial difference and the 'inferior' nature of 'oriental' people. Such opinions were formed and circulated through networks linking people within the British Empire, through widely read popular and scientific publications and through the media. All these contributed to the white settlers' belief that they were superior to the Chinese settlers. Competition for jobs, trade and land between whites and Chinese people in the emerging city also helped to bolster these anxious assertions of perceived racial superiority. The white settlers were not averse, as we saw, to using their privileged representation on the council (the Chinese could not vote) or their greater numbers to ensure that they kept the best opportunities for themselves.

In this example, relative power was used by one racial group to segregate another. The powerful actors and institutions which grew up with the city – the local council, law-making and law-enforcing bodies, notable citizens and aggressive mobs – were racially exclusive and were used to preserve racial privileges and racial separation. A similar example has been recorded by Sue Parnell in Johannesburg, South Africa, in the 1920s. Here white residents in the city were lobbying local and national governments to remove African people from an area known as the 'Malay location', which was situated adjacent to some popular white working-class neighbourhoods (Parnell, 1991). Initially the area had been set aside for Malay or coloured (mixed race) people. However, under pressure from a rapidly increasing African population who were desperate for accommodation close to town, and with the attraction of a lucrative income from dividing up homes for boarding houses, landowners had taken in African tenants and evicted coloured families. The local white working-class population – at the time, a powerful political force as the national government was struggling to keep Afrikaans- and English-speaking whites together in a ruling coalition – agitated strongly against the African presence in the area. Part of this was once again to do with competition for jobs, as African trade union organizations based in the Malay location were threatening white workers' positions.

From the local council's point of view, though, things were rather more complicated. They were interested in having closer control over the area, which was becoming something of a political power base for African people. But an earlier outbreak of influenza in 1918 had heightened their sense that 'public health' was also threatened by the abysmal living conditions in the locations. No running water or sewerage, unpaved streets and extreme overcrowding made the area dangerous not only to its inhabitants but, in the view of the Medical Officer of Health, it was also a serious threat to the health of the city. On the whole, any attempts to improve the living conditions in the area were opposed by white residents of nearby Vrededorp and Mayfair, who wanted the area dismantled, not improved. However, removing the residents meant that there had to be alternative places for them to live and the council was not willing to spend its own money on building houses for Africans. Racial segregation was promoted as an 'ideal' that would respond to the demands of racist agitators for removal, facilitate political control and enable public health concerns to be met (by allowing the provision of new, modern and sanitary conditions in segregated areas). But the council needed both more money and more staff before they could do anything. It was only much later that African people from the Malay location, as well as those living in areas such as Sophiatown, were able to be removed to more distant, state-built segregated townships.

Both of these examples point to two different but closely related groups of powerful actors who were involved in shaping the differentiation of urban space in former settler colonial contexts: on the one hand, the ordinary white citizens; on the other, the institutions and agents of the local government. The local white citizens attempted to use their influence and right to vote to remove less powerful racial groups to segregated neighbourhoods. The local council, however, had many different things to consider in their efforts to shape urban space. They had to be accountable to their electorate – the whites – but they also had to maintain public health, improve the way the city worked, and keep political order among the dominated population. These two sets of concerns did not always match up.

The concerns and practices of government have had very significant consequences for the spaces of cities.

- The need to control and order what happens in society – for example, the need for *surveillance* – has reshaped cities, as was shown in Chapter 3. This involves opening up previously inaccessible spaces, such as disorderly squatter settlements or slum areas, for inspection as well as restricting what can and cannot happen in certain spaces.
- The spaces of the city have also been radically altered by attempts to *improve* the conditions of living in cities, as was suggested by the Johannesburg example, discussed in this section.

Together with racist attitudes among dominant populations, both of these imperatives of city government – surveillance and improvement – have contributed historically to the racial differentiation of urban space in former settler colonial contexts, creating racially defined areas that are relatively homogeneous and separated from the rest of the city. It is quite ironic to think that attempts to improve living conditions in cities might contribute to divisions and separations. Let us explore this further in a different context.

2.2 IMPROVEMENT OR SEGREGATION?

An important strategy for improving living conditions in cities has been 'slum clearance' and redevelopment – the wholesale removal of houses and populations from run-down, usually central areas of the city. Associated with this has been the provision of public housing in new neighbourhoods, creating differentiated urban spaces as part of efforts to improve the quality of city life for poor and working class people. The middle decades of the twentieth century saw a spurt of 'slum clearance' initiatives in many countries around the world, including the UK. The clearance of inner city areas contributed to the process of suburbanization, as new housing was often provided on the outskirts of cities – although, as Mooney discusses in the case of Glasgow (Chapter 2), a number of suburban estates on the outskirts of cities have deteriorated and are no longer seen to be desirable locations. Another feature of this process of urban improvement in the UK was its contribution to racial segregation in Britain's larger cities. Susan Smith (1987) summarizes the events in Extract 4.2.

EXTRACT 4.2
Susan Smith: 'Residential segregation: a geography of English racism?'

Ironically, almost every major Housing Act ... has had the consequence of sustaining racial segregation and reducing the residential options open to black households.

One of the earliest policy changes with such an effect was the shift from slum clearance and redevelopment to *in situ* improvement, introduced in the 1969 Housing Act. Between 1958 and 1968, slum clearance in England and Wales decanted some 160 000 to 180 000 people per year from inner-city slums into peripheral estates and high-rise flats. Most of those removed were white. The 'middle ring' Victorian and Edwardian apartment houses and terraces which then accommodated the majority of black households would have been the next to go had clearance policies remained in place. By the mid 1960s, however, there were many pressures working to change the emphasis of housing policy towards gradual *in situ* improvement. These pressures included local opposition to the scale and organization of redevelopment; a massive repairs problem in housing *not* scheduled for demolition; and government concern to control public spending in the face of economic decline ... An additional pressure, though, was perceived social resistance to the rehousing of black people onto new council estates. Reginald Freeson raised this issue in Parliament in 1966, asking about the extent to which 'redevelopment plans are being held up by local authorities because they do not wish to accept responsibility for rehousing immigrants living in twilight zones of major city areas' (Hansard 1965–6, v. 725, c. 239). His question was dismissed as unfounded. Yet it was an allegation considered in a Political and Economic Planning Survey in 1966 ... and ratified in Rex and Moore's classic study of Sparbrook (1967). ... The *timing* of the shift away from comprehensive redevelopment had the consequence, intended or otherwise, of retaining the black population in those relatively highly segregated areas of cities to which the migrant labour process had originally drawn them.

Reference

Rex, J. and Moore, R. (1967) *Race, Community and Conflict*, Oxford, Oxford University Press.

Source: Smith, 1987, pp.29–30

Susan Smith traces the underlying racist assumptions and practices which led various individuals and government bodies to contribute to the emergence of racial segregation in British cities in the course of their efforts to improve the quality of housing in cities. Quite often this was a result of a laissez faire approach on the part of the government, which ignored evidence of racism in the private housing market and in the allocation of public housing. The government might have assumed that the problem would go away in time, but they were also unwilling to act against racism. The persistence of what Smith calls 'white supremacy' (or racism) in many spheres of British life – from the formal political arena to the private decisions of landlords and housing managers – ensured that racial segregation became a feature of British cities. By the 1991 census, levels of segregation seemed to be declining, although Ratcliffe (1997, p.98) concludes that there are still differentials in housing quality which are not 'explained' by class or economic factors. Discriminatory factors, he suggests, have not disappeared, despite the introduction of legislation to address racial discrimination and substantial changes in practices of allocating public housing.

Smith uses the term 'white supremacy' to mark the historical connections between the racism which shaped racial segregation in colonial and settler cities, and the fate of immigrants from former colonies in the racially discriminatory housing market in Britain. In this case, attempts to improve the city worked in concert with racist practices to create racially differentiated spaces in cities. David Theo Goldberg makes an even stronger connection between the practices of segregation in colonial contexts and cities in the West. He argues that planning ideas used in New York to reinforce segregation in the course of attempts to improve the city were similar to those used in South Africa to create the apartheid city.

ACTIVITY 4.2 A number of writers – such as Mike Davis (1993) – use the term 'apartheid' to describe the experience of racial segregation in the USA. Read through the extract from David Theo Goldberg (Reading 4A) and make a note of the ways in which the practices of planning in South Africa and other colonial situations seem to have influenced planning for segregation in the USA. ◆

An important aspect of Goldberg's argument is that ideas about how to improve and plan cities travel the world. Thinking spatially about the history of these practices within cities makes us aware of the connections outwards, the links with practices and ideas applied and developed elsewhere. Many colonial attempts to improve sanitary and health conditions in cities were learned in the colonizing country – either directly by people who subsequently went to work in a colonial context, or else indirectly, as planners in these countries looked to the latest innovations and laws in the West (today we might call this, learning from 'international best practice'). South Africa's apartheid city planning, for example, drew heavily on British and other planning ideas. However, Goldberg is concentrating on the connections which imply the movement of planning ideas in the opposite direction. The dominant practices of segregation developed in South Africa, Goldberg alleges, have also influenced contemporary Western town planning.

For example, Goldberg shows that the 'buffer strip' – cornerstone of the apartheid city – added a new dimension to otherwise benign elements of the city: parks, freeways, commercial or industrial areas came to be seen as convenient separators between different ethnic groups. The differentiated structure of the city was reinforced by these methods, he argues, and boundaries between different groups were strengthened.

Attempts to improve city life have therefore:

- contributed to the creation of socially differentiated and spatially separate neighbourhoods, for example, through slum clearance;
- meant that both racist attitudes and associated spatial strategies for separating racial groups have influenced contemporary planning practices; and also
- provided an opportunity for reinforcing racial separation.

So far we have emphasized the importance of political processes (racial politics and the state) in shaping the differentiation of city spaces. Many of the differences we observe between different city neighbourhoods are also associated with class, or income: the inner city and the suburb, for example, are often distinguished from each other on the basis of the relative wealth of their residents. The next section considers how these kinds of economic differentiations come about.

2.3 ECONOMIES OF DIFFERENTIATION

Many differentiations in land use in cities could be understood intuitively to be an outcome of different income levels and different abilities to pay. Certain types of location are more desirable for certain types of activities and a person's or firm's ability to pay, or bid, for use of the land which would most suit the activity in question would determine the nature of land use in that place. Some neighbourhoods may be very desirable – perhaps they are well located, have attractive houses, are close to good schools and have clean and pretty streets which feel safe at night. Some people may be prepared to pay a premium for a house there and would bid up the house prices in the area. This kind of intuition has been formalized into a 'bid-rent curve', that is, a mathematical model showing how different land uses come to be distributed across the city (see Figure 4.3).

This model leads us to expect that those land users able to pay the most for land in one area of a city – for example, the centre of town for retail traders who wish to be accessible to the greatest number of people – would be able to dominate that area. This would then appear as a distinct zone in the city, say the central business district. Other users would be excluded from buying or renting property in this area as it would be too expensive. Residential users, for

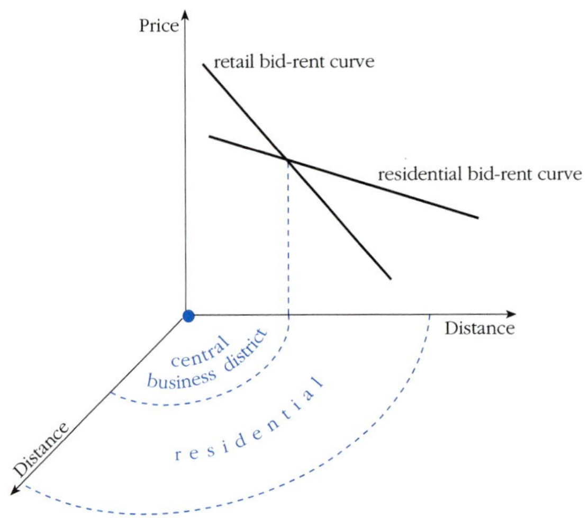

FIGURE 4.3 *Bid-rent curve model of the distribution of retail and residential land use in the city*

example, would only be able to pay competitive prices further from this central, commercial area and would be confined to the outskirts of the city (as shown in Figure 4.3). This is part of the thinking behind the Chicago School's division of the city into zones (see Figure 1.1 in Chapter 1). Formally, we could say that the market in land is partly responsible for the distribution and differentiation of the uses of space across the city.

The market, however, is a social creation. It is the product of many different actors and is specific to the context in which it operates. The urban property market is a particularly good example of this. David Harvey puts this point very well when he notes that in order to explain residential differentiation we cannot look only to the workings of the anonymous capitalist market. In addition to the rhythms of the capitalist economy which influence the fluctuations in investment in different parts of the city, he argues that

> we still have to turn to the examination of the activities of speculator-developers, speculator-landlords and real estate brokers, backed by the power of financial and government institutions, for an explanation of how the built environment and residential neighbourhoods are actually produced.
>
> (Harvey, 1989, p.121)

In the purchase, sale and use of urban land, for example, we find not only the buyer and the seller involved, but also the estate agent who negotiates the deal, the local council who control use of land, neighbours who affect the value of the property, and transport authorities who determine the accessibility of the area. We find, too, a whole range of socially specific sentiments about the design of the property and the neighbourhood, banks and finance houses who are probably involved in calculations about the individual purchaser and the property, as well as planners, community organizations and governments who may be busy negotiating future plans for the area. If we stop to consider how each of these actors is shaped by social processes ranging from gender relations to international flows of money, we realize that the 'market' constitutes only a very first step in an explanation of the differentiation of urban space.

As we have already seen, the state plays an important role in shaping the urban environment through direct investment, such as in housing construction, providing infrastructures and controlling the quality of buildings erected. The state has also contributed to the differentiation of urban space through less visible means. This includes using planning procedures and laws to 'zone' land for different specified uses. This has been especially important in creating the kinds of homogeneous uses of space which characterize the suburb, for example, and in allowing the slum clearances and redevelopment projects discussed in the previous section.

The very possibility of there being a widespread market in residential property based on private home ownership has in many contexts depended upon state intervention. Unwilling to meet a growing housing shortage through direct construction, the US government in the 1920s set out to create conditions in which ordinary middle-class and working class Americans would be able to purchase their own houses through the private sector (Boyer, 1983). Since a house would cost such a large amount of money compared to an individual's income, potential homeowners had to be convinced that the huge sums of money involved in the purchase would be secure. This was achieved through the implementation of *land use zoning* – setting aside certain areas of land for particular land uses and forbidding others.

The urban historian Christine Boyer describes the implementation of zoning legislation in US cities in her book, *Dreaming the Rational City.* Zoning enabled the state to campaign successfully for the spread of private housing developments through home ownership by ensuring the stability of the investments involved. As Boyer explains it,

> to draw this money into such investment areas and to guarantee the establishment of mortgage credit would require an insurance policy that the single-family home owner's investment would be protected in stable neighbourhood communities through zoning, economy of land uses, and the intelligent planning of improvements. 'Many ... men although willing to acquire homes, were afraid to do so lest they later ruined their investment if an apartment, stable, laundry or public garage were built next door ... Big industries and businessmen therefore have good reason to work for the establishment of protected residential zones, as a definitive encouragement to home ownership and to more stable labour conditions.'
>
> (Boyer, 1983, p.148)

The uniformity of suburban spaces was secured as much through state zoning practices as it was by the pricing structure of the market. So far, then, we have seen that

- the differentiation of cities into relatively homogeneous areas or zones has been a product of the workings of the market, but that
- the political actions of states and other agents have also been centrally involved in shaping how this market works.

If we press the matter further, we will see that an even wider range of actors have been involved in shaping the differentiation of city space. This is especially so if we consider that suburbs are differentiated both on the basis of income and also frequently on the basis of race.

Estate agents and bankers figure highly in explaining this – both have encouraged racialized perceptions of what a 'stable' suburban community might look like. Banks refuse loans in areas which don't fit these criteria and estate agents encourage 'white flight' from neighbourhoods in order to benefit from the premium that black families are willing to pay for access to good homes in a racially restricted market. 'Redlining' (a process whereby banks literally or metaphorically draw a red line around certain neighbourhoods where they are unwilling to give mortgages) has had a profound effect in shaping the potential for development in these areas and has been shown to be closely linked to racial criteria. Individual white homeowners have also taken various actions to create and protect their racially exclusive environments. These have included racially restrictive clauses in deeds and neighbourhood covenants prohibiting sale to other than white owners (both now illegal in the USA), movement to new municipal areas, and even violence to force out unwelcome black neighbours. Douglas Massey and Nancy Denton give some examples of this in their book, *American Apartheid*:

> In Philadelphia, for example, an interracial couple made national headlines in 1985 when [they] moved into a white working-class neighbourhood and [were] met by an angry mob and fire bombs. When Otis and Alva Debnam became the first blacks to buy a home in an Irish neighbourhood of Boston, they experienced a sustained campaign of racial intimidation, violence, and vandalism that culminated in a pitched battle with white youths on the eve of the nation's bicentennial in 1976. In New York City, an Italian American told the sociologist Jonathan Rieder about the treatment he and his friends gave to Puerto Rican and black families who invaded their turf: 'we got them out of Carnarsie. We ran into the house and kicked the shit out of every one of them'.
>
> (Massey and Denton, 1993, p.91)

There has been an astonishing persistence of very high levels of racial residential segregation in large American cities. In considering the reasons why this should be, Massey and Denton (1993) ask whether this is not simply to do with the over-representation of African–American people in the poorer classes, a coincidence of class and race, as was discussed in Chapter 2. They find themselves unable to agree with this suggestion, however; Table 4.1 sets out the evidence they had to confront in this regard.

ACTIVITY 4.3 Look at the data in Table 4.1. Do the data suggest that racial segregation in US cities is closely associated with poverty? ◆

TABLE 4.1 *Segregation [indices*] by income in thirty [US] metropolitan areas with the largest black populations, 1970–1980*

		Income category	
Metropolitan area	**Under $2,500**	**$25,000–$27,500**	**$50,000+**
Northern areas			
Boston	85.1	83.9	89.1
Buffalo	85.2	80.0	90.0
Chicago	91.1	85.8	86.3
Cincinnati	81.7	70.9	74.2
Cleveland	91.6	87.1	86.4
Columbus	80.3	74.6	83.4
Detroit	88.6	85.0	86.4
Gary-Hammond-E. Chicago	90.6	89.5	90.9
Indianapolis	80.8	76.6	80.0
Kansas City	86.1	79.3	84.2
Los Angeles–Long Beach	85.4	79.8	78.9
Milwaukee	91.3	87.9	86.3
New York	86.2	81.2	78.6
Newark	85.8	79.0	77.5
Philadelphia	84.9	78.6	81.9
Pittsburgh	82.1	80.6	87.9
St. Louis	87.3	78.4	83.2
San Francisco–Oakland	79.9	73.7	72.1
Average	85.8	80.7	83.2
Southern areas			
Atlanta	82.2	77.3	78.2
Baltimore	82.4	72.3	76.8
Birmingham	46.1	40.8	45.2
Dallas–Ft. Worth	83.1	74.7	82.4
Greensboro–Winston Salem	63.2	55.1	70.8
Houston	73.8	65.5	72.7
Memphis	73.8	66.8	69.8
Miami	81.6	78.4	76.5
New Orleans	75.8	63.1	77.8
Norfolk–Virginia Beach	70.1	63.3	72.4
Tampa–St Petersburg	81.8	76.0	85.7
Washington, D.C.	79.2	67.0	65.4
Average	74.4	66.7	72.8

Source: Denton and Massey, 1988, p.811

* Ranging from 0–100, where 0 is perfectly integrated and 100 completely racially segregated

Table 4.1 seems to indicate that quite independent of income category, levels of segregation of African-American people in most of the thirty major metropolitan areas were high. When Massey and Denton combined this information with survey data concerning the preferred racial mix of neighbourhoods, it seems that these levels of racial segregation were not chosen or desired by the average African-American respondent. White residents, however, indicated that they would choose to leave neighbourhoods as the numbers of black residents increased.

In addressing the question of how city spaces come to be differentiated, we have seen that:

- the operations of the *land use market* and its management by the *state*, such as in zoning laws and planning, have created and reinforced class and racial divisions in the city;
- *powerful groups* in society, such as white people or banks, have either directly or indirectly used force or legal authority to enforce segregation in some cities;
- various strategies for trying to *improve urban life*, such as concerns with sanitation, disease or slums, have contributed to segregation.

Nonetheless, even in the most separated of cities, people usually travel across the boundaries of their neighbourhood to arrive at or look for work, or to make use of amenities in another part of the city, to visit relatives or friends, to find food, clothes, or useful items for building a shelter, or perhaps to sell goods they have made or found. The flows and rhythms of the city mean that patterns of separation and division do not capture the whole meaning of living in cities which are differentiated. For just as different neighbourhoods that are juxtaposed, or in the same city, are separate from one another, they are also *next* to each other, *connected in some way by their proximity*, through being part of the same city. Furthermore, they are separated and connected differently depending on the particular histories and geographies of the city and neighbourhood concerned.

In some ways this is quite paradoxical: just as cities bring people together at the same time as they also seem to separate people out, the differentiated spaces of cities do the same. They separate people in differentiated spaces, but also shape new possibilities for mixing and interaction – boundaries in cities are crossed and recrossed all the time. The following section tries to pin down what difference spatial divisions make to peoples' lives in the city. In order to answer this question we certainly need to think about how differentiations and separations impact on city life by keeping people apart. But we also need to think about the extent to which these divisions also involve specific kinds of crossings, flows and movements through different parts of the city; the extent to which they entail links and connections across boundaries and beyond localized spaces of social difference.

3 *Living in the differentiated city*

ACTIVITY 4.4 Turn to Mzwanele Mayekiso's account of walking along the streets of Alexandra township (Alex) in Johannesburg, Reading 4B. We will be returning to consider elements of this reading later in this section. For the moment, as you read it think about what the township feels like for those who live and work there. Take note of the various historical forces which have shaped the area, and the extent to which the lives of people in this segregated space are connected to or disconnected from the rest of the city. ◆

Like many South Africans, Mzwanele, now a political activist and scholar, moved from his family home in a rural area to Johannesburg, South Africa's largest city. At the time, he was fleeing political harassment for his work as a student activist. He joined some family members in Alexandra township, one of the oldest and most overcrowded of South Africa's segregated townships. In many ways, Alexandra township has a very similar history to Sophiatown (see Chapter 1), except that for various reasons it was never removed. Once in Alex, Mzwanele became deeply involved in local community politics in the township. In the reading he explains how the area's history and its present circumstances are part of wider social and political power relations, such as state action (and inaction), white racism and housing provision.

You may have noticed, though, that life in Alex is about much more than the external forces which have been responsible for shaping its past. Perhaps one of the most striking things in the reading is the vibrancy and diversity of life in Alex – notwithstanding the extreme poverty of many people who live there. Despite the difficulties and expense of transport, people in the area are connected to other parts of the city, the country and the world through travel, visitors or cultural pursuits. The separateness of Alex, though, is also very apparent, perhaps most obviously in terms of those who find it too expensive or difficult to travel, whose efforts to make a living are confined to this very poor township, or for whom there are no alternative places to live.

This section will consider what effect the differentiation of city spaces has on life in the city.

- Did segregation, for example, work as its various architects had hoped: to separate lives, impose power and entrench privilege?
- Or does the separation and differentiation of spaces in cities not only keep people apart, but also bring them together in new kinds of ways?

3.1 SEPARATE LIVES?

Differentiated and segregated spaces shape the lives of city dwellers. One of the first ways in which this happens is through the impact (on people's lives) of the process of achieving segregated cities. A common aspect of creating differentiated cities has been the need to move people from one place to another and, whether this is in the name of urban improvement or racism, the consequences of being uprooted from your home, neighbours and community are considerable.

John Western spent some time in Cape Town during the 1970s interviewing and talking with people who had recently been removed by the South African government from the settled middle and working class neighbourhood of Mowbray, which was quite racially mixed, to 'purified' 'group areas', set aside for the use of racially defined groups on what was known as the 'Cape Flats' (see Figure 4.4 for a photo-map of the areas).

The government had defined most of the people Western spoke to as 'coloured'. Their family histories were in fact mostly quite 'mixed' – many coloured people are the descendants of children of cross-racial affairs or marriages, and some can trace their family histories back to important Dutch settlers, or more recently to white parents and grandparents. A number of coloured families are descendants of Malay slaves, brought to the Cape during Dutch rule, and many trace their

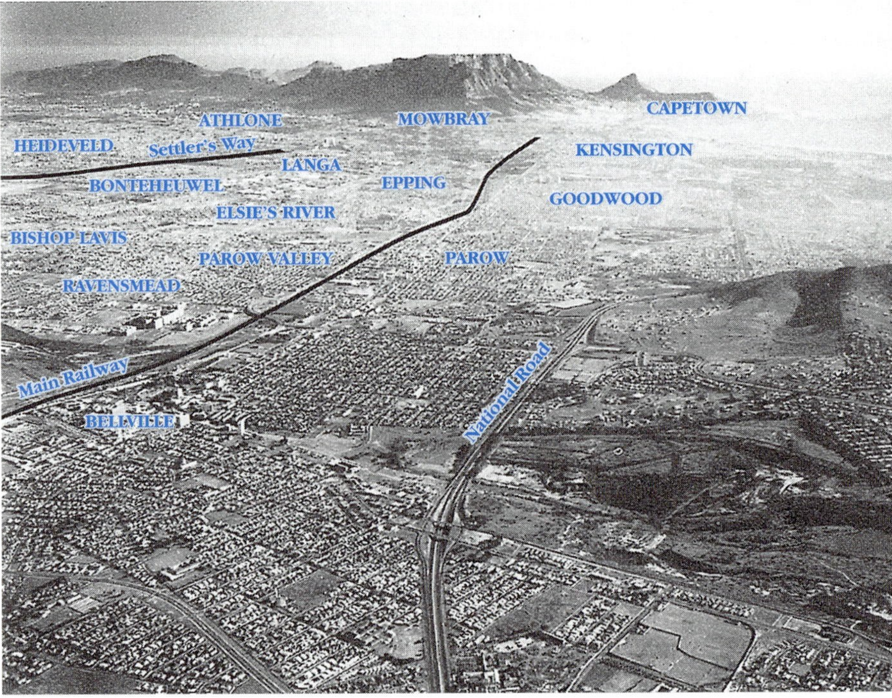

FIGURE 4.4 *Cape Town from the East: the northern suburbs are in the lower left on one side of the rail line, the Cape Flats are on the other side, and the southern suburbs are in the upper left*

ancestry back to indigenous Khoi-San groups. In many senses, then, this was a very diverse community, and often government attempts to 'classify' people as coloured were haunted with indeterminacy. Several of Western's informants were able to 'pass' as white – a precarious and often nerve-racking process, albeit with obvious material rewards, such as being able to stay in Mowbray.

Nonetheless, separate residential areas were created and a whole generation of Capetonians experienced the effects of segregation and removal very directly and very personally. Unlike those removals in which poor living conditions and overcrowding were some of the main reasons for the painful process, the upheavals experienced by the people of Mowbray were entirely a product of the government's policy of racial division. The impacts on people's lives, though, were in some respects similar to those experienced elsewhere.

One of the chief results of the removals was to break up the family and friendship networks which had been built up over the long decades of residence in Mowbray. As families moved out at different times, they were allocated state housing in different, more distant townships on the Cape Flats. Old people, who had lived their whole lives in Mowbray, became depressed and ill, perhaps dying sooner than they might have as a result of the grief. People with jobs often had to travel much further to their place of employment after the move – as Western writes:

> The daily, inevitable journey to work and back is probably the most important statistic ... it is the largest tangible imposition stemming from group areas removals from Mowbray. Basically, the poorer are now farther from their work than before; this same change has also occurred in other cities, for example in Detroit and Philadelphia ... However, in these North American metropolises, it is the jobs that have been suburbanized while the Black poor have remained trapped in their ghettos. In Cape Town, the Coloureds have been involuntarily moved to suburban ghettos and, for the present at least, their jobs have remained in the inner city ... For ... two-thirds of the principal wage earners of the households, the average journey time each day was ... 1 hour and 56.4 minutes. Of this *1 hour and 7 minutes of the journey extra each day was thanks to the Group Areas Act.* Common sense surely tells us that such erosion of otherwise discretionary time will likely have a negative effect on the worker's family relationships.
>
> (Western, 1996, pp.221–3)

Another difficult aspect of the move was leaving a settled and relatively safe middle-class environment for the racially 'pure' but mixed-class environments of the new townships. Western writes about how being from Mowbray had given people a certain status within the coloured community. For many people, the move completely changed that. The new environments quickly became unsafe: recent evidence suggests that the government allowed gangs to flourish, as a perverse way of keeping anti-apartheid politics out of these areas. The fruits of this today include extremely dangerous gang warfare, increasingly drug-related and with a complex relationship to religious differences. Residents removed from Mowbray reported trying very hard to keep their children indoors to prevent them from mixing with 'rough' types (Western, 1996, p.241). The new townships came to signify a very different social standing from that implied by Mowbray, the place they had been forced to leave. Western reports one person's experience of this:

> A conventionally respectable Muslim, a florist, summed up his feelings about Bonteheuwel, to which he was forcibly removed from Mowbray and from which he was trying to escape into a home-ownership area: 'As far as I'm concerned, Bonteheuwel's just a roof over my head'. Mr. Gierdien went on to try to refute the stigma of his address: 'No, when you say you're from Bonteheuwel, some of them look at you down their noses, people from Fairways or Wynberg. I tell them there are decent people in Bonteheuwel, too. We've got a social apartheid amongst ourselves, as well as the big thing'.
>
> This statement is a striking example of a spatial expression given to a social differentiation that was only immanent when people were still resident in Mowbray. Certainly, then, space is mirroring society in Cape Town, but the very visibility of space – the different housing areas with their various names – underscores more clearly than before the social divisions. Space *enhances* societal distinction; social structure in a sense mirrors space'.
>
> (Western, 1996, p.253–4)

Western is arguing that

- the spaces produced by segregation, in turn, shape people's experiences of city life;
- they ascribe new meanings to people's identities as the new differentiations of the spaces of the city help to make and mark social distinctions.

There are many people around the world who have lost their homes through efforts to 'improve' the city, provide better housing, or clear slums, improve transport or stimulate the economy – efforts which have resulted in cities being divided in various ways. The racial character of these experiences in South Africa and other contexts exacerbated the psychological and economic losses involved. While there are many situations in which people might be enthusiastic about moving to new homes and neighbourhoods (even if they are in segregated areas), common to many experiences of differentiation and separation in many cities are the problems they bring in terms of relocation, access to work and income opportunities, services and the amenities of urban living. These have both a material and an emotional content – separated from many parts of the city, unable to travel too easily, you may feel as if the city does not belong to you. The stigma of living in certain neighbourhoods is also common and can have very material effects, for instance, in some situations employers simply refuse to hire people from certain neighbourhoods. Many commentators note that the disadvantages of living in deprived areas can compound and intensify the position of people who are already poor.

David Harvey makes the general observation, borne out by many studies, that:

> Residential differentiation in the capitalist city means differential access to the scarce resources required to acquire market capacity … For example, differential access to educational opportunity – understood in broad terms as those experiences derived from family, geographical neighbourhood and community, classroom, and the mass media – facilitates the intergenerational transference of market capacity and typically leads to the restriction of mobility chances.
>
> (Harvey, 1989, p.118)

Obviously this is far too general a statement to cover all cases, but has a particular resonance in American cities where white flight from central cities has often been associated with attempts to avoid racial mixing in schools (Cox and Jonas, 1993). Wealthier white people move to suburbs in order to have access to better educational opportunities there, leaving poorer and black children concentrated in underfunded inner city schools.

The story so far has been one of the losses and hardships imposed through separation and differentiation in cities. We can't end the story here, though, at the borders of neatly separated areas. In many cities 'segregation' is not a very prominent feature – as we saw for example in the maps of ethnic communities in London (Figures 4.1(a)–(h)) – and many of these experiences would be felt less starkly or not at all. However, even where cities are characterized by relatively high levels of differentiation and separation, they still bring people and resources together in a set of relations characterized by mixing, movement and flows as much as by division. How does thinking about this aspect of cities change what separated and differentiated spaces might mean? To the extent that juxtaposed spaces are as much *next to one another* as they are *separate from one another*, what does it mean for different kinds of neighbourhoods to be *adjacent* to one another or at least caught up in some of the same kinds of dynamics and rhythms?

3.2 BORDER CROSSINGS …

ACTIVITY 4.5 Reflect upon the extent to which the contemporary, post-colonial city of Dar es Salaam, described by Richa Naga in Reading 4C, still bears traces of its colonial past. What elements of social and spatial separation remain pertinent? How do these segregations shape people's daily lives? ◆

Two aspects of the reading about Dar es Salaam are interesting to me. *First*, racial patterning in the city has persisted from colonial times, but is by no means intact or complete. People of different racial groups are still more or less clustered in different areas, but many live among people with different backgrounds from themselves or often travel to parts of town with a quite different ethnic character. *Second*, the article shows that social or racial separations can be maintained even where separations in space are not complete. Where the segregations in space have broken down, individuals and communities can still maintain separateness in new ways. The high level of interactions associated with the religious and communal life of Asian people in the city mean that for some of these people their paths seldom cross those of people from other backgrounds. This could be especially true for women living in the concentrated Asian and trading area, whose lives revolve very much around the home. Later on in the article, Nagar discusses how, even in these tightly knit and spatially distinctive religious and communal circles, quite severe social separations are maintained between people of different races, classes and genders. Nonetheless, the spatial segregations which have been imposed in the past continue to play a part in shaping today's cities, sometimes reinforced, sometimes attenuated, by new processes.

This is a good time to return to the reading about Alexandra township (Reading 4B) and to notice how Mzwanele's walk along the streets leads him to reflect upon the diversity of the township's population. He comments on the different African cultures which are found in the area, such as the Zulu, Xhosa and Pedi traditional dancing in backyards at weekends. He also remarks on the shebeens, informal drinking places, that he is considering visiting, each of which specializes in different types of popular music – jazz, disco, funk, rap. Alex is made up of people from all over the country (and increasingly, the continent) whose enthusiasms and cultural pursuits are drawn from all over the world. Divisions in Alex are also present, and even among fellow churchgoers, some people are poorer than others. Women and men have quite different experiences of the difficulties shared in the community; crime and danger further divide the residents of this tightly packed area, and there is an ongoing ethnic-political conflict between Zulu-speaking hostel-dwellers and African National Congress (ANC) and civic supporters. We could recall the Chicago School's comments that the segregated 'ghetto' is a truly *cosmopolitan* place (see **Pile, 1999**, section 2.5)

Not only is the space of the segregated neighbourhood connected in many ways to places and experiences from all over the world, the separations from adjacent areas locally are also not as complete as our mappings of places might indicate. Let us take Alex again. Mzwanele talks about the neighbouring industrial areas and white residential areas, separated from Alex by a 'buffer strip' (as described by Goldberg in Reading 4A). One of the nearby roads even has a high concrete fence down the middle of it so that Alex residents cannot cross it and enter the industrial/commercial area. Neighbouring wealthy Sandton has fenced itself in and employed private security guards to recreate the level of control over black people's movements in the area which characterized the apartheid era (see Reading 3B in Chapter 3). White people in these neighbourhoods clearly feel there is a significant possibility that 'dangerous' black people will find their way into the area – along with those black people they are happy to see there, such as domestic workers. Indeed, if you were to drive or walk around almost any white suburb in South Africa, very often the only people you would pass on the street will be black – people looking for work, visiting relatives who work in the area, selling goods, looking for some help or money, or on their way to or from work. In this way, white suburban streets become (black) meeting places, places for friendly encounters and for socializing, especially for black people employed or living as domestic workers in the neighbourhood.

In these examples we see that

- efforts at segregating are often a response to a reality of connections and crossings;
- social segregations can also take place without spatial differentiation, as in Dar es Salaam;
- segregated places are connected to other parts of the city and to other parts of the world; they are also cosmopolitan, heterogeneous places.

At the same time as they keep people apart, then, separations also encourage certain kinds of crossings, or connections. The opportunities for work which segregated spaces create is often a cause for drawing people across the boundaries of separated spaces, as we saw in the relation between Alex and Sandton in Johannesburg. Extract 4.3 is from Lee Jellinek's harrowing account of the life of 'Sumira', a street trader in Jakarta in Indonesia (see also Reading 2B, Chapter 2). Jellink's account describes Sumira's experiences when she left her poor squatter neighbourhood, or *kampung*, to work in an expatriate home as a domestic worker.

EXTRACT 4.3
Lee Jellinek: 'Saga of a Jakarta street trader'

Even though she was not a good cook and had never worked in a Western household, I was able to find [Sumira] a job amongst my expatriate friends. At first she was overwhelmed by the prosperity of the household where she worked. She could not get over the spaciousness of the house or the size of its garden. Her entire *kampung* neighbourhood would have fit in the garden. Yet only six people lived there. Sumira was amazed by the enormous refrigerator and the pantry stocked with tins of imported food for months in advance. Like other kampung dwellers Sumira had always bought just enough food to last for the day. Despite her urban background, she had never used a telephone or seen a shower, a Western-style toilet or a swimming pool. The unlimited supply of hot and cold water in the three bathrooms and kitchen was a novelty. The automatically opening car-port door was a source of astonishment. Her employer's noisy teenage children gulped down bottles of milk and gorged themselves on chocolate cake. Yet, she observed, when a little sugar or spice was missing from the pantry there was inevitably an uproar. More than anything else Sumira was amazed to see the family's children shout at their parents. The family was wealthy and secure, but Sumira noted that did not seem to make them any happier or more contented with their lot.

Though Sumira had at last found a measure of security at least for as long as her employers remained in Jakarta, she felt imprisoned behind the high

> walls of the expatriate compound. She missed the ceaseless activity of her *kampung's* crowded lanes, the familiar faces, the gossip, the smell of *kampung* food and, above all, her family. Sumira worried about leaving her aged mother and young daughter to fend for themselves. Her job required her to live where she worked. She was given one day off a week and that was her only opportunity to see her family. Sumira's mother was in her late sixties and had become accident-prone. ... Rahmini, now 8 years old, was left to care for her grandmother ...
>
> As a domestic servant in 1979, Sumira was paid in cash about one-fifth of what she had earned as a food-trader in 1972, if inflation is taken into account. It was just enough to feed herself, her daughter and her mother. She did obtain medical care, clothing accommodation, and some food for herself and occasional presents for her daughter and mother, but she was cut off from her family and her roots, and forced to live in an alien and lonely world. She longed for the interaction with people, her freedom, and the independence of petty trade ... Despite the drawbacks of her new job, the lack of freedom, and the realization that it offered little scope for material improvement, life as a domestic gave her economic security. For Sumira that was worth a lot.
>
> Source: Jellinek, 1997, pp.146–7

When her mother died, Sumira gave up her job to take care of her daughter. By this time they were living in a new government-built apartment block to which they had been moved from the *kampung*. The networks of neighbours on whom she relied before were no longer accessible. Sumira's story is very sad and such experiences in precarious circumstances are quite common in many poor cities. The account of her feelings about crossing the borders between the *kampung* and the wealthy suburb where the expatriates lived is a poignant reminder of the meaning of the social divisions we have been describing. Cut off from her family, her life was lonely and isolated and she herself died in impoverished circumstances.

The experience of being in the home of a wealthy family had provided many benefits for her, but it had also brought her face to face with the kinds of assumptions these people had about their employees – that they were inclined to steal from them, for example – and probably with various kinds of racial assumptions as well. Sumira's life moved across the boundaries which various social processes in the city had produced, but in many ways it was a difficult and painful crossing.

A quite different example of how separations entail a measure of bringing together is found if we go back to the experiences of colonial and apartheid separations. Both of these systems of rule depended upon state employees to oversee the colonized, or dominated, populations. In South Africa, the township manager, responsible for governing the segregated African township, was usually a white man fluent in an African language, claiming familiarity with

African culture, and often someone who had grown up in a situation in which he had a lot of contact with African people – on a farm, or a mission station perhaps, or even in a township. Many stories about these white men stress their complicity with apartheid rule, their racism and their distance from the communities in which they worked and often lived. Yet some of these men tried to be 'paternalistic' towards 'their natives': they wanted to help them, like fathers might help children. There are deep ambivalences at work here, and of course the nature of the relationships in question depended very much on the individual concerned. Many administrators may have been kindly men, despite their racist assumptions about the people they were governing; but some were combative and aggressive, generating considerable opposition to their style of administrations (Sapire, 1994; Robinson, 1996).

Administrators physically crossed the boundaries which were set in place to separate people, but we could perhaps extend this sense of ambivalence in relationships between colonizer and colonized to include ordinary settlers in colonial contexts. Although the idea of 'natives' living near their homes frequently horrified them (as we saw in the examples of Vancouver and Johannesburg), colonizers' homes were nonetheless often full of objects drawn from indigenous cultures. Their own cultures (language and music, for example) were also shaped by those of the people among whom they lived, but in many ways lived separately – except that most children of colonizers were raised by women from native societies doing work very similar to that of Sumira in the previous extract. Her story reminds us, though, of the different quality and meaning of the crossings involved here. Some are more difficult and painful than others.

However, in all sorts of ways, lives and cultures cross boundaries, even in separated societies; in the geographies of our minds, places and people separated from us in space are brought very close to home in our imaginations, fears, fantasies, dreams, and perhaps even in our politics as we try to change the cities in which we live.

So far we have seen that:

- segregation causes suffering and loss as people are moved from their homes or are disconnected from other parts of the city;
- the spaces of segregation in turn shape social life, reinforcing new identities in new environments;
- differentiated spaces in the city are also connected to one another through physical crossings and cultural and imaginative interactions.

The impacts of segregation are ambivalent, causing social separation but also undermining it. If the initial ambitions of those implementing segregation and differentiation were intended to keep different people apart, then they have only been partially successful. However, as we saw in section 2 above, differentiation and segregation were also associated with power relations in cities. Powerful institutions and actors enabled these patterns to emerge in cities because they thought they would enhance their capacities or advance their interests in some way. We turn now to evaluate whether these ambitions were achieved.

3.3 SPACES OF POWER?

In order to asses whether segregation has served the interests of the powerful, let us return to the two examples, Vancouver and Johannesburg, discussed in section 2.

As you will recall, Chinese migrants to Vancouver had been forced to take up residence in one part of town, a swampy rather marginal area. Today this is next to downtown Vancouver and has become one of the city's major tourist attractions (Figure 4.5). It is also adjacent to the poorest neighbourhood in the city – certainly not a salubrious environment. But many Chinese people, especially traders and restauranteurs, have made use of this spatial segregation to improve their economic situation. Until the mid 1930s, Anderson writes, Chinatown in Vancouver was subject to all sorts of negative stereotyping by the white majority. The area was seen as run-down, a source of disease and ill-health, a site for evil and dissolute practices, including opium dens. Agitation to prevent Chinese people from moving into the suburbs was combined with national initiatives to bar further Chinese immigration into Canada altogether: 'If leakage from the ghetto could not be directly plugged, at least flow into Chinatown could be stemmed at the source' (Anderson, 1988, p.139).

However, things started to improve as white Vancouver residents were affected by a more romantic vision of the Orient – ironically, just as Chinatown was becoming more Westernized in appearance! These more positive attitudes, despite still involving a form of stereotyping of Chinese people, formed the basis for marketing the area as a tourist attraction. Prominent Chinese residents joined in this process of commodifying and Orientalizing their neighbourhood, seeing benefits to this new form of racial representation:

> With time, some Chinatown residents came to identify interests of their own in the implementation of the dominant view. Some of their receptivity was economically based, geared to the tourist trade; for others, the courting of Chinatown offered a form of power and status as community advocates, and a pride in the Chinese experience in Canada.
>
> (Anderson, 1988, p.145)

An important point that emerges from Anderson's account of the history of Vancouver's Chinatown is that *European conceptions of a 'Chinese race' and the making of a local Chinese identity were both enhanced by the clear demarcation of the space of Chinatown* and the construction of a place that visibly gave testimony to a form of 'Chineseness'. If the segregated space enhanced ethnic identity – a perhaps anticipated outcome of segregation – it also provided a basis for a form of community empowerment, which was perhaps a less expected result.

One of the unexpected consequences of spatial separation in many contexts, therefore, has been the way in which it has *enabled new forms of political action.* Segregated areas often separate people along racial lines, for example,

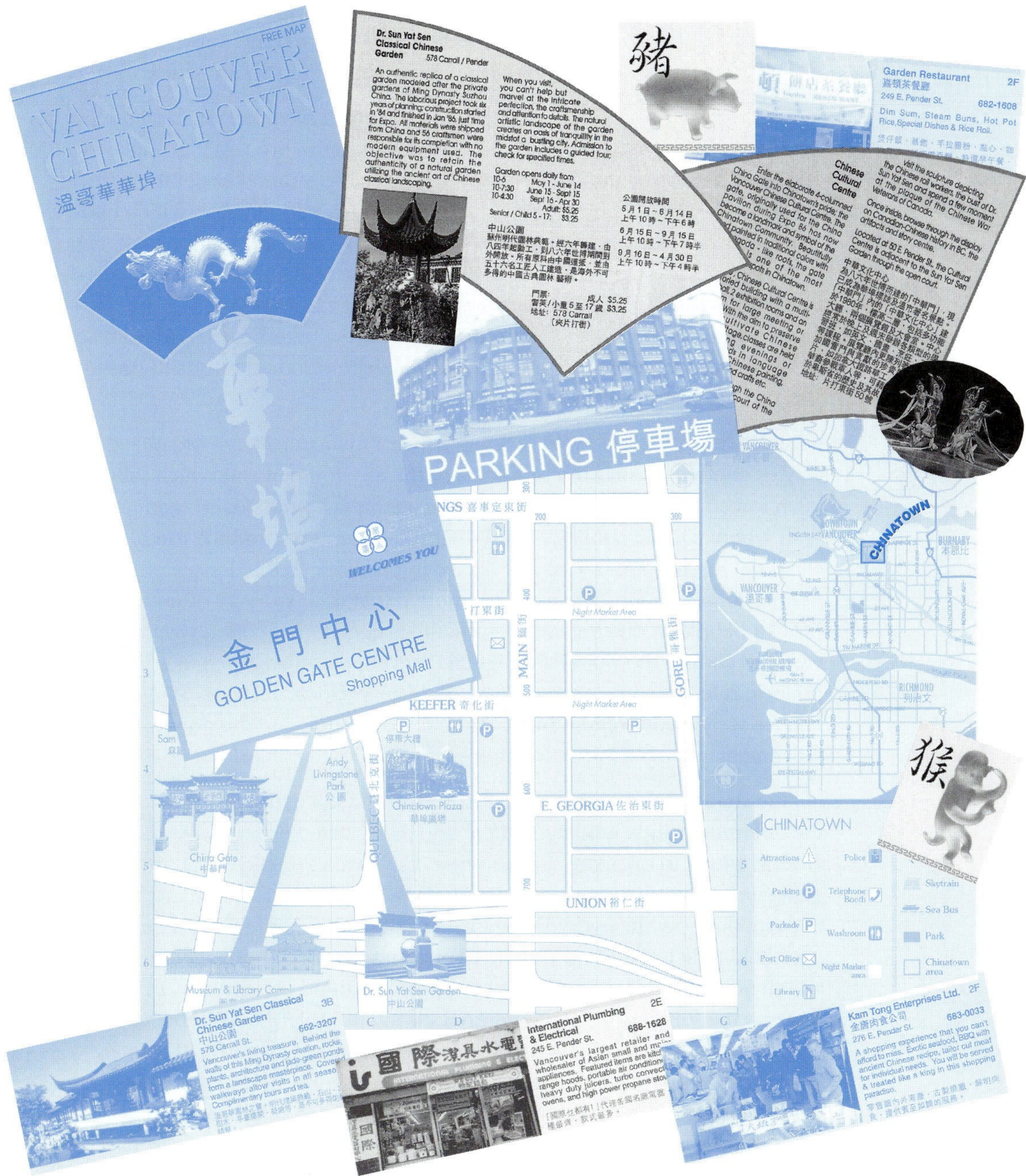

FIGURE 4.5 *Vancouver's Chinatown – a major tourist attraction*

but in doing so they bring people together across many other divisions, such as class or gender. American women, for instance, have argued that the home, far from being a place where power relations between men and women are central, is a site where women and men can receive support and protection from a wider racist environment. Harlem, as we saw in Chapter 1, has been seen both as a haven and a hell, both a disconnected neighbourhood and a diverse city within a city connected to places around the world, including South Africa.

In South Africa's townships racial segregation may have been more effectively enforced than anywhere else in the world, but in the process it brought together people across class and ethnic barriers. This made it much easier to organize political resistance against the very system of racial domination which had created those divisions. Mzwanele Mayekiso discusses the existence of 'civic' organizations in Alexandra. More recently, these organizations have been drawn into battles with rival ethnically based political groupings (such as the Zulu-based movement, Inkatha), but initially they drew together people from across the township, regardless of ethnic or class background, in efforts to protest about living conditions in the township.

For decades, African people in South Africa's townships have organized in one way or another around the problems of the environments in which they have been forced to live, and also around the attempts by the government to control both where people could live and their movements around the town. It is very clear that forcing African people to live in one area enabled more effective political organization. The state had hoped that their strategy of putting all African people in one location – sometimes even fenced off – would make it easier to control the population and to prevent the emergence of a threat to their political rule. To some extent this worked, as political alliances across racial divides were undermined, but it also helped to stimulate the political movements which contributed to the downfall of the apartheid government.

So, the strategy of segregation 'backfired' as the township brought together all the different groups who lived there and gave them very specific grievances around which to organize. In this sense, the spatial practice of apartheid generated contradictions which led to its downfall. It was very easy for activists to point out how the everyday grievances of township life were a direct result of apartheid policies, and from there to encourage a large movement to contest apartheid based in many different townships around the country. In the 1950s this was the form adopted by the ANC during the celebrated Defiance Campaign, in which individuals defied newly enforced apartheid laws by entering racially restricted areas (Lodge, 1983). In the 1980s, continuing in this tradition, civics such as those in Alex joined together with other anti-apartheid organizations around the country to form the United Democratic Front which spearheaded the growing resistance to the apartheid government during these years. Some townships were rendered ungovernable and some civics tried to assume responsibility for governing and servicing these areas. The government's ability to govern African townships was substantially undermined through such community-based politics, as well as through growing financial difficulties and problems of state capacity and overcrowding in the limited housing stock.

The segregated spaces of apartheid both enabled racial domination and facilitated resistance to that domination. We could make a case that both the way in which cities were planned and governed under apartheid, and the urban movements which opposed this, made a substantial contribution to the downfall of apartheid. The spaces produced by apartheid society worked not only to enforce division and domination, but also contributed to the ending of apartheid rule.

- Cities may be divided and people living in them may be forced to live lives in separated neighbourhoods for a whole variety of reasons – but these same *people can also re-make these divided spaces.*
- In small ways, people in the city *cross the boundaries* which are either drawn on a map or cast in concrete fences – both in their everyday lives and perhaps also in their imaginations.
- But sometimes in much more noticeable ways, the spaces of the city are remade through *collective political action.* The example of civic and community organizations in South Africa, which have successfully contested the racially divided apartheid city, is only one among many such political movements as we will see in Chapter 5.

The outcomes of segregation have been ambivalent, then, both enhancing the capacities of the powerful and enabling new forms of political activity. Nonetheless, the negative effects of segregation and differentiation remain, and along with them the question as to whether the spaces of segregation and differentiation in cities should be changed in some way. The following section will explore the different ways in which South Africa's cities are changing as the society tries to shake off its apartheid past. This will enable us to think about whether segregation and differentiation in cities should be abolished, and if so, how this might be achieved.

4 *What should be done about segregation?*

Even as segregation and differentiation are persisting and perhaps being reinforced in many cities around the world, the archetypal segregated city, the apartheid city, is experiencing fundamental change. Most commentators on contemporary South African cities are keenly aware of the destructive effects of past urban policies on city life. Not only is there a need to overcome the racial basis for the allocation of urban space, but government and community groups agree that there is a need to break free from the planning ideas that underpinned the organization of South Africa's cities in the past.

Ideas about how to reshape these cities have coalesced around two sets of visions. The first concerns the idea of '*one city*', as opposed to the racially divided city of the past. This is an attempt to address the *political* fragmentation of South Africa's cities, uniting all the different areas and racial groups in cities which were previously governed by many different political bodies into one municipality. The second vision is more concerned with the divisive planning principles common to South Africa and many other cities; these principles have contributed to urban sprawl, poor accessibility and a fragmented environment and have been compounded by the practice under apartheid of locating segregated residential areas for the poorest groups on the peripheries of cites. This second vision imagines producing a '*compact city*', a city which is both *physically more integrated* and more densely settled, therefore more sustainable (see Chapter 6). David Dewar (1995, p.413) summarizes below some of the strategies which would need to be adopted in order to develop compact cities characterized more by connection and integration than by disconnection and separation:

> – Compaction rather than sprawl (the use of densification and infill projects is clearly central to the achievement of this);
> – the integration, overlap and mixing of urban land uses, activities and elements, as opposed to their separation;
> – the continuity of urban fabric as opposed to its fragmentation;
> – linearity as the dominant structural geometry, as opposed to point-focused approaches;
> – the 'extroversion' of more intensive activities and facilities towards dominant transport routes (particularly public transport routes), as opposed to their 'introversion' or embedding them in local areas.
>
> (Dewar, 1995, p.413)

Dewar has been influential in shaping thinking in South Africa about how to achieve a more compact city form – as opposed to the sprawling, divided and unsustainable city which apartheid produced. This would be a city which celebrates heterogeneity, proximity and interaction, rather than separation, uniformity and suburban sprawl. One suggestion he makes is to fill in the many empty spaces which were produced under apartheid, either through removals or through the excessive buffer zones required between different land uses. He also suggests encouraging people to come together from adjacent, separate neighbourhoods to meet in 'activity corridors', where shopping, employment and some housing will offer an opportunity for racial divisions to be broken down – in effect, for juxtaposition to be seen as a basis for interaction rather than separation. For future developments, he suggests a linear emphasis rather than the segmented, mosaic structure encouraged by apartheid planning.

ACTIVITY 4.6 Read again the suggestions in the Dewar quotation above. Do you think they will work to overcome the problems generated by a segregated city form? ◆

One of the biggest problems remains the isolation of poor African residential areas from work opportunities and services. African townships have historically been very poorly provided with these facilities, and long journeys to other parts of the city continue to be necessary; as Mzwanele notes, this is the case even for residents in Alexandra, a relatively well located township. Building new housing for poorer people closer to the centre of town seems important. The alternative would be to provide better services and facilities in existing residential areas. But, as the Gauteng draft White Paper on urban development puts it:

> - will investment in townships exacerbate segregationist planning?
> - will a failure to invest in townships isolate disadvantaged communities from goods and services?
>
> (Gauteng Provincial Government, 1997)

In addition to this, problems of land availability and costs have meant that new housing for low-income people is still being developed far from the city centres, fairly close to already existing townships. Changing the geography of the apartheid city could be a very long and difficult task. Certainly, there are some obvious and very real constraints on the extent to which we can hope to transform the apartheid city. John Western expresses it well:

> To actively dismantle the apartheid city which is so evidently and intendedly reflective and indeed constitutive of [racial] inequality would cost a great deal. The money it would take to pursue social engineering significant enough to somehow rescramble the racial and/or ethnic groups (as formerly defined) seems to make such a prospect utterly remote.
>
> (Western, 1996, p.xix)

Some elements of the segregated apartheid city are unlikely to disappear – and we may want to ask whether it really is important to change the physical spaces in question. Would the important thing, perhaps, be to change the types of connections and flows which contribute to shaping the opportunities and quality of life for people in these segregated areas? Instead of changing the spaces of the city, we could try to influence the spatial relations and networks into which they are connected.

ACTIVITY 4.7 As you read through the case study from Los Angeles below, consider what implications you think this might have for plans to change South Africa's cities. ◆

Case Study: Los Angeles

Oliver, Johnson and Farrell (1993) consider the case of South Central Los Angeles, a segregated black neighbourhood. Over the last two decades the area has become increasingly ethnically mixed as South and Central American immigrants have boosted a more settled Latino population and the previously Jewish-owned shops have been taken over by Koreans (see Figure 4.6). Unemployment in the area is high, and the mostly Latino workers earn low wages in insecure jobs. Most African-American men are

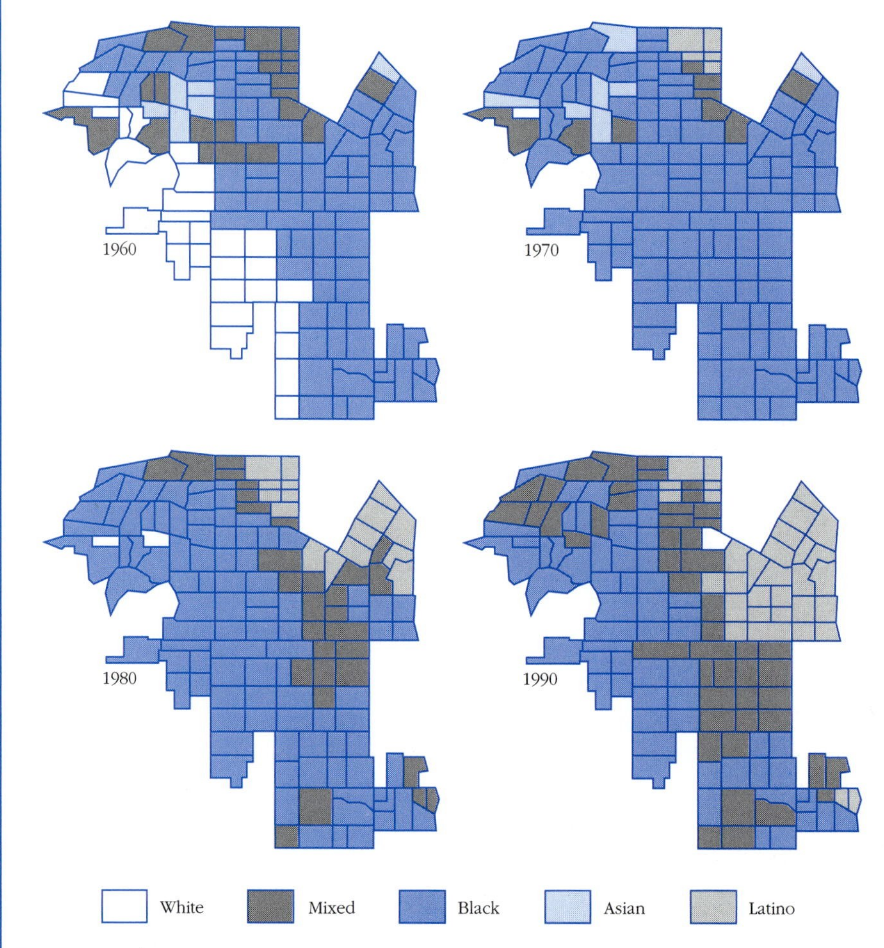

FIGURE 4.6 *Ethnic/racial composition:: South Central Los Angeles, 1960–90*

unemployed and the neighbourhood bears the brunt of Los Angeles's notoriously brutal and racist policing.

It is these people who responded most to the uprising which took place after policemen accused of beating up a black man, Rodney King, were acquitted despite video evidence which seemed to indicate that they were guilty. In their assessment of the causes of the rebellion, Oliver and his co-writers argue that more should be done to change the economically depressed and socially conflictual situation in the neighbourhood. They hold that the economic and social conditions there are at least partially responsible for the uprising. They write that, in these circumstances, 'The three ethnic groups – blacks, Latinos, and Koreans – have found themselves in conflict and competition with one another over jobs, housing and scarce public resources' (Oliver *et al.*, 1993, p.121). Increasingly isolated from job opportunities, cut off from the rest of the city and providing shelter to some of the poorest sectors of the city's population, this area seems to be experiencing some of the worst effects of segregation in the city.

And yet, Oliver *et al.* believe that a different future is imaginable. They write in favour of a programme of public action that would:

> [First] create meaningful jobs that could provide the jobless with skills transferable to the private sector. Second, it would rebuild a neglected infrastructure, making South Central Los Angeles an attractive place to locate for business and commerce. Finally, and most important, by reconnecting this isolated part of the city to the major arteries of transportation, by building a physical infrastructure that could support the social and cultural life of this richly multicultural area (e.g., museums, public buildings, housing), and by enhancing the ability of community and educational institutions to educate and socialize the young, this plan would go far in providing a sustainable 'public space' in the community. For it is our contention, that only when South Central Los Angeles is perceived as a public space that is economically vibrant and socially attractive will the promise of this multicultural community be fulfilled.
>
> (Oliver *et al.*, 1993, p.135)

We have learned much in this book about the vibrancy and potential creativity of many ethnically segregated and poor neighbourhoods. It is this vitality which Oliver *et al.* wish to see supported and encouraged, and not dispersed or diffused. The problems they foresee in implementing this plan are mostly to do with a lack of public investment in infrastructure and services, the racist employment and policing practices which isolate especially young black men, and the disconnections from the rest of the city which deny access to the opportunities and interactions which other, perhaps more affluent, parts of the city offer. They don't want to change the ethnic character of the neighbourhood, but rather to alter the economic and social circumstances which isolate the area and which lead to high levels of deprivation.

- *Thinking spatially* about the problem of segregation means paying attention not only to the problems of borders and boundaries, but also to the potential of connections and crossings.

But while planners and government try to find ways to rearrange the city or to encourage reconnection and integration, city dwellers are remaking urban space at a rapid rate! Faced with a restricted local market in poor townships in South Africa, many informal traders have relocated to central city areas. The streets of downtown Johannesburg, for example, are covered with informal stalls set out in the shadows of the tall skyscrapers, selling everything from fruit to handbags, cosmetics and clothes, as well as haircuts, shoe repairs, photographs and telephone calls! The major retail stores continue to trade behind the facade of street-sellers. Some traders now live in inner city neighbourhoods (see Crankshaw and White, 1995), while others find it necessary to stay overnight on the streets with their goods – in some places street settlements have sprung up, allowing easier access to the centre than is possible from distant townships. There are also a number of more central areas where squatters have seized the opportunity to build more permanent shelters in a well located area, rather than on the remote, if more abundant, land on the edges of the city where most newcomers and homeless people are forced to seek shelter.

However, much as this indicates a process of deracialization in that spaces which had been reserved primarily for white people are now accessible to black people, these changes also reflect a process of reracialization, as white people commonly avoid areas which are now predominantly black and retreat into new, privatized segregated spaces. Shops and services previously located downtown are now relocated to suburban shopping malls with private security guards and protected parking spaces. Even traders on the Johannesburg Stock Exchange, located on Diagonal Street – the Johannesburg equivalent of Wall street or the City of London – are now starting to move out to Sandton, the wealthy suburb neighbouring Alexandra township. Those businesses still choosing a central location, mostly back office functions of banks and insurance companies, design their buildings to enhance security for staff, creating the familiar defended spaces of the walled city (see Chapter 3).

There is another dimension to efforts at urban integration though, and that is *political*, as captured in the vision of *one city*. Here, in place of the racially fragmented municipal governments characteristic of the apartheid city, the entire metropolitan area is to be governed by one, democratically elected municipality. All revenues will be shared out across the municipal area – unlike in the past when African areas had to be self-financing and white residents were able to benefit from the taxes earned from business activities located within the 'white' city. Integrating the city in this way has not been easy; Sandton residents, for example, initiated a boycott of local rates when they went up significantly as the new Johannesburg Metropolitan government starndardized the rating system across the city. Without this income, the possibilities for meeting the needs of areas long deprived of infrastructural investment will be much reduced. The politics of integrating the city is still fraught with tension.

'Race' remains a key element in local urban politics and the city remains divided – albeit in new ways – along racial lines. The racial differentiation of city space may be much less clearly evident as city dwellers have created new ways of relating to urban space, drawing the city together and transforming the racially exclusive nature and use of streets and public spaces. But there are also many changes which are recasting barriers and replacing divisions, sometimes with a quite different pattern of separation or along new lines such as class or immigration status.

We have noticed some paradoxical characteristics of segregated spaces: their vibrancy as well as their disadvantage; their links to other places as well as their disconnection from parts of the city; their persistence, despite the best efforts of planners, as well as the rapid changes which can ensue as people remake and reuse city spaces. All of these paradoxical features make it difficult to find an answer to the question of whether we should try to do away with spatial segregation. There seem to be so many negative consequences of certain kinds of segregation, yet at the same time spatial differentiation in general is a very common feature of city life. The diverse and spatially differentiated character of cities has the potential to be one of their most appealing and attractive features. However, in particular social circumstances certain forms of differentiation are produced in concert with disconnection and disadvantage and seem to demand some form of intervention to try to change the situation.

Just as the felt intensity of city life can be both exhilarating and overwhelming, spatial differentiation, too, has ambivalent potentialities.

- It could be enabling and empowering, facilitating political opposition to racial domination, for example; or
- it could be a source of power for dominant groups, who might be resisting change.
- Segregated neighbourhoods can be dynamic and a source of potential and promise, but
- segregated spaces can also be places of despair, poverty and danger.

The imperatives identified in the Los Angeles case study – to *reconnect* segregated spaces to the wider city context, to empower segregated communities politically and to change the negative stereotypes and attitudes of dominant social groups and institutions – are also important in ideas about how to reshape the apartheid city. In this case, the ideas include building links between different parts of the city and making sure that all areas have access to services and facilities. Changing the physical infrastructure of the city (building new roads and transport links and shopping centres) is designed to facilitate more, and stronger, connections between different parts of the city.

We have already seen, though, that these plans have almost been overtaken by other changes in the city; defensive spaces have been created by white people to defend racial privilege, and black people have already remade certain areas, such as city centres and inner city neighbourhoods. In addition, we should note that the segregated spaces of the city are already connected up in different ways to other parts of the city and to other parts of the world. Concentrating too much on the physical form of the city itself – looking only within the city – can mean ignoring the wider connections which are already shaping city spaces. And once again these connections can be deeply ambivalent. The links which draw places on opposite ends of the world into a similar creative urban culture are perhaps links which should be fostered in whatever way possible. Links which mean work and opportunities for people are also productive: immigrants, for example, can contribute to developing new types of economic activity. But there are many connections which might actively undermine the well-being of people in different parts of the city and make those places more dangerous and difficult to live in.

Criminal networks, including drug trading, car theft, arms smuggling and perhaps even selling child labour, can find a safe space in poorly policed, racially segregated areas and extend across continents and even around the world. One of the emergent concerns for public safety in South African cities since the ending of apartheid has been the way in which different parts of the city have been drawn into a continental and international drugs trade. Arms smuggling connections set up during and since the liberation struggle now feed criminal elements, and car thieves make good use of the easy links out of cities across international borders. Clearly, some connections are positively harmful to the safety of city dwellers and some people have more power than others to forge and encourage certain kinds of connections.

Networks which stretch beyond the city are themselves a source of power for different groups within the city, who can then influence the opportunities for change (see **Allen, 1999b).** Those whose power and income depend on maintaining poorly policed and poorly serviced segregated spaces – perhaps both druglords at one extreme and informal sector traders or landlords in shack areas at the other – are able to undermine efforts to transform the city (Edwards, 1994). The politics of the city, then, can limit the possibilities for remaking the meaning and opportunities of differentiated spaces. If ambitions for transforming segregated city spaces are to be achieved, citizens will need to organize and governments and businesses will need to take effective action to improve social and economic opportunities and to reinforce the positive changes which planners and city dwellers are actively creating.

5 Conclusion

City spaces may be produced through dynamic flows and movements, but in this chapter we have seen how changing processes can involve the emergence of certain patterns of settlement which persist over long periods of time. The differentiation of city space sometimes hardens into segregation as communities become disconnected from the city and isolated from economic opportunity. The causes and extent of differentiation and separation differ from place to place, but are often connected to processes associated with managing city life in its diversity and complexity and with trying to improve the physical conditions of urban living. The ways in which different parts of the city are imagined (dangerous, insecure or unpleasant) can compound their separation from other parts of the city and strengthen their disconnection from certain city networks.

Nonetheless, even areas which seem disconnected from the city in some ways are also shaped through various links and connections to other parts of the city and to other parts of the world. This reminds us that the effects of segregation can be ambivalent, in that they may be positive as well as negative. Segregated areas may offer opportunities for crossing boundaries and for forging new kinds of connections as much as they can keep different kinds of people and activities apart. Thinking spatially about addressing some of the problems of segregation means bearing both these kinds of spatialities in mind – borders and boundaries as well as crossings and connections. If it is more and better connections with other parts of the city, country, continent or world that need to be encouraged as a way of changing the spaces of segregation in cities, then we will have to be aware of the many different kinds of connections that are possible – physical and imaginary, productive and destructive. We will also need to recognize that new kinds of connections produce new kinds of power relations and new kinds of spaces. The dynamism of the city can make our efforts to improve city life somewhat unpredictable!

The next chapter considers in more detail how politics shapes city life. Segregation, of course, is only one of the many different sources of possibilities and problems that accompany the diversity and dynamism of city living, shape city politics and pose a challenge for city management.

References

Allen, J. (1999a) 'Worlds within cities' in Massey, D. *et al.* (eds).

Allen, J. (1999b) 'Cities of power: settled formations?' in Allen, J. *et al.* (eds).

Allen, J., Massey, D. and Pryke, M. (eds) (1999) *Unsettling Cities*, London, Routledge/The Open University (Book 2 in this series).

Anderson, K. (1988) 'Cultural hegemony and the race-definition process in Chinatown Vancouver: 1890–1980', *Society and Space*, vol.6, pp.127–49.

Boyer, C. (1983) *Dreaming the Rational City*, Cambridge, Mass, MIT Press.

Cox, K and Jonas, A. (1993) 'Urban development, collective consumption and the politics of metropolitan fragmentation', *Political Geography*. vol.12, no.1, pp.8–37.

Crankshaw, O and White, L. (1995) 'Racial desegregation and inner city decay in Johannesburg', *International Journal of Urban and Regional Research*, vol.19, pp.622–38.

Davis, M. (1993) *City of Quartz*, London, Verso.

Denton, N.A. and Massey, D.S. (1988) 'Residential Segregation of Blacks, Hispanics, and Asians by Socioeconomic Status and Generation', *Social Science Quarterly*, vol.69, p.811.

Dewar, D. (1995) 'The urban question in South Africa: the need for a planning paradigm shift', *Third World Planning Review*, vol.17, no.4, pp.407–520.

Edwards, I. (1994) 'Cato Manor: cruel past, pivotal future', *Review of African Political Economy*, vol.61, pp.415–27.

Gauteng Provincial Government (1997) *Draft White Paper: Vusani Amadolobha. Urban Regeneration and Integration Plan for City, Town and Township Centre*, Department of Development Planning and Local Government, Gauteng.

Goldberg, D.T. (1993) 'Polluting the Body Politic': racist discourse and the urban location' in Cross, M and Keith, M. (eds) *Racism, the City and the State*, London, Routledge, pp.45–60.

Harvey, D. (1989) *The Urban Experience*, Oxford, Blackwell.

Jellinek, L. (1997) 'Displaced by modernity: the saga of a Jakarta street-trader's family from the 1940s to the 1990s' in J. Gugler (ed.) *Cities in the Developing World: Issues, Theory, and Policy*, Oxford, Oxford University Press.

Lodge, T. (1983) *Black Politics in South Africa since 1945*, Johannesburg, Ravan.

Massey, D., Allen, J. and Pile, S. (eds) (1999) *City Worlds*, London, Routledge/The Open University (Book 1 in this series).

Massey, D. and Denton, N. (1993) *American Apartheid*, Cambridge, Mass, Harvard University Press.

Mayekiso, M (1996) *Township Politics: Civic Struggles for a New South Africa*, New York, Monthly Review Press.

McDowell, L. (1999) 'City life and difference: negotiating diversity' in Allen, J. *et al.* (eds).

Naga, R. (1997) 'Communal places and the politics of multiple identities: the case of Tanzanian Asians', *Ecumene*, vol.4, no.1, pp.3–26.

Oliver, M., Johnson, H.J. Jr., and Farrell, W.C. Jr. (1993) 'Anatomy of a Rebellion: A Political-Economic Analysis' in Gooding-William, R. (ed.) *Reading Rodney King – Reading Urban Uprising*, London, Routledge.

Parnell, S. (1991) 'Sanitation, Segregation and the Natives (Urban Areas) Act: African exclusion from Johannesburg's Malay Location, 1897–1925', *Journal of Historical Geography*, vol.17, pp.271–88.

Pile, S. (1999) 'What is a city?' in Massey, D. *et al.* (eds).

Ratcliffe, P. (1997) 'Race', Housing and the City' in Jewson, N. and MacGregor, S. (eds) *Transforming Cities: Contested Governance and New Spatial Divisions*, London, Routledge.

Robinson, J. (1996) *The Power of Apartheid: State Power and Space in South African Cities*, Oxford, Butterworth Heinemann.

Sapire, H. (1994) 'Apartheid's "Testing Ground": Urban "Native Policy" and African Politics in Brakpan, South Africa, 1943–1948', *Journal of African History*, vol.35, pp.99–123.

Smith, S. (1987) 'Residential segregation: a geography of English racism?' in Jackson, P. (ed.) *Race and Racism: Essays in Social Geography*, London, Allen and Unwin.

Starkey, M (1994) *London's Ethnic Communities: One City, Many Communities. An Analysis of 1991 Census Results*, London, London Research Centre.

Western, J. (1996) *Outcast Cape Town* (first published 1981), Berkeley, University of California Press.

READING 4A
David Theo Goldberg: 'Racism, the city and the state'

It seems clear that concerns of race have played some part in the unfolding of urban planning rationale. Consider the contemporary history of slum clearance. The racial dimensions of this concept were set at the turn of the century by colonial officials fearful of infectious disease and epidemic plague. Unsanitary living conditions among the black urban poor in many of Africa's port cities were exacerbated by profiteering slumlords. Concern heightened that the arrival of the plague, which devastated the indigenous population, would contaminate the European colonists. As fast as the plague spread among the urban poor, this 'sanitation syndrome' caught hold of the colonial imagination as a general social metaphor for the pollution by blacks of urban space. Uncivilized Africans, it was claimed, suffered urbanization as a pathology of disorder and degeneration of their traditional tribal life. To prevent their pollution contaminating European city dwellers and services, the idea of sanitation and public health was invoked first as the legal path to remove blacks to separate locales at the city limits, and then as the principle for sustaining permanent segregation. When plague first arrived at Dakar in 1914, the French administration established a separate African quarter. This was formalized by colonial urban planning as a permanent feature of the idea of the segregated city in the 1930s. The urban planner Toussaint formulates the principle at issue: '… between European Dakar and native Dakar we will establish an immense curtain composed of a great park'. In the post-war years, active state interjection in urban development of Western cities, as of colonial cities, was encouraged, by means of apparatuses like nuisance law and zoning policy, to guarantee the most efficient ordering and use of resources. The principle of racialized urban segregation accordingly insinuated itself into the definition of post-colonial city space in the West, just as it continued to inform post-independence urban planning in Africa.

Thus, administration of racialized urban space in the West began to reflect the divided cityscapes produced by colonial urban planning. The massive urban renewal and public housing programmes in the US in the late 1950s and 1960s started out explicitly as the exclusive concern with slum clearance. This is reflected in the titles of the bureaucracies directing the programmes: in terms of the heralded Housing Act of 1949, urban renewal was to be administered by the Division of Slums and Urban Redevelopment; the country's largest urban programme in New York City was originally headed by the Slum Clearance Commission, and in Chicago by the Land Clearance Commission (Glazer 1965: 195, 200). The experience of the Philadelphia Housing Authority is typical. The federal Public Housing Authority rejected slum locations in the 1950s as the sites for (re)new(ed) public housing projects, yet they did little to generate available alternatives. Strong resistance to encroachment by white neighbourhoods, a strict government unit-cost formula, shrinking federal slum clearance subsidies, and high land costs (caused in part by competition from private developers) left the Housing Authority with one realistic option: to develop multi-storey elevator towers on slum sites (Baumann 1987: 176). The effects were twofold: on the one hand, reproduction of inner city racial slums on a smaller but concentrated scale, but now visible to all; on the other hand, massive removal of the cities' racial poor with no plan to rehouse them. Inner city ghettos were centralized and highly rationalized; the larger proportion of the racialized poor had to settle for slum conditions marginalized at the city limits. The first effect turned out to be nothing short of 'warehousing' the racially marginalized; the second, no less than 'Negro removal'.

This notion of 'slumliness' stamped the terms in, and through, which the urban space of the racially marginalized was (and in many ways still is) conceived and literally experienced by us (the other's racial and class other). The slum is, by definition, filthy, foul smelling, wretched, rancorous, uncultivated and lacking care. The racial slum is doubly determined, for the metaphorical stigma of a black blotch on the cityscape bears the added connotations of moral degeneracy, natural inferiority and repulsiveness. The slum locates the lower class, and the racial slum the *under*class …

In terms of structural formation, then, the planning prototype of project housing and slum reproduction for the racially marginalized throughout the West, is, I want to suggest, idealized in the Group Areas Act of the apartheid polis. This hypothesis will be considered by many to be purposely provocative and obviously over-generalized; by others, it may be thought trivially true.

So I should specify what I do *not* mean by it. First, I am emphatically not claiming that urban planners and government administrators in the West have had apartheid-like intentions. Indeed, although there may have been the rare exception, primary intentions appear to have been to integrate neighbourhoods along class lines. Second, the planning effects under consideration in the West have not been formalized or instituted with anything even closely resembling the precision of the South African state; urban movement and racial displacement in the West have more often been responses to the informalities of market forces than the function of legislative imposition. Third, I do not mean to suggest that project housing (or ghettoization for that matter) ever was, or now is, considered a single residential solution to 'the Negro problem' or to 'the problem of the underclass'. Fourth, and most fundamentally, my aim is not to exonerate apartheid morally by normalizing *it*, that is, by rendering it in terms analogous to common (and so seemingly acceptable) practice in the West. Rather, I am concerned, by invoking the comparison, to condemn segregated housing in the West, the practice of reinventing ghettos (whether formally or informally), and the peripheral dislocation – and thus reproduction – of the racially marginalized. Finally, I am not claiming that all elements of the apartheid idea of Group Areas are manifest in the practices outlined above, only that they embed key elements of the apartheid structure.

The key structural features of the Group Areas Act of 1950 that I wish to emphasize here include:

(a) A residential race zone or area for each racial group.

(b) Strong physical boundaries or barriers to serve as buffers between racial residential zones. These barriers may be natural, like a river or valley, or human constructions, like a park, railway line or highway.

(c) Each racial group should have direct access to work areas (industrial sites or central business district), where racial interaction is necessary, or to common amenities (government bureaucracies, airports, sport stadiums, etc.) without having to enter the residential zone of another racial group. Where economies in furnishing such common access necessitate traversing the racial space of others, it should be by neutral and buffered means like railways or highways.

(d) Industry should be dispersed in ribbon formation around the periphery of the city rather than amassed in great blocks, to give maximal direct access at minimal transportation costs.

(e) The central business district is to remain white controlled.

... An example of (b) is the division of Harlem from south-west Manhatten by Central Park and Morningside Park, as well as by double-lane, two-way traffic cross-streets (110th and 125th Streets; most east–west streets in Manhatten are one way). The South Bronx is divided from Manhatten by river, and from the rest of 'respectably' residential Bronx by a steep hill. Black public housing in the racially split and discriminatory City of Yonkers is all [to] the west of the Saw Mill River Parkway, the railway line and a large reservoir park; white middle-class housing is all to the east. The strong buffer zones of item (b) ideally include space for each racial residential zone to expand. In the urban metropolises of the West, upon which the residential race 'problem' was foisted and where space is at a (costly) premium, this ideal has not been an option. It is replaced in the scheme of things by a testy area of racially overlapping, common-class residential integration (as, say, in South Philadelphia).

Examples of (c) include the West Side Highway and the east River Drive along the sides of Manhatten, the 195 and Schuykill Expressways in Philadelphia, and Chicago's Lake Shore Drive ...

I do not mean to suggest by my illustrations of (b) and (c) that city parks, highways or reservoirs are developed for the purpose of dividing urban space along racial lines, or to deny that given (racialized) communities have their own internal logics of formation. Obviously, the historical determinations of urban structure are multiple and complex, and there are, as I have suggested, urban areas which remain racially ambiguous, resisting or escaping idealized racial (self-)definition. But once in place, these urban facilities were often used or, at the very least, had the effect of reifying racialized city space.

References

Baumann, J. (1987) *Public Housing, Race and Renewal: Urban Planning in Philadelphia, 1920–1974*, Philadelphia, Temple University Press.

Glazer, N. (1965) 'The renewal of cities', *Scientific American,* vol.213, no.3, pp.195–204.

Source: Goldberg, 1993, pp.48–51

READING 4B
Mzwanele Mayekiso: 'Township politics: civic struggles for a new South Africa'

A welcome to Alexandra: Alexandra as freedom dawned (early 1990s)

Welcome to Alexandra Township, just north of Johannesburg ... I will provide a walking tour of the township as it appeared in the early 1990s, just prior to our transition to democracy ...

View from the street

You always remember your first sight of Alexandra Township, it is so striking. Visitors expect to see many high-rise buildings, since around 350,000 people live in this small area of less than two square miles.

Instead, you are immediately surprised as you leave the Pan Africa taxi rank on the outskirts of Alex. You must dodge cows, goats, horses, chickens, and mangy dogs that no owner would claim, all roaming the streets, scavenging in litter pits for food. But you quickly conclude that this is not, by any means, a rural scene.

Alex is unlike any city in the world. No South African township is as densely populated, as well developed politically, as decimated by unemployment and economic despair, or as socially tense.

If you have come from the Wynberg suburb into Alexandra, you will have seen the main commercial area on First Avenue. The taxi rank is central, with thousands of people going to and fro, except when there is a taxi war in progress. There are some major shops such as Sales House, OK, and Checkers, but commerce is also controlled by smaller shops owned by Indians and some whites.

The prices charged here are higher than in downtown Johannesburg, due to the ability of shop owners to exploit customers who must otherwise pay an expensive taxi for a trip to buy cheaper goods. Hawkers outside the shops have slightly better prices. A few African-owned shops are on the Alexandra side of First Avenue.

There is also a good deal of industry on the outskirts of Alexandra. Marlboro, Wynberg, and Bramley View are filled with small factories. And there are white residential areas nearby as well, separated from Alexandra by a buffer strip of small industry in Bramley View and Kew, as well as by open fields.

The most interesting things happen within the township. Take a walk through these streets with me and you will understand more about these contradictions. It is always wise to have a guide on your visits to Alex, if you are an outsider. When I first came here from the Transkei, when I was in my late teens, my sister Nobantu showed me around. Things have become much more complicated since the early 1980s.

This is obvious when you talk to the people of Alexandra. Looking to your right, you find small groups of unemployed youth. They often spend time gambling with dice and cards, with some winning or losing up to R1,000 a day. By the way, if you have a lucky streak and win that much and then try to walk away, it might cost you your life. Some of the players stole their families' savings to play this game, and do not take losing well.

As an outsider, there is a fair chance you will be mugged if you walk these streets carelessly. Thugs carry knives, and guns are now becoming the fashion. Ordinary criminals have begun a virtual arms race here. One result is that you will quickly learn of the organized gangs that operate here.

Not all our street-life economic activity is criminal, of course. The informal sector is buzzing. Coming up this road here is comrade Tsediso, who is unemployed, a school drop-out, and now a peanut seller. He sells these nuts in little packets the size of your palm for R1.50.

Tsediso takes in about R30 each day, and in the early hours of the next morning he must spend R15 of that getting more nuts. Tsediso starts selling to workers near a factory at 6 a.m., moves his stall around the township during the day, looking for concentrations of people, and ends up at night near the factory gate. Twelve hours is not unusual, as Tsediso manages one sale every half hour. It is tragic that a person must spend time in such a manner.

The R15 profit Tsediso brings home must feed eight people. He barely survives. Other vendors are competing for Alex residents' scarce money. Some sell the heads of sheep, and also the internal organs. Very little goes to waste here, since people are so desperate.

There is a small crowd of youngsters, even though it is still school hours. Some are also regulars at the shebeen, the equivalent of a corner tavern, but located within a house. Some youngsters are to be found in houses and shacks, seduced into having sexual relations with adults who have money. It is not formal prostitution, but just reflects how the social fabric has decayed.

Further along this road is a garbage dump. Homeless people are living there, moving around the township scavenging for food in dustbins. They are social rejects, and must live as human garbage in these conditions with makeshift cardboard covers or even just in the open.

But we must be careful, and not just about muggers: a kombi (minibus) taxi is approaching fast. These taxis can be very dangerous, racing through the streets to beat the competition. Transport is the best example of the success of the informal sector, the experts say. There is no railway linking Alex to Johannesburg, and the bus company went bankrupt. Transport is a big issue, and the entire community now depends on taxis.

A trip from here to Johannesburg and back costs R3.40, which is a large amount for people who earn less than R1.000 per month. Worse, they must still wait as long as two hours during the morning and evening rush hours for space in a kombi. The kombi that nearly ran us down belongs to Alexandra-Randburg-Midrand-Sandton Taxi Association. Their politics are different, but their drivers are just as crazy. You must pray if you are a passenger, or even just crossing the road.

And for quite some time, you also had to pray if you went anywhere near Roosevelt Street, one of our main arteries. Not even the taxis used this street. Why, you ask? Look up the hill, and you see the answer: that brown, military-style barracks. That's Madala Hostel, the oldest hostel in the township, and therefore called Old Man's Hostel, or M1.

There was a good chance of being a sniper victim if you proceeded along Roosevelt near the hostel; it was best never to go closer than a few hundred meters. The life-long community residents who used to live next to the hostel, in what we call Beirut, were forced to move to churches and community halls and even the offices of the hated Alexandra Town Council ...

There is a women's hostel, however, which is safe to visit. The 2,800 women in this prison-like structure live in overcrowded conditions, four or five women in tiny rooms with no space for relaxation. They cannot have their husbands or children even visiting them here. Life is hard, but they keep their spirits up.

For women elsewhere in the township, life is also very hard. They must get up earlier than anyone to take care of the family and go to work. Their pay is nowhere near the man's, for even the same job. They come home at night, clean up, cook, do laundry, and prepare the children for bed. The man of the house is usually out visiting friends, or sitting in the corner, relaxing and reading a newspaper.

Outside, the streets are dangerous. I think it is time to move into a shebeen nearby. In fact, every third house in Alex is a shebeen and there are many types of alcoholic drinks to chose from. Some shebeens specialize in jazz music, some disco, and some funk. The music in this one, rap from the United States, is extremely popular. But it is too loud for us to carry on the conversation for long.

Out on the road again, we must cross the little river that divides Alexandra, called the Jukskei. Children play there when it is hot. The water is filthy, a health hazard. The river banks are not well kept, and indeed they sometimes collapse when people walk on them. You can see that the shacks built right up to the edge of the banks are in severe danger of falling into the water, which they sometimes do. If, on the other hand, you visit the same Jukskei River in Sandton, you will see how clean it is and how the banks are well fortified ...

Let's keep walking. These shack areas we're passing have little kiosks known as 'spaza shops' interspersed throughout. This is another informal sector activity, where people can earn a few rand without a license, selling sugar, the staple corn meal, bread, rice, matches, and other essentials. Over there is a typical 'corner shop', which is a real shop with a license. Prices are more expensive in both kinds of shops than in Johannesburg. Some of the owners have begun to climb the ladder, but with such intense competition between them, and because of robberies, the owners are not really prospering ...

Life is very different if you go just three kilometers to the west. It is unbelievable, but there you find one of the richest suburbs in all of Africa, even the southern hemisphere. Sandton is a dream to the blacks of Alexandra, in terms of living standards, a place to raise a family in peace and quiet.

And if you look into the distance you will see Sandton City, the high-rise, luxury office, and shopping center. Our people travel there to work as security guards, cleaners, or menial laborers. They see the wealth all day. At night, the confusion and despair of Alexandra welcome them back. It is a tragic tale of two cities.

A popular joke in Alex is that when Harry Oppenheimer, former boss of Anglo American Corporation, flew in an airplane over Alex, he thought it was an old, disused scrapyard for junk cars, set in the middle of a beautiful suburban scene dotted with swimming pools and leafy trees, and immediately thought about redeveloping the scrapyard for profit. As one *Star* newspaper report has it, 'Approaching Alexandra by air shocks the sense more than it would on the ground. One moment you are gliding over large beautiful homes set in lush gardens amid tennis courts, stables and sparkling swimming pools, then the earth below you suddenly turns brown and scabrous as if it had died'.

It may seem like an abandoned pile of junk to the very rich, but life in Alexandra is always vibrant at the street level. If you return to Alex on a weekend, what you see in the backyards will amaze you. Each ethnic group reverts to traditional cultural recreation. The Pedi people come out with their beautiful dances, as many people watch in admiration. The Zulu people have their dances too, as do the Xhosa. And you can go from one to the other, enjoying the richness of African culture. Then you visit a stokvel, or burial society, which offers a mix of saving money and socializing.

The churches are also very well attended on Sundays. But from the Catholic church to sects such as the Zionists, it is older people who are the loyal core of churchgoers. They pray, sometimes with great energy, running in circles in small, unventilated areas. But you can feel the tension between the mainstream churches and the sects. And there is a real class difference between churchgoers, which you can see from the quality of the clothing and the way they walk to church.

Alexandra is a very splintered place, you see. What nearly everyone shares, though, is a sense of insecurity. It is difficult to see this at first glance, because of the tradition of hiding the problems. But once you meet people and gain their trust, you hear more and more of the problems that poor and working-class South Africans must face.

Source: Mayekiso, 1996, pp.17–27

READING 4C
Richa Naga: 'Communal places and the politics of multiple identities'

Communal places in people's maps

The centrality of communal places in the lives of Asians emerged prominently in the sketch maps of Dar-es-Salaam drawn by some informants. The maps [discussed] here were drawn by women and men from diverse backgrounds – middle and lower classes, educated and illiterate, living in both the relatively prosperous Asian-dominated city centre and the relatively poor, racially mixed neighbourhood of Kariakoo. While these maps emphasize how people experience, interpret, and represent their communal places, they also suggest that social space is not merely personal. Individual experiences of places are shaped by social and spatial structures rooted in existing hierarchies even as individuals constantly enact, reproduce, challenge, and sometimes modify these structures. [Figure 1 is an 'actual' map of Dar-es-Salaam showing geographical locations of the life-historians.]

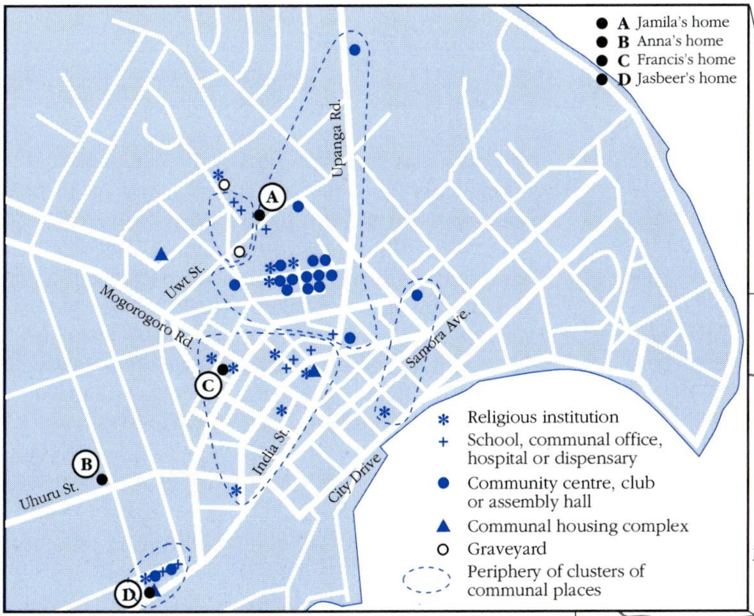

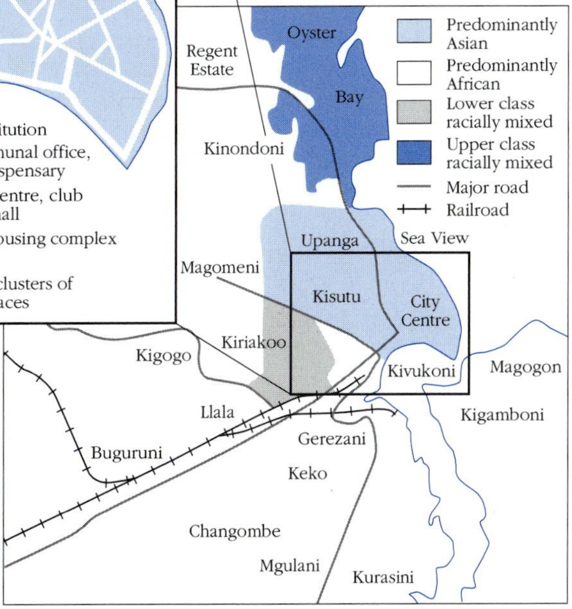

FIGURE 1 *Dar-es-Salaam showing locations of individuals who made mental maps*

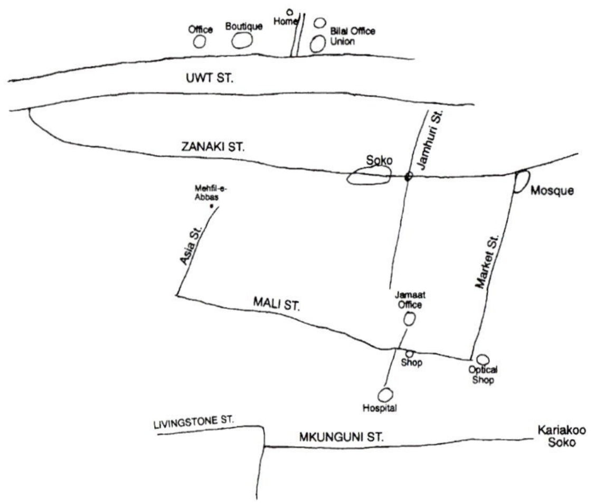

FIGURE 2 *Mental map of Jamila*

The first mental map (Figure 2) was drawn by Jamila, a 46-year-old working-class Ithna-Asheri woman. Jamila has to depend on financial help from the Ithna-Asheri community to maintain her family, which includes her husband and four children. She works informally as a part-time sales agent for various small-scale Ithna-Asheri importers. The centrality of the Ithna-Asheri Jamaat (the formal body that organizes the religious, social, and economic affairs of the Ithna-Asheri community) and community-based networks in Jamila's life is reflected in her map. She lives in a subsidized apartment owned by the Jamaat, located close to other Ithna-Asheri buildings such as the Bilal Muslim Mission and the Ithna-Asheri Union. Most of Jamila's day is spent in an area comprising a few blocks, which is dominated by Ithna-Asheri communal places and businesses. On her map, Jamila pointed out the Ithna-Asheri mosque, the Jamaat office where she frequently goes to borrow money, the Ithna-Asheri charitable hospital where she collects free medicines, some Ithna-Asheri-owned shops, and Mehfile Abbas, where weekly religious gatherings of Ithna-Asheri women are held. Beyond this Ithna-Asheri-dominated area, the only place that appears on Jamila's map is Mkunguni Street, a border between the Asian-dominated area and Kariakoo. Although Jamila gets cheap vegetables from here, she prefers to not venture into the interior of Kariakoo because it is 'too African'.

The meaning of Kariakoo is very different for Anna, a 44-year-old middle-class Goan woman. Anna lives in a racially mixed section of Kariakoo with her Goan husband. Kariakoo forms the centre of Anna's life. While drawing her map (Figure 3) she remarked:

> I will start from Kariakoo because even though it is on the periphery of the town, it is my home. I love Kariakoo. Here's my home, and here's our colourful Kariakoo market ... My mother lives in that building behind us. I have several friends close by.

Anna's world is made up of people from various communities. Close to her home live an Arab friend, two Hindu friends with whom she converses in Gujarati (not Anna's mother tongue), and a close friend, Moona, who is racially mixed. Outside Kariakoo, the only places which appear prominently in her map, are the Roman Catholic church in the city centre where she often goes for the Sunday mass, and the United Nations office where she works as a secretary.

Figure 4 was drawn by Francis, a 45-year-old working-class Goan man who mainly earns his income by driving Asian children to their schools. He also works as a part-time taxi driver and motor mechanic. Francis is married and has one son. He lives in the city centre but, being a taxi driver, he feels that he knows 'every bit of this city'. Francis's map begins from his home and the streets close by from where he picks up his school kids, and the schools to which he drives them. The Roman Catholic church, which Francis refers to as 'the Goan church', is a place he visits every

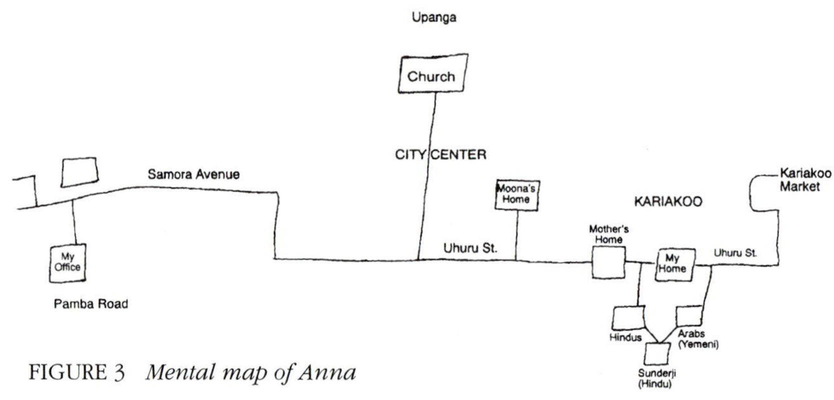

FIGURE 3 *Mental map of Anna*

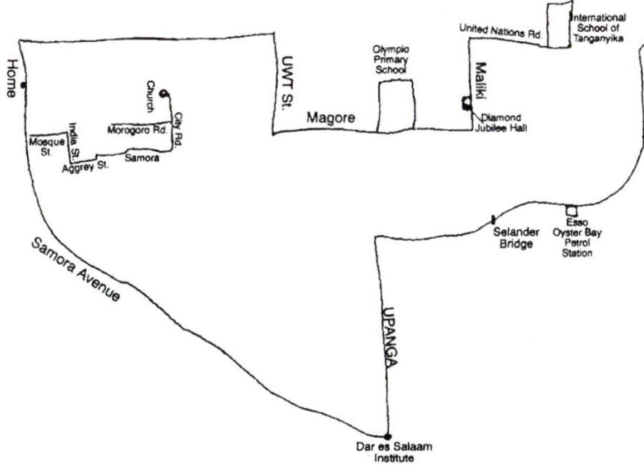

FIGURE 4 *Mental map of Francis*

morning. The Goan club (Dar-es-Salaam Institute) is another place that Francis visits frequently. Beyond the Asian-dominated city centre and Upanga, Francis's map includes the Drive-in Cinema, where he worked as a gatekeeper and ticket-seller during his schooldays.

The last map (Figure 5) was drawn by Jasbeer, a 35-year-old lower-middle-class Sikh man who migrated from Punjab twelve years ago and works as a small-scale building contractor. He lives in the Sikh Trust building with his wife and two children. His map shows his home next to the Sikh Temple, his son's nursery school nearby, the shops on India Street where he gets his building materials, the Daily News office where he goes to place advertisements in the newspaper, and the construction sites in Upanga and

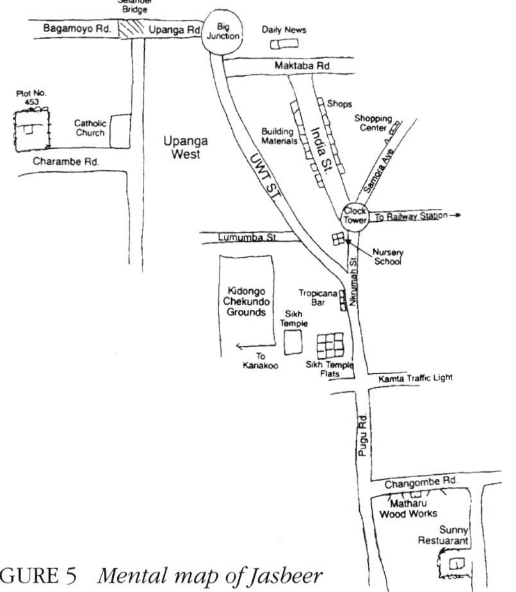

FIGURE 5 *Mental map of Jasbeer*

Changombe where his projects were going on at the time he drew this map.

These maps represent the activity spaces of the map-makers and the manner in which they perceive their city … These maps also reflect that different individuals assign different meanings to the same places and that there can be no one reading of any given place. The meanings and interpretations that people assign to social places depend on the social relations and structures of power in which people are embedded. For example, the significance of communal places for middle-class women such as Anna lies primarily in the religious or spiritual meanings of these places, and in the opportunities for social interaction that they provide. For poor women such as Jamila, however, dependence on communal places is a matter of survival, physically, economically, and socially.

Despite the multiplicity of interpretations, however, the common thread tying all the maps together is the rootedness of people's identities and sense of place in their communal places and neighbourhoods (see 'tales of the field' 1 and 2). Communal places and neighbourhoods not only define each informant's sense of social space, they become salient points of reference in each person's perception of the city irrespective of their gender or class.

At the same time, the maps are also characterized by an important gendered difference that cuts across class and community. The maps of Jamila and Anna (Figures 2 and 3) indicate that Asian women's perceptions of their city are associated more intimately with their homes, communities, and neighbourhoods than are those of men (see the maps of Francis and Jasbeer, Figures 4 and 5). Women's maps cover a considerably smaller area than men's. This gendered contrast in maps results from the difference in the activity spaces as well as the physical mobility of women and men. The gendered nature of racial, religious, caste, and class boundaries, which often imposes severe restrictions on women, plays a significant role in shaping women's mobility and activity spaces.

Finally, these maps reflect an 'Asian-centric' view of Dar-es-Salaam where the perceived limits of the city coincide with those of Asian residential and business areas. Anna clearly voices this perception when she describes Kariakoo as 'the periphery of the town', even though it is central to her own life. African areas and institutions are conspicuously absent, or appear only marginally in these maps. The marginal position of Africans in these maps represents the social and spatial distance between Asians and Africans. It also indicates the role that Asian communal places and residential areas indirectly play in intensifying racial segregation and stereotyping by circumscribing people's social activities and lives primarily around their religious, caste and sectarian affiliations. The rootedness of people in their communities and the racially segregated pattern of social interaction are also reflected in my 'tales of the field' below:

Tales of the field (1): Social interactions and neighborhood communities (impressions recorded on 20 June 1993)

People define their places at every step. I see residents of Mtendeni Street playing volley ball and badminton with their neighbours in the courtyards of their apartment buildings. All the sounds which play on loudspeakers in the downtown area – Hindu devotional songs in the mornings, Hindi movie songs during the rest of the day, prayer calls from the mosques, and the sounds of satellite telecasts of Ithna-Asheri gatherings during Moharram – give the city centre an atmosphere which not only looks and smells 'Indian' but is also full of Indian sounds.

Social contacts, visits, lunches and dinners are defined primarily along communal lines. Ithna-Asheris socialize with Ithna-Asheri friends and relatives, Hindus with Hindus who are often from their own caste, Goans with Goans, and Sikhs with Sikhs. However, this intra-communal contact is almost always with people of the same economic class. Cross-class socializing is rare. Some people also have diverse neighbourhood communities. For example, Nargis's friends are mostly from her Ithna-Asheri family community but she also has a neighbourhood community which includes a Sunni and three Hindu families. Neighbourhood communities are more predominant in the city centre and Kariakoo, less so in Upanga, and almost nonexistent in wealthy areas such as Mikocheni, Msasani, and Oyster Bay, where people are connected by phones and cars with their friends and relatives, rather than through personal, random and spontaneous interaction that characterizes the neighborhoods in the town and Kariakoo ...

Tales of the field (2): Hindu Neighbourhood and communal scene (impressions recorded on 4 Feb. 1993)

The Hindu communal scene in Dar-es-Salaam is one dominated by middle-class, self-employed shopkeepers with their homes, shops, places of worship, and socializing clustered in a small area. Men leave their shops to their assistants or partners, and frequently drop in and out from the temple and halls as and when they feel like it. Middle-class housewives and old women gather with women from their age group at about 3.30 to 4 p.m. Lower-class women from Bhoi, Rana, Vanand, and Divecha castes from Kumbharwada and Kariakoo escort their children to Pathshala [Hindu religious school] and gather every day for about two hours while their children study religion and Gujarati. Well-off Hindus normally drive with their spouses to the temple in the evenings after 6.

With the exception of African domestic servants, vendors and employees who work in Asian shops and businesses, the majority of middle- and upper-class Asians have minimal social contact with Africans. In fact, a quick glance at the Asian communal scene can gain one the impression that the only Africans present in Asian communal places are servants who clean Asian temples, mosques, and community halls. Although it is true that the majority of Africans are marginalized from the mainstream social life of Asians, racial boundaries are not as clear-cut as they might seem. Notions of racially 'pure' communities are continuously challenged and interrupted in Asian communal places by the presence of groups such as African-Asian Sikhs, Seychelloise Roman Catholics, and African Muslims who have embraced the Ithna-Asheri faith.

Source: Naga, 1997, pp.7–13

CHAPTER 5
City politics

by Sophie Watson

1 Introduction

The politics of a city can be understood in many different ways. We can think of politics as being about the way a city is administered and governed – as we shall in Chapters 6 and 7 – or we can see politics as being about the struggles and conflicts in the city over resources and services. We can also see politics as being about daily life and the possibilities or constraints that enable or disable city-dwellers in living the ways that they want. The differentiation of urban space itself produces struggles over resources and services of the city as we began to see in the previous chapter. Some areas have good housing, transport or open spaces while others do not. Some of these resources are tied to specific places, like schools and hospitals, while others are not, such as clean air or knowledge.

This differentiation of urban space operates at a number of levels. Materially some city-dwellers have access to good amenities and locations while others do not. In this sense cities are spatially differentiated. They are also differentiated socially, different groups live in different places so some urban spaces are characterized by high levels of unemployment and poverty, while others may be the enclaves of the wealthy. Third, cities are differentiated symbolically, which is to say that they carry different meanings and significance depending on who we are and what we need and value. This will vary according to race, age, gender and any number of other factors – social, economic and cultural – which affect peoples' perceptions and experience of the world. City politics is, therefore, about how, and whether, people can represent themselves in the city, as well as about whether they can gain access to the resources they need and want.

Cities are constituted in part by everyday interactions and conflicts as well as by struggles around resources and provision, which the urban sociologist Manuel Castells, in his book *The Urban Question* (1977), called struggles around 'collective consumption'. Castells used this term to describe the processes whereby city-dwellers are all engaged in consuming services like housing and education in the city, and this shared activity from which some gain and others lose, he argued, is a significant force in shaping the social relations of the city. This argument was important in that it highlighted the ways in which divisions in the city between the rich and poor are not only a result of the productive relations in the city, that is in terms of where people are located in relation to work and the ownership of capital, but are also a product of *processes of consumption*. For access to home ownership, good education or community services is as significant as the relation to the means of production which was the more classic focus in the Marxist thought at the time. Where the notion of collective consumption was less useful was in analysing women's role in providing services in the city as unpaid workers (*International Journal of Urban and Regional Research*, 1978) and women's unpaid provision of services. The point here is that the city works because women provide all sorts of services for free – from rearing children to looking after the sick or dependent. To look at collective consumption processes in the city without recognizing these hidden aspects of women's work is to tell only half the story.

Cities are also constituted by struggles that may be less immediately observable, or perhaps are observable in a different way. These are struggles around questions of representation, meaning and identity which have profound impacts on cities as we shall see. How particular spaces are defined and controlled may just as much derive from cultural practices as from economic ones. One example here might be who represents themselves through public art in the city or through graffiti. Another such example is spaces designed for one use but used for another, like underpasses or car-parks which are used for skate-boards or roller-blading (usually) by young men, where anyone else will enter at their peril.

A high population density in cities intensifies the struggle between people for limited resources and amenities. The intense juxtaposition of people also means that where conflicts arise they are likely to be more explosive than elsewhere. Or, more positively, proximity between city-dwellers can also provide opportunities for community networks and collaborative initiatives at a local level. So the intense social relations can be both a source of conflict or a source of community, in other words the city can be a disabling or an enabling force. Conflicts often arise when compromise and accommodations cannot be found, that is when the intensification of social relations brings together incommensurable differences for which there is no obvious resolution.

The intensified social relations in cities can take a myriad of forms. People from many different places, backgrounds and cultures live side by side and above and below one another. Each of these people are connected with different places of work, different relations with others in the city, in Britain and overseas, and each bring different memories and experiences to bear on daily life (see **Amin and Graham, 1999**). In the context of this differentiation of urban space, socially and spatially, people have to learn, at least to some degree, how to live together with people who are different from themselves. They are forced to confront each other's differences which can be avoided more easily in rural and suburban areas. City politics can be intense.

Urban politics arise in a context of the heterogeneity of cities. As we will see, this means that city-dwellers are constantly negotiating the 'unassimilated otherness' of other people as Young (1990, p.301) proposes; as a set of multiple and differentiated public arenas (Fraser, 1990), where some groups are included and others excluded (see **McDowell, 1999**); or as a site of mutual withdrawal, indifference and lack of acknowledgement, as feared by Sennett (1994, p.357) (see **Allen, 1999a**). Cities can also be places of division and segregation, as we saw in the previous chapter.

Crucial in determining who has access to the resources of the city and who does not, and who can move freely through urban space and who cannot, are power relations (see **Allen, 1999b**). In this chapter we will explore how who is included and who is excluded is a dynamic process which shifts over time and space and depends on a group's location in the networks of power. Overall, as we shall see, the picture is one where the poor, women, minority racial and ethnic groups have had less access to city resources or to defining and controlling city space.

In this chapter we shall be

- exploring *struggles around resources in the city* and how these are produced by, and produce, differentiated urban space,
- looking at how people's interests are represented in the city and whose voices are heard, and
- considering who is excluded and who is included, and how and why, from access to resources or from finding a voice in the city.

This will involve looking both at more traditional analyses of urban politics and power as well as more contemporary understandings of city politics, new urban social movements and cultural practices. We will explore what underlies the collective consumption processes and how these are connected with networks of power. In this chapter a central strand of the argument is that city politics is about everyday experiences and conflicts as much as it is about more formal notions of urban governance and party politics. This is to shift the focus onto the daily conflicts and interactions of city-dwellers away from the Town Hall and sites of official political representation, committee meetings, and local and city government.

ACTIVITY 5.1 Read the extract from Elijah Anderson's – a black American sociologist's – fascinating account of daily interactions in the streets of Greenwich Village in New York: *Streetwise* (1990) (Reading 5A). It is a text which illustrates how the intense juxtaposition of people in urban spaces causes people to adopt coping strategies to make themselves feel safe.

As you read the text, jot down the ways in which a sense of safety was established by each of the people he describes. How was this gendered? Why does he argue that black people cannot hold the eyes of white people for too long? ◆

When I read the piece I found the metaphor of the overcoat (which both protects and repels) at the end a very poignant one. For me this reading illustrates very clearly how living amongst strangers in close proximity makes us develop all sorts of strategies to protect oneself from unwanted interactions with another person, particularly if we think that a person different from ourselves is threatening. I laughed to find that my own spatial practices as a woman are clearly common ones. Another way in which women and other people who feel vulnerable to attack in public protect themselves is not to enter a railway carriage unless at least two other people are in the carriage. Did you pick up how important power is in the unspoken dynamics of interaction in the street? The notion of power, which has been explored in the context of networks of cities (**Allen, 1999b**), will be a recurring theme in this chapter.

In the extract we were looking at the relations between black and white people in the street. But these strategies can be seen in all sorts of other social relations of the street. Think, for example, of the way in which people often avert their gaze away from the homeless or beggars. One of the important successes of the magazine *Big Issue* is that it gives the homeless sellers of the magazine a more active engagement with the public which can also provide a forum for conversation, interaction and an exchange of information and ideas. In this way the magazine not only provides an opportunity for the homeless to gain some income (a proportion of the street price goes to the seller), it also returns dignity and agency to the homeless – a group who have been highly marginalized in the streets of our cities.

In this introduction I have highlighted several important themes:

- City politics are as much about the everyday as they are about the big picture of city governance and administration.
- City politics are about access to resources in the city, and where these are distributed unequally conflict is likely to arise.
- City politics are also concerned with spaces of representation and the possibilities for expressing oneself in the public spaces of the city.
- City politics are constituted in a field of power relations where some people have power and others do not.
- The intense juxtaposition of different activities and different people in the city constitutes the site of city politics – a site that can sometimes be explosive.

Understanding the city as a site of tensions – in some instances productive and in others negative – which arise from living close to one another in often limited space helps us to think about all the many forms of politics in the city. Let us look now at a case study of a struggle over resources and urban space.

2 *Case study: Sydney – the Better Cities Programme*

We have all seen photographs of Sydney's spectacular harbour glinting in the sun. Who is it that actually gets to live on the water's edge and enjoy the beautiful views and access to the water? You might well imagine that it is only the wealthy home owners that can afford this luxury. In many parts of the harbour you would be right, but one area – Pyrmont Point – which is close to the harbour bridge and the city centre, tells a different story.

Before Captain Cook arrived in 1788 to set up a British colony in Australia, Aboriginal peoples lived across the continent in different tribes moving freely from one place to another. Early settlement by the British, who took over the land using convict labour, led quickly to the destruction of the Aboriginal peoples and their culture through outright slaughter and the introduction of disease. The Aboriginal population in the coastal regions, where Sydney was established, was decimated by disease and brutality. Sydney, in the ensuing years, saw the development of patterns of slum housing juxtaposed with the occupation of prime sites in the city by the wealthy. For the first 150 years of white settlement most migrants to Sydney could afford a single dwelling on a quarter-acre block close to the city centre and foreshore areas. Following the Second World War, with the expansion of the migration programme and pressure on land, suburbanization shifted further west as prime-site locations were only accessible to higher income groups – with some exceptions.

This process was exacerbated as Australia became increasingly affected by global processes of restructuring, which meant that Australia's control over its own economy, wage levels, employment practices and property market was

FIGURE 5.1 *Sydney harbour*

radically diminished. Global restructuring (see **Thrift, 1999**) forced a reduction in tariffs to increase international competitiveness, leading to huge increases in overseas investment in Australia – from Japan and Hong Kong particularly. These forces impacted on the housing market, leading to a massive increase in house prices pushing poorer people into the outer suburbs. This is an excellent illustration of global and local interconnections impacting on daily life where global financial decisions directly affected the housing choices of local people. The result was urban sprawl, a deteriorating environment, traffic congestion, and suburban isolation where those without cars – often women, youth and older people – became trapped at home. A heated debate in urban policy arenas ensued in the 1970s and early 1980s over whether the Australian dream suburban home had become a nightmare (Kemeny, 1980). Did continued suburban development represent egalitarian access to individual home ownership on a quarter-acre block, as many of you will have seen on the Australian television programme *Neighbours*, or an environmental and social disaster?

The 1983–95 Labor government was determined to address suburban inequality through its Better Cities Programme, whose main objective was to address spatial and social inequalities, particularly between the outer and inner suburban areas. Part of the strategy was to move towards a more consolidated city, bringing people back from the suburbs and increasing density in the inner-city areas. The federal government thus encouraged developers to build mixed developments of private and public housing at varying densities so that a range of households could find affordable housing in a good location. The policy met resistance from local governments and local residents in low-density suburbs who feared increased housing density would lead to diminishing house prices and decreased amenity in their local areas. Academics from varying perspectives engaged in fierce debate, some defending free choice as egalitarianism while others favoured strategic intervention. The following Liberal government of 1995 was quick to dismantle the Better Cities Programme, advocating a return to the free market and a diminished role for interventionist urban policy.

ACTIVITY 5.2 Read the following extract from the *Australian*. Why, according to Ward, does Patrick Troy object to the Better Cities Programme? What do you think of Troy's argument? ◆

EXTRACT 5.1
P. Ward: 'A view to the future'

Million dollar views are going for a song [relatively] in Sydney and it doesn't mean the property bubble has burst. They are part of the legacy of the Federal Government's abandoned Better Cities Programme [BCP] 1991–6 which had equity and amenity in inner city housing as one of its key objectives. Residents in the NSW [New South Wales] Department of Housing's new Bowman Point are accordingly now enjoying a splendid outlook as well as living just up from Darling Harbour ... Bringing community and public housing into Pyrmont helps fulfil a primary aim of

> the BCP to create a precinct that contains a full range of socio-economic groupings. It is an attempt to stop the place becoming a yuppie ghetto … The question, however, is will those for whom such projects were made – government agencies, local authorities, private developers, urban planners and researchers – look and learn that it's impossible to mix dense public and private housing and create useful, affordable, civilized and safe neighbourhoods? … So how is it, in the midst of the political ice age, when only the financially fittest are supposed to survive, that egalitarianism has broken out on Pyrmont Point? …
>
> Patrick Troy in his book 'The Perils of Urban Consolidation' [argues that] denser housing than the standard detached house and block is un-Strine [un-Australian-English]. Perhaps the hapless folk on Pyrmont Point don't know what's bad for them.
>
> The book erects a critique for every assumption of BCP planners and designers. Their demographic figures are flawed, he says, the infrastructure and housing cost estimates are skewed, assertions of housing preferences and size are incorrect, environment imperatives have been misinterpreted, and ideas of efficiency and equity have been consistently misapplied.
>
> Not only that, motor car dependency while causing some pollution is also socially liberating … In the end, in proposing an alternative urban policy to that pursued by the BCP planners, he begins on a base which finds little to fault with the 'traditional' form of Australian cities. 'The way households vote through the market expressing preference for traditional forms of housing could be taken as prima facie evidence that it best meets their needs for most of the time' he says.
>
> … But does not [this] quote encapsulate a central problem that the BCP sought to address? Of course households 'vote' through the market, but that market, and all the vested political and financial interests that operate it and speculate up and down the greenfields, has traditionally offered a very narrow series of housing choices.
>
> Of course people choose the location that best meets their needs – especially their need to service their housing debt. Cost, not comfort, forces a vast number of people to vote for the cheapest, most distant, most environmentally problematic housing developments.
>
> Patrick Troy is critical of what he sees was the BCP's attempts to coerce, through administration and planning instruments, households to accept forms of housing they would not want. Yet that cuts both ways. A planning regime that trammels households towards the so called traditional form of Australian cities – that is the detached house on the average 700 sq m greenfield block – is just as coercive.
>
> Source: Ward, 1997

There are two points that need to be highlighted here. As the writer points out, choice operates in the context of constraint. Thus we are mistaken if we argue that people necessarily choose their place of residence because it is exactly what they want. People choose from what they can afford, from what is on offer, and from what they know and they make decisions as to what best suits their needs given these constraints. In the article Ward points out that households in Sydney have a very narrow series of housing choices and are also constrained by housing costs. Second, the article refers to the long tradition of suburban housing as the predominant form of housing development in Australian (and American) cities. This housing form, Ward suggests, has been promoted by the planning system, long before the BCP stepped in to address the problem of unfettered suburban development. Though this may have suited the country in a period when the nuclear family was dominant, where there was full employment and not every household had a car or even two cars, new socio-economic conditions raise new questions.

One reason that the suburbs functioned successfully in the post-war years was the fact that many married women stayed at home to carry out domestic responsibilities (Watson, 1988). Thus, women's days were taken up with conveying different family members between dispersed activities: school, shopping, the doctor, the train station and so on. As more women have entered the workforce, their double burden of paid and unpaid work has been increasingly hard to carry out in these low-density suburbs where employment opportunities, transport, and community and child-care facilities are not interlinked. For single-parent families accommodating the complex and multiple tasks of everyday life is increasingly difficult. Many women have no car and experience isolation, and there are similar problems to those found on the peripheral housing estates in Glasgow (Chapter 2). The dispersed city form has become unviable for another reason: as more people overall own and drive cars, people spend a much greater proportion of their time travelling between home and work – which exacerbates problems of pollution and congestion.

What we see here is:

- A government programme (the BCP) attempting to address the unequal distribution of resources (in this case housing), changing needs and circumstances through new forms of mixed-density housing which is, perhaps, more appropriate to cities of the future, where the traditional nuclear-family model is no longer dominant.
- That city politics are complex and there is often no one view of what is the best solution to a problem. The politics around urban space – sprawl versus equity – are ambiguous. Troy, who is himself a committed social reformer, sees the older suburban form of housing as egalitarian, whereas the new BCP politicians have a different view.
- That even if the earlier suburban form of housing worked well when it was first developed, over time it has led to new problems – the isolation of women and traffic congestion – which has led to new political struggles and interventions.

Thus we can see that city issues and the politics that arise are in a constant state of flux and change. Nevertheless,

- it is usually the poor and the least powerful who lose out in city politics.

3 Power and politics in the city

City politics cannot be detached from an understanding of how power operates (**Allen, 1999b**) and the wider networks of power within which cities are embedded. To develop this argument, we will examine closely the connections between power and politics in the urban landscape. In this context power can be understood in a number of ways:

1. *Who has the power to define what is an urban problem?* For example, until the mid–late nineteenth century ill health, slum conditions, lack of sanitation and so on were not perceived as urban problems to be addressed by government. The social reformers and philanthropists identified and defined these issues as an urban problem and pressed for reform. It was not the people who were living in poor conditions themselves at this time whose voice was heard. Similarly, in the post-Second World War period, urban redevelopment took little account of the people whose lives were being dispersed. A famous study in Bethnal Green in East London (Young and Willmott, 1962/1986) in the 1960s explored the break up of family and kinship networks that was brought about when old working-class housing was knocked down to make way for large public housing estates. There was little notion then, or earlier, of citizen participation which has become more prevalent in urban governance recently.
2. *Who has the power in a situation of conflict?* When an urban problem becomes manifest what matters is who has the power to intervene to find a solution, on whom is power conferred by others, and what are the mechanisms at work. Sometimes who has the power is very elusive or hidden. For example, an issue may simply be written off the agenda so that it does not enter public discussion and debate. In the case of a property developer having local political influence, the agenda may be set which only lays out certain options for development and not others. In this way power is exercised behind the scenes and cuts out alternative methods of approach. Thus, power is exercised without obvious coercion or force and because there appears to be no alternatives than the ones presented, people are unaware that they are being manipulated to agree to only a limited set of options (Lukes, 1975).
3. *Who can maximize their advantages at the expense of the weaker?* Power affects different peoples' access to resources and this power is held by particular groups – those who can take advantage of the resources and amenities of the city, or those who can have their needs met in the city. This might be visible in terms of place of residence: in Los Angeles, for example, many of the rich and famous live in huge mansions on Beverley Hills. Or it may be visible in terms of a household being able to afford to move close to well known successful state schools where housing is more expensive, in order to give their children an advantage. For example, houses in the

FIGURE 5.2 *Local residents fight cuts in expenditure to their public library*

catchment area of the Cherwell comprehensive school in north Oxford are at least 10–20 per cent more expensive than similar houses nearby but outside of the catchment area.

So power in the city may be constituted by a person or group's ability to have their interests represented and their voices listened to, thus preventing, for example, the building of a new road through their property or a development against their wishes. These may be people who are visible in local politics, or less visible, in that they can influence and control the local agenda through covert means and influential connections. Those with power may be able to resist local developments which are likely to improve the lives of other local citizens through both overt and covert means of resistance. *Resistance* can take many forms and be both regressive and progressive, but romantic notions of local politics tend to focus more on the weaker sections of the community protecting and fighting for their interests, rather than on local élites maintaining their positions of privilege. Once again we find a different nexus of powers and exclusions.

ACTIVITY 5.3 Think of a conflict around a proposed development (for example, of housing, a community centre, hospital, road) that you have experienced directly or have read about in a local paper. What was the conflict about and which constituency was resisting the proposal? Would you consider that the forces of resistance to the proposed development were progressive or not? Now read the following extract by Doreen Massey about the estate where she grew up.

Think back to your own example, how does it compare with the issues discussed by Massey about the development of Wythenshawe, Manchester? ◆

EXTRACT 5.2
Doreen Massey: 'Living in Wythenshawe'

The spatiality of my parents' lives is negotiated within a lattice of differentially powerful spatialized social relations.

Some of their confinement we regularly put down to 'Them'; to 'capitalism' or 'the Tories'. The meagreness of state pensions, the low level of social services and the difficulties of public transport (think what 'high tech' could do for the mobility of the old, the frail, the infirm, if only it were differently directed), the broken paving stones. All these things entrench a rigid framework of constraint: they restrict your movement. Literally close down your space, hem in that less tangible sense of spatial freedom and ease. Their weight is undeniable.

But things are also more complicated than that. The very creation of this estate was the result of a battle. Moreover, it was a battle in which were ranged against each other a powerful local state (the city of Manchester) and the local people of rural north Cheshire. Planners against the people. The state against private citizens. The classic terms of so much current debate slide themselves easily into place: domination versus resistance; strategy versus tactics; the system versus local people.

That romanticized classification/identification would here be quite misplaced. The state, the planners, the system, were here a collection of socialists and progressives battling to win more, and healthy, space for the city's working class. The 'locals' combined a relatively small number of villagers, a high proportion of people who commuted into Manchester to work, and a group of large landowners. The commuters depended upon Manchester for their livelihood but wanted nothing to do with the consequences of their large incomes – the higher taxes of the city, the necessity of living amongst the poor. A poll taken in three of the parishes central to the 'local struggle' showed that 82 per cent of the parishioners wanted to resist Manchester's advances; yet nearly half of them worked there. The landowners had extensive spreading acres, could often trace a lineage back through several centuries, and lived still at the apex of a set of (spatialized) social relations which had even now more than a touch about them of feudal settledness and expectation of deference. There is a tendency in recent literature to glorify 'resistance', to assume it is always ranged against 'domination', to accept without further consideration that it is on the resisters' behalf that we should organize our rhetoric. Maybe it is because today we feel ourselves so relatively powerless. Whatever is the case it is an assumption which avoids thinking about the responsibilities of 'power' as necessarily negative. And it is an assumption which can lead to the misreading of many a situation.

Here in north Cheshire, the resisters against the state, clad in the mantle of 'local people', were defenders of the local ways of life of property and privilege. There are few, if any, abstract or universal 'spatial rules'. Local people are not always the bearers of the most progressive values, 'resisters'

> though they may be. Battles over space and place – that set of sometimes conflicting embedded socio-spatial practices – are always battles (usually complex) over spatialized social power. Personally, I'm glad this lot of locals lost, and that the Wythenshawe estate was built.
>
> Source: Massey, 1999

The extract from Massey gives us an illustration of the workings of power in an urban context. A traditional analysis of the battle, as Massey points out, would be to talk of the local state, or the system, as having power, as opposed to the local people who have none. In this instance the distribution of power is much more complicated; the local state consisted of progressives who were battling to improve the position of the city's working class and people resisting had a great deal of social power and the ability to influence planning decisions because of this. So it is not always easy to see where power lies and city battles often present a much more fluid picture than one of 'domination versus resistance', as we began to see in the Sydney case study.

There are two main ways of thinking about power: *power as possessed* and *power as exercised.* The more traditional models of power relevant to thinking about city politics have tended to see power as *possessed.* In this formulation power is usually conceived as repressive, coercive and negative. Marxist or political economy approaches saw power in the city as residing with the capitalists – financial and industrial – or the bourgeoisie who make the city work in their interests (Harvey, 1973). While in a different approach, deriving initially from Max Weber's work, power is seen to reside in institutions and with the urban managers of the city who have the power to control and distribute resources. These are the gatekeepers of the city (Pahl, 1975), such as housing officials, loans officers and social workers.

The notion of power as *exercised* potentially offers us a more fluid concept where power is seen to be imminent in, and constitutive of, all social relations. In its earlier formulations this was the power of different interest groups in the city who theoretically could influence decisions and outcomes through rational debate and democratic means. This pluralistic view of power and city politics, though descriptive of the ideals underpinning a democratic city politics, failed to acknowledge either structural interests, such as those of capital, or the ways in which power tends to be concentrated in the hands of specific individuals, such as city councillors or key political agents who could determine outcomes (Dunleavy, 1980). Early studies of community power were conducted in the USA which set out to analyse who were the key figures in the local community and their strategies of decision-making (Dahl, 1961).

More recently this idea of power as exercised was extensively developed in the writing of French philosopher Michel Foucault (1978, 1980). Here power is a productive force – a multiplicity of force relations – rather than a negative or repressive force. It is relational rather than possessed or seized and operates in a capillary fashion from below. Power, Foucault (1978, p.92) argues, must be understood 'as the process which, through ceaseless struggles and confrontations, transforms, strengthens, or reverses them; or on the contrary, the disjunctions and contradictions which isolate them from one another; and lastly, as the strategies in which they effect'.

Foucault also links power with knowledge, and power with resistance. These ideas provide a different way of thinking about city politics where we can see the importance of information (as one aspect of knowledge), and the lack of it, as well as recognize that there are many potential sites of struggle and confrontation. This focus highlights the importance of the local and the specific in city politics. Some writers have developed Foucault's ideas to suggest an understanding of urban politics that foregrounds ideas of fragmentation and difference, and the emphasis is on questions of difference, identity, the local, and the specific (see Watson and Gibson, 1995; Keith and Pile, 1993). In the context of intense social relations where different people from different groups are living close together, questions about how people form political allegiances – despite their differences – are emphasized. An emphasis on the local and the specific construction of political alliances allows for a city politics that does not try to find grand solutions to urban problems but which attempts instead to act locally and strategically to influence wider circumstances.

Another way of thinking about city politics and power is to ask the question 'who has symbolic power in the city?' or 'who is imagined to have power?' This might throw up quite different people and sites depending on the age, gender, sex, race or cultural group with which an individual identifies. If we look at young men as an illustration: in a particular locality the dominant social relations may take the form of a football club, or to take a more extreme example, a gang (Davis, 1990). Those seen to be powerful may well be those who cut the appropriate image, who bear the appropriate marks such as tattoos, who are seen to be cool, or even to command sexual attention. Powerful groups of this kind are also able to secure spaces of the city as their own, be it a particular street or park or building, from which others are excluded.

There are several things we can now say about power in the city:

- Power is crucial to understanding who has access to the resources of a city.
- Power is constituted in a number of ways: such as, the ability to define a problem, to set the terms of the agenda, to influence others, to have one's needs met, to determine the distribution of resources.
- Power relations are spatialized. Given the intense juxtaposition of different groups of people in the city different groups will exercise power at different times and in different places.
- Power is exercised by individuals, by groups and/or by institutions.
- Power can have a cultural dimension.
- Power can be elusive and hidden.

Together, these points suggest that power relations are dynamic, part of an ever-changing urban scene.

4 City politics take many forms

There are many different political terrains in the city. We will look at three areas here.

4.1 PRIVATE–PUBLIC PARTNERSHIPS: URBAN GROWTH COALITIONS, CITY MARKETING

With the dominance of conservative governments in the USA and the UK during the 1980s and early 1990s came the growing perception that local governments were inefficient, wayward and wasteful. In the urban policy literature there was an increasing emphasis on marketization, privatization, competitive tendering and private–public partnerships. In this scenario key actors in the private sector are mobilized outside the state structures to develop structures for renewal and regeneration. This in turn has led to a new terrain in urban politics. The following chapter will return to this theme so I shall only discuss it very briefly here.

Regime theory was developed in the 1980s to make sense of the growth of private–public partnerships (Logan and Molotch, 1987). These were initiatives where local governments, city governments or other public sector bodies joined forces to address a particular problem in an area or to create a new development such as a housing estate or leisure facility. Some of these consortia were very large, such as the London Docklands Development Corporation which oversaw the Canary Wharf development on the Isle of Dogs. Regime theory emphasizes the relationship between government and non-government actors in dominant coalitions, exploring how different interests achieve the results they desire through a combination of government and non-government means. What is important here is the recognition of the importance of private business in local governance (see also **Thrift, 1999**). Rather than seeing urban decision-making as coming from the top down, this analysis stresses that urban decision-making is not hierarchical and is fragmented (Stone, 1989).

A related strand of urban decision-making is the trend for city governments to increasingly engage in their own self promotion and to sell themselves to the outside world. Here cities engage in urban regeneration and other forms of urban transformation – often producing a new image of the city through brochures and other forms of promotional marketing. These strategies are used to attract inward investment or the siting of international exhibitions, conferences or international events such as the Olympic games. The hope is that these events will bring money and jobs to the city, though the impact of such events on the lives of citizens – particularly the poor – has been shown to be far from unequivocally positive.

In his discussion of the then forthcoming Olympic games in Atlanta, Rutheiser (1996, p.264) tells how the city and the Olympic games committee joined forces to exclude the homeless from a place in the public realm. And it has happened elsewhere. In 1984, the organizers of the Seoul Olympic games also sought to beautify the city by eradicating evidence of slums (Kim and Choe, 1997). Slum clearance projects were established using a private–public partnership which was fine in theory but a nightmare in practice. The profit motive of the powerful major corporations (*chaebol*) involved, coupled with the lack of political foresight by the municipal government, meant very few – 10 per cent – of the original low-income residents affected could afford to buy the new units constructed. Instead they were forced to relocate from the slum areas – breaking up social and family ties – to often impermanent housing in an unfamiliar area. As a result social tensions emerged in the city and the city government failed to intervene to prevent a widening gap between the interests of the private developers and the general public whose collective power to resist was undermined by relocation. And all in the name of turning a bright shiny face to the world. Similarly, Murphy and Watson's (1997, p.162) study of city marketing in Sydney to attract the Australian Motor Cycle Grand Prix away from Victoria was a financial disaster with millions of dollars wasted.

There are two points here:

- First, the growth of private–public partnerships has led to a new and shifting terrain of city politics. In each case different players will exercise power, sometimes the lead may come from the private sector, sometimes it may come from government. In different instances, different individuals, groups or institutions will be powerful. At the same time other groups will have no say in the decisions that are made and will be marginalized.
- Second, in the case of cities marketing themselves to the outside world to boost their image and attract investment, we again see that there are winners and losers. Often it is the poor and homeless who lose, while the wealthy – such as property developers – are the ones to gain.

4.2 NIMBYISM

Let us turn now to a different kind of terrain of city politics. This is the terrain where communities are involved with local struggles around issues that affect them in their home or neighbourhood. These struggles often arise from urban decision-making concerning developments over which local residents have no say. These decisions might be made by institutions of urban governance, like the road authority, or by private developers, or by a local community organization. Decisions are often made which affect local people with little concern as to what the effects might be. In the case of a new road, for example, the road authority may argue that it will ease congestion for commuters, while ignoring the local pollution and deteriorated environment that will result from its construction. The struggles which ensue over developments such as these are frequently passionate and intense.

The term given to the resistance to proposed developments in a local neighbourhood to which residents object because of perceived negative impacts is *nimbyism* – the 'not in my backyard' syndrome. We saw in the earlier extract from Massey (Extract 5.2) how these struggles over space are constituted in networks of power relations and interconnections. We saw also how acts of resistance represent different interest groups which will in some instance be arguing for progressive ideals, while in others, dominant class or other group interests are being preserved. Nimbyism is similarly ambiguous.

As I have stressed throughout, cities are constituted by the intense juxtaposition of different activities and people in the city. The fact that people live close together and different and conflicting land uses exist side by side, exacerbates the intensity of conflict and tension. A key role for urban planning has been to manage and regulate the different uses of urban spaces. This is done through the zoning of areas whereby one area is zoned for residential use, another for commercial and another for industrial. Sometimes these areas exist in close proximity and sometimes urban areas are highly segregated. Territorial-based conflicts can often arise when urban space and buildings are put to uses side by side which those living in the area see as incompatible. This may be as a result of what are called 'negative externalities' – for example, when a car repair workshop is located in the midst of a residential area causing noise and parking congestion, or when a factory's emissions produce bad air to breathe for the local residents. When these conflicts arise the usual mechanism for resolving them is through the planning system. But often, though not always, it is those with more power – economic or social – who get their way.

FIGURE 5.3
A terrace house is turned into a mosque

Another form of conflict can arise when a building is located in a mixed race neighbourhood which serves the needs of one specific cultural group and not others. Mosques and temples built in residential areas of Western cities have often given rise to conflict. In Finsbury Park in north London there was a lot of hostility to the building of a mosque in a local street mainly inhabited by white gentrifiers and private tenants. Hostility to such developments may be because they are visually distinct and different, give rise to local traffic problems or are used for social functions as well as religious ones which mean different patterns of use than those expected. The antagonism expressed may reflect covert or overt racism from the dominant racial group. This may be legitimated through mobilizing a contravention to planning regulations which themselves tend to reflect the social and cultural patterns of dominant social groups.

ACTIVITY 5.4 Read the following extract from *Surface City* which concerns the building of mosques and temples in the Sydney suburbs. As you read jot down what were the major objections to the proposed buildings. Do you think the same issues would have arisen if a new McDonald's was placed in a residential area? ◆

EXTRACT 5.3
Peter Murphy and Sophie Watson: 'Surface city: Sydney at the millennium'

In several cases a great deal of conflict arose during the site selection processes, with council officers suggesting sites on open space land away from residential areas. These sites are contentious for several reasons. The architectural forms are distinct and the patterns of usage are different from those of the Christian norm. Parking becomes an issue because of regional drawing power, and poor public transport exacerbates the need for cars. It was also argued that the noise impact of mosques was greater than churches since mosques and temples are social meeting places as well as sites of worship. Controversial cases arose where groups were unaware of what was, and was not, permitted. As recently as the late 1980s, local politicians were not embarrassed to express racist attitudes: 'the temple will not be an asset. There are no Vietnamese people living there. You are hoping to put in a complete foreign body in anticipation of people coming to use it'. [The body metaphor is interesting here: Anglo-Australia here is the 'pure' body in danger of contamination.]

On some occasions buildings were refused simply because they were different or 'other'. An increase in the height of the mosque spire at Smithfield in the Fairfield locality was refused by the council 'on the grounds that it was likely to spoil the amenity of the area'. The council report stated that 'the visual privacy of adjoining properties would be reduced. It is out of character with the surrounding areas and the height is considered to be excessive.' There were no similar instances where a church tower had not been permitted. More often than not refusal was legitimated on the grounds of inadequate parking provision.

In 1991 a Greek bishop submitted an application for a private chapel on his property in Penrith. The local council objected on the grounds that it was too close to the house and that its 'design was not in keeping with the area'. New plans were submitted and fifty nine objections were received – mostly concerning crowds, noise and traffic. Some residents also complained that 'the fence was not in keeping with the area, and that trees

> have been cleared and a track cut through the land'. This was followed by claims that the bishop was not an authentic bishop of the Greek Orthodox church. The council then deferred its decision until it received a statutory declaration that the chapel would only be for private use. Opponents of the site said that the aldermen were 'fence sitting to avoid being labelled racist'. Eventually permission for the private chapel was granted but on condition that the bishop furnish a letter of intent. It was recognized that a refusal would constitute an infringement of his freedom of religion which could be taken to the Anti-Discrimination Board. One alderman shouted during the debate: 'I'll stand here and say Germans are better than bloody Greeks'. Another, who was speaking on behalf of the majority of residents, pointed out that it would be difficult to provide evidence should the bishop break the ruling at weekends. In cases such as these, planning norms and regulations are invoked to neutralize objections which derive from a different source.
>
> Source: Murphy and Watson, 1997, pp.27–8

What struck me about this situation was the dominant concern by local residents that their area had a homogenous image that they wanted to retain. The idea of something different being placed in the locality appeared to be very threatening. Perhaps it is for this reason that a new McDonald's would be less likely to be resisted by local residents even though it might be equally disruptive. McDonald's restaurants are usually built in a fairly standard and conservative style which is familiar to most city-dwellers (see Massey, 1995).

Another form of nimbyism derives from the trend in social policy towards community care and long-term residential housing for marginal groups in the so called 'community'. As part of a strategy to reduce public expenditure on the one hand, and part of the ideology of community-based care as preferable to institutional care on the other, people from mental institutions and homeless hostels are increasingly housed in residential areas. Housing associations buy and convert dwellings, with funding from the housing corporation, for voluntary organizations and other such groups for long-term housing. Here conflict arises when local residents object to having people who they regard as disruptive or disreputable in their street. Similar concerns are expressed when sex offenders are released back into local communities.

The hostility that many residents feel can be understood in terms of fear of the unknown 'other'. This is the idea that those unlike ourselves are unknown, are 'other' and are thus threatening. In his work on gypsies David Sibley (1995) suggests that people are marginalized because they are feared and created as 'other'. Powerful groups 'purify' and dominate space to create fear of minorities and ultimately exclude them from having a voice. This seems very much to be the case with the Vietnamese temple in Sydney where the metaphor is used of the temple as a foreign body disrupting a purified space and allowing new and different voices to be heard (Extract 5.3). Sibley (1995) argues that often in local conflicts in the city a community will represent itself as normal and is threatened by those who are perceived as different. Fear of attack, harassment or burglaries has led to the establishment of gated communities for the wealthy in order to keep out strangers (**Amin and Graham, 1999**). Fears and anxieties are expressed in stereotypes which logically could be challenged by greater knowledge of, and interaction with, the unknown others, though such a move could have limited success if done half-heartedly. This argument also resonates with issues raised by **McDowell (1999)** and in Massey and Jess (1995).

A more positive alternative to Sibley's rather negative scenario is Iris Marion Young's (1990) ideal of city life as a vision of social relations as affirming group difference or as 'openness to unassimilated otherness' or a 'being together of strangers'. For Young the unpredictability and heterogeneity of the city are the ingredients for a sense of openness to others and mutual affirmation of difference. Richard Sennett's (1970) work on the uses of disorder follows a similar track. Sennett rails against the suburban neighbourhood as socially homogeneous and enclosed. Such environments, he suggests, encourage the persistence of the myth of a purified community of neighbourliness and security (although divisions may lurk under the surface: remember Chapter 1?). They also perpetuate a form of parochialism based on a naïve child-like psychology about other communities. In the Sydney example, the myth was that no Vietnamese people were living in an area despite census statistics which revealed a relatively high proportion of Vietnamese in the locality. Sennett argues that the only way to break down these purified notions, the only way to develop full maturity, is to encounter other groups and situations (as we will see later in Reading 8A). This happens in the inner city with its mixed communities in a form of politics of encounter. In Sennett's vision there would be no such thing as nimbyism because people could *enjoy* being next to 'unassimilated others'.

As I write from my study in Kentish Town, London, I laugh at the irony of Sennett's unambiguous celebration of difference. The house next door is owned by the housing association St. Mungo's and houses long-term middle-aged male alcoholics. Through many evenings of the summer months the noise of fights and shouting disturb the peace of this quiet terrace. Yet I support the provision of long-term housing for marginal groups in the community and worked in a similar organization for the homeless in the 1970s. What do I do? Do I join my neighbours' struggles to get them out?, do I negotiate with the men themselves?, do I meet with the community workers to sort out the problem?, do I move? What would *you* do? Living in the city requires negotiation and accommodation and learning to live with difference in the ways suggested by Young and Sennett's work.

Yet sometimes people simply cannot tolerate certain activities in or close to their backyard and conflicts cannot be negotiated. In these instances an outside mediator may be necessary. One such instance is noise from neighbouring houses. Local councils, as a response, have initiated codes of practice to address this problem and action can eventually be taken if the problem is not resolved.

The difficulty, though, is that sometimes the solution moves the problem elsewhere onto someone else's patch.

This discussion of nimbyism illustrates the very intense and often difficult political issues which arise out of different people with different needs and circumstances living cheek by jowl in the city. To summarize:

- Nimbyism, like other struggles over space, is constituted in networks of power relations.
- Nimbyism arises in sites of intense juxtaposition of people, conflicting needs and conflicting land uses.
- Nimby resistance may or may not be progressive. Nimbyism is ambiguous.
- Nimbyism can result from the fear of an unknown 'other'.

4.3 COMMUNITY-BASED CITY POLITICS AND URBAN SOCIAL MOVEMENTS

Let us move now from city politics at the neighbourhood or local level to consider wider communities based on an identification with a particular social group or political struggle. This kind of city politics is sometimes referred to as community politics. There is a long tradition of community-based city politics organized outside of the mainstream institutions of governance both public and private. In the UK urban community-based politics had its heyday in the 1960s and 1970s. Though the very form and essence of community politics is marked by its diversity and heterogeneity, there are some common features.

FIGURE 5.4 *Anti-racist campaigners demonstrate against racist attacks*

Community groups in the city usually come together to represent interests that have been marginalized or excluded in some form – in terms of lack of provision or services to meet their needs, or experiences of prejudice, harassment or discrimination, or lack of access to urban space such as housing. They are not usually allied to any political party although there is often a clear political strategy inherent to the project which is typically organized around social justice goals. Another feature of community groups in the city, particularly in the earlier period in the 1970s, was an ambivalence towards institutions of the state (London Edinburgh Weekend Return Group, 1978). A desire to retain autonomy and not be co-opted by institutional forms, procedures and bureaucracy sat uneasily with the need for funding and provision. Drawing on ideals of mutual co-operation, and an antipathy to hierarchy and leadership, which were particularly important to the early women's liberation movement, many of the community groups operated as collectives with shared and revolving responsibilities. Given that a further aim was to empower members of the group and encourage participation, the people whom the organization addressed were motivated to take on positions of responsibility as soon as possible.

Three case studies can be used to illustrate community-based politics in the city.

4.3.1 The squatters movement

In the early 1970s many local authorities sought to expand their public housing programme through purchasing private property in mixed neighbourhoods from landlords and owners. The objective was to rehabilitate the properties and rent them to applicants from the housing waiting list, thereby creating mixed areas of public and private housing and breaking down the ghettoization of public tenants in large housing estates. Cut backs in public expenditure in 1974 led to severe restrictions on local authorities' abilities to press ahead with the programme with thousands of empty properties, in the Greater London area particularly, as a result. This combined with a growing number of young single households moving to the city for work, many of whom were committed to an alternative lifestyle and politics, together with growing numbers of homeless households.

FIGURE 5.5 *Squatters protest about their eviction when council properties remain empty*

In the 1970s, many public properties were thus occupied by squatters who were prepared and able to move in and fix up the dwelling to a livable standard. In some cases the squatters had to move on within months, while in others households occupied the same place for several years. In various parts of larger cities entire streets or squares were squatted, such as Tolmers Square near Euston in London, Chapeltown in Leeds and Montpelier in Bristol, weekly meetings were held, food co-operatives and shared facilities were established. These squatters' groups were very politically active in a number of ways. First, they drew attention to the large number of public properties that were left neglected and empty while thousands of people were registered on housing waiting lists. Pressure was put on local authorities to rehabilitate dwellings, as well as to rehouse non-traditional households such as single people and special interest groups. This was the seeding ground for the growth in housing co-operatives and housing associations which were expanded in the latter part of the decade.

Second, the squatters' groups were involved in local community politics campaigning around a plethora of issues from the provision of play areas for children to the provision of local community law centres. This was a vibrant and diverse political movement which changed the face of large metropolitan areas throughout the 1970s. By the end of the decade the squatters had been moved on, and the dwellings they had occupied had been renovated and allocated to applicants from housing authority or housing association waiting lists and households that were defined to be in housing need. Similar trends emerged in European cities like Amsterdam and Berlin and squatters came together through informal networks across national borders at festivals, political events and local gatherings.

4.3.2 Women's refuges

Refuges for women who were victims of domestic violence emerged at a similar political moment in several large cities, though by the end of the 1970s some refuges had been established in smaller towns in rural areas. A refuge for battered women was established in Chiswick in 1971 by Erin Pizzey bringing attention to an issue which had been hidden from the public and political agenda. Feminists were quick to voice the significance and extent of hitherto unrecognized domestic violence and they established groups throughout the country to pressure local authorities to provide accommodation and financial support. To increase political pressure groups joined together in 1974 to form the National Women's Federation to lobby government for changes to the law and public awareness. Women's aid groups, like the squatters, were locally based and organized non-hierarchically. In some areas local authorities were quick to respond to demands for a dwelling and funding. In other areas groups had to run refuges on voluntary contributions and support. Workers in the refuges encouraged the women who were residents to become involved in the running of the refuge and in the national political campaigns. This was fuelled by a politics of empowerment where those in receipt of services were encouraged to define and articulate their own needs. A desire to retain autonomy and control often meant that limited public financial support was available.

A sense of the political impact on the city of women's aid groups is hard to recreate 25 years on. Issues of women's safety in the home and the city, and the need for long-term housing for women on their own with their children, amongst others, entered the arena of city politics in a way that had never happened before. In 1998 Gill Hague summed up the situation:

> There are now more than 300 refuges throughout the country, including a variety of specialist refuges for black women and women from minority communities. These include specific projects, most commonly for Asian women, but also for African and Caribbean women, and Latin American,

> Turkish, Irish and Chinese women in various areas. The first Jewish women's refuge was opened in London in 1997. However, many of these specialist projects are in a shaky financial position. Most refuges are affiliated to the Women's Aid federations in either England, Scotland, Wales or Northern Ireland which coordinate the provision of refuge, support and advocacy services and work to promote the interests of abused women and their children nationally. Some refuges remain independent of the federations, usually those associated with church groups, or increasingly in the mid-1990s with housing associations (which frequently provide housing services only, with little attention to support, care and advocacy issues). Independent refuges also include some black women's organizations and *Refuge*, the successor to Chiswick Women's Aid (and Chiswick Family Rescue), which maintains a high profile.
>
> (Hague, 1998)

We see here how a political struggle for resources to be allocated to a previously marginalized group has been extremely successful. Yet at the same time the gains remain shaky and many women's refuges have not managed to secure a stable financial base.

4.3.3 The Green Ban Movement in Sydney

Environmental movements have played an increasingly important role in city politics over the last 20 years, attracting in particular young people and students. There are many different groups from bicyclists pressing for bicycle routes in the city, to groups campaigning against waste dumps, factory emissions, water poisoning and traffic pollution. One of the most interesting strategic alliances around environmental politics in the city was the Green Ban Movement which was active in Sydney from the early to mid 1970s.

FIGURE 5.6 *Bicycle campaigners take to the streets*

In these struggles, the builders' labourers' union (the Builders Labour Federation) joined together with residents and tenants to protect public parks on the foreshore of the harbour, to prevent the destruction of old working-class housing and to resist rampant private-property development. The strategy was a simple one. The residents or tenants resisting a development called on the union to support their campaign by not permitting their union members to work on the site. The movement was very effective in protecting environmental resources and low-cost housing in the city for a number of years. One of the key reasons for the success of these campaigns was the strength of the building industry at the time and the high demand for builders' labourers, who were able to exercise a lot of power. Another reason was that there was a progressive federal Labor government in power (under Gough Whitlam) which was committed to urban policy reform.

ACTIVITY 5.5 Read the extract from Jakubowicz on the Green Ban Movement. Consider the following questions as you read. How does Jakubowicz account for the success of the Green Ban Movement period? What does he cite as the main reasons for its final demise? ◆

EXTRACT 5.4
A. Jakubowicz: 'The Green Ban Movement: urban struggle and class politics'

The major implication of the green ban movement period is that effective defence of working-class 'urban' interests is a most difficult goal, impossible indeed without organised power in the industrial sphere, and a state, either local, regional or national, with some commitment to redistributive politics. Yet the period also indicates that without 'grass roots' local action aimed at securing those interests, industrial and political 'power' is insufficient in itself to guarantee urban equity.

For as the advances of the working class in Australia in the mid-1970s met increasing resistance, and urban or collective consumption crises increased, the strategies adopted by capital and the ruling classes changed. These changes were composed of two major modes: that of political co-option and incorporation ('corporatism'); and that of cutting the social wage and reducing real direct wages ('recommodification'). The struggles over the extension of social provision that were part of the early 1970s push had by the 1980s become defences of the small gains already made. Thus, the possibilities of the momentary advances wrested by the working class from capital and from/through the state, depend on the confluence of rising class power at the local level, over collective consumption and its social provision, in the sphere of production of those consumption commodities, and over the allocative mechanism, or the state. The working-class green bans depended for their success on all these factors, and on the active complicity of the state. Indeed key 'bourgeois' bans, such as those in Centennial Park, Hunters Hill and on the figtrees at the Opera House, were successful against a conservative state government.

The green bans were smashed by three connected processes, again the outcome of a 'restructuring' of capital, but this time in the face of the onset of the current international depression. Finance and development capital resourced another political faction in the BLF [Builders Labour Federation] to 'roll' the green ban leaders. The massive insecurity affecting the building industry in 1975 aided this move in the wake of deregistration proceedings launched by the developers against the NSW branch of the BLF …

> The final 'reason' was the massive attack on the working class that preceded, was expressed in, and followed upon the dismissal of the Whitlam government in 1975.
>
> Source: Jakubowicz, 1984

One of the points that came across strongly to me was the importance of strategic alliances between the different players in the city – the state, the unions and the community groups – in achieving urban reforms. When these break down it is much harder to effect change. I was also struck by how fragile these alliances can be and how quickly the green bans were smashed. Again what we see is how city politics can shift and change over time in unpredictable ways with different groups exercising power and making gains at different moments and in different places.

We have seen from these examples (and there are countless others) that community politics have been an important force in shaping our cities. The squatting movement saved many empty houses from dereliction and destruction and paved the way for new groups such as housing associations and co-operatives. It was also linked to other community-based politics in the city, such as campaigns for open play spaces and adventure playgrounds for children and facilities for other special needs groups. The presence of women's refuges in cities provided not only a safe environment for women who were battered but also brought to public attention the previously hidden issue of male violence towards women. One of the ways this was done was through graffiti.

In a period where community-based politics are a less visible force in the city, the many and various impacts of these movements on the urban fabric and on the distribution of space in the city can be underestimated.

Before moving on we need to return to the work of Manuel Castells, which was mentioned briefly at the beginning of this chapter, who developed an important notion – *urban social movements* – to describe the multiplicity of social and political groups based in the city pressing for change. Castells (1977) argued that urban social movements were framed by conflicts within the sphere of processes of collective consumption – that is the sphere of struggles over resources, services and provision in the city. As we have just seen, these movements have had significant impacts on the provision of services in the city. There are also many examples of the ways in which local community politics have contributed to new notions of citizenship and democracy. The squatters, for example, formed associations which in many areas were able to find a voice through the local council. More recently some of these initiatives have involved a shift away from the association of democracy with only the formal political institutions of the state towards new forms of democratic participation (Chapter 7 discusses this further). Soon, city-dwellers may voice their views on local issues to the council via the internet. Already, city-based groups, environmental groups and others, as well as large political movements like the Zapatistas in Mexico, use the internet as a way of communicating and organizing campaigns.

FIGURE 5.7 *Women bring male violence to the public's attention*

In earlier analyses urban social movements tended to be seen as a progressive force. Yet Castells (1997) ambiguously defined social movements as 'purposive collective action whose outcome, in victory as in defeat, transforms the values and institutions of society'. Urban social movements can be formed within any political and social context and are thus not necessarily progressive forces. In the examples of community-based politics discussed above I focused on movements such as these. In *The Power of Identity*, Castells (1997) draws our attention to fundamentalist groups, both Christian and Islamic, whose popularity, he suggests, can be located in growing uncertainties around work, the traditional patriarchal family, gender roles and so on. In many parts of the world there are frightening signs of fascist and racist groupings emerging in cities which mar the more positive image of urban social movements as vehicles for progressive change.

From a different perspective, the notion of urban social movements has been used by world systems theorists like Wallerstein (1974) to suggest that people are beginning to organize politically on a global basis. A critique of this notion lies in its assumption that people across the world are increasingly connected. It is certainly the case that cultural images, environmental problems, the decisions of multinational corporations, international politics, world trade, etc., can affect the lives of ourselves and our neighbours at the same moment as someone right across the world. But the notion implies a similarity of experience and a cosy homogeneity across space. What is ignored are the spatially differentiated effects of global processes and these are important when we think about urban social movements and global and local identity politics. What is at play in urban social movements are connections between people at different levels which may be intensely locally experienced while being also connected into wider global movements which, moreover, can take a different form in another place.

To summarize:

- Community-based politics over many years have had a profound impact on our cities from the provision of new forms of housing, to the protection of threatened public services and goods, such as public parks.
- Community-based politics differ from institutionalized politics in that they emphasize the importance of mutual co-operation, empowerment and autonomy from the state.
- Community-based politics have played a key role in bringing to public awareness a range of social and political issues which are often ignored.
- Community-based politics wax and wane with different issues coming to the fore at different times.
- Community-based politics have given rise to political alliances in the city which Castells has termed urban social movements.
- Some urban social movements have connections across the world bringing together different groups in cities who share similar concerns.

5 The politics of identity, meaning and representation

In the last decade there has been a shift from a discourse of urban social movements to a discourse of the *politics of identity* (Castells, 1997). The move marks a recognition that most people occupy a variety of political identities at one time. Thus, someone might take up, at different times, contradictory positions as they define themselves as a woman, a mother, an unemployed person, an environmentalist, a tenant and so on. In cities people are connected in complex ways across different issues and sites of identity, but many interpretations of urban social movements imply a homogeneity of interests or positions which end up obscuring the heterogeneity underpinning different strategic alliances, alliances that also change and shift frequently.

With the fragmentation of the traditional left we have seen a worldwide growth in different forms of identity politics and community mobilization. With the rise of the Civil Rights movement in the USA placing race politics on the political agenda from the early 1960s, followed in the latter part of the decade by the formation of women's and gay liberation movements in many countries, there has been a shift from class politics to a politics of difference. This is not to say that class divisions have gone away, rather that the claims and voices of other marginalized groups have increasingly come to be heard. Stuart Hall (1991, p.57) describes the politics of difference as 'the politics of recognizing that all of us are composed of multiple social identities, not of one. That we are all complexly constructed through different categories, of different antagonisms, and these may have the effect of locating us socially in multiple positions'.

5.1 THE POLITICS OF MEANING AND REPRESENTATION

We have been considering the differentiation of urban space, conflicts over resources, and city politics in a number of ways throughout this chapter. Another way in to thinking about city politics is to consider the social and cultural significance of city spaces as opposed to the more obvious material understandings where urban analysts have hitherto tended to focus their attention. Political identities are not only constructed in the spheres of the visible or the material. The meaning of place to an individual can matter as much as its visible or structural attributes. Let us look at some contemporary feminist urban writing as an illustration of this point.

Early feminist urban analysts (e.g. McDowell, 1983) viewed the process of suburbanization as subjugating and controlling women, separating them from the arenas of power, work and politics in the city and reinforcing their domestic position. As mothers and carers of children, dependants and husbands, as well as workers in the paid labour market, women have to take complex and time consuming journeys each day to carry out their many responsibilities. Where women try and combine paid work and domestic responsibilities their opportunities for full-time employment are limited. Feminists (e.g. Watson, 1988) thus have argued that suburbs are not simply the havens from the hurly-burly of the city as they are sometimes portrayed. Though they may be constituted in this way for the male city worker, the suburbs were seen to limit women's choices because of their dual roles. The political meanings of suburban life were different for different women. We are in a position now to extend the analysis begun in Chapter 1.

In her study of migrant women in the Western suburbs of Sydney, Susan Thompson (1995) tells a story which has implications for a feminist city politics. For the migrant women Thompson interviewed, the suburbs provided an opportunity for the appropriation of personal power, both as women and as representatives of a minority ethnic group living in white Anglo-Australia. The suburban home was seen as a tangible way for atoning for the losses in the migration process. It served as a symbol of success in the adopted country and as means of maintaining cultural, religious and personal links with the past. This is done through the placing of a Buddhist altar in the sitting room, or through growing vegetables that are unavailable in the local grocery stores, and other cultural practices which preserve some sense of continuity with the country that has been left behind.

Public spaces also carry different meanings and resonances for city-dwellers and these can take on political significance. In the Aboriginal context this takes the form of sacred sites – a particular rock on Sydney's harbour shore where there are rock paintings may constitute a place where land claims can be made. In Australia Aboriginal peoples have been thrown to the margins of society, often living in poverty and disease, and their voices have been little heard. This began to change with the 1972–75 Labor government under Whitlam, and the more recent (1983–95) Labor government under Hawke and Keating. The new discourse was one of reconciliation between white Australians and Aboriginal Australians. There are many strands to this politics of reconciliation, but the important point for our argument here is that the reclamation and establishment of symbolic spaces have played an important part. Thus in Australian cities, sites that were once inhabited by Aboriginal people have become subject to land claims by Aboriginal people, and areas of the city where there are concentrations of Aboriginal people have become important sites for the building of political strength and solidarity.

Struggles which take place to preserve a building, a pub, a local park make sense in the context of their significance as spaces of pleasure, a sense of freedom, or a site of the construction of one's social identity away from the private sphere. Walking tours in cities often map sites that are key to different interest groups; for example, there is a Stonewall tour in New York which takes visitors along the route of the gay struggles that took place there, and in many

British cities there are walking tours which mark political histories of earlier inhabitants. Key city spaces bare the traces of earlier political struggles, such as Trafalgar Square in London from the suffragette demonstrations to England's battles with France under Nelson. Trafalgar Square is certainly an ambiguous site representing, as it does, both Empire and resistance, since many political demonstrations have taken place there.

So city spaces can have different meanings and importance to different groups and individuals depending on who they are. Let us now think about what places matter in people's everyday lives.

ACTIVITY 5.6 Read the two extracts from Virginia Woolf's *The Years* and *Three Guineas.*

Take some time to reflect on what spaces matter to you and your sense of your own identity. Do you find that your sense of yourself can shift as you enter different parts of a city or town? What is it about these places or spaces that matters to you? ◆

EXTRACT 5.5
Virginia Woolf: 'The years'

If the bus stopped here, Rose thought, looking down over the side, she would get up. The bus stopped, and she rose. It was a pity, she thought, as she stepped on to the pavement and caught a glimpse of her own figure in a tailor's window, not to dress better, not to look nicer. Always reach-me-downs, coats and skirts from Whiteleys. But they saved time, and the years after all – she was over forty – made one care very little what people thought. They used to say, why don't you marry? Why don't you do this or that, interfering. But not any longer.

She paused in one of the little alcoves that were scooped out in the bridge, from habit. People always stopped to look at the river. It was running fast, a muddy gold this morning with smooth breadths and ripples, for the tide was high. And there was the usual tug and the usual barges with black tarpaulins and corn showing. The water swirled round the arches. As she stood there, looking down at the water, some buried feeling began to arrange the stream into a pattern. The pattern was painful. She remembered how she had stood there on the night of a certain engagement, crying; her tears had fallen, her happiness, it seemed to her, had fallen. Then she had turned – here she turned – and had seen the churches, the masts and roofs of the city. There's that, she had said to herself. Indeed it was a splendid view … She looked, and then again she turned. There were the Houses of Parliament. A queer expression, half frown, half smile, formed on her face and she threw herself slightly backwards, as if she were leading an army.

'Damned humbugs!' she said aloud, striking her fist on the balustrade. A clerk who was passing looked at her with surprise. She laughed. She often talked aloud. Why not? That too was one of the consolations, like her coat and skirt, and the hat she stuck on without giving a look in the glass. If people chose to laugh, let them. She strode on. She was lunching in Hyams Place with her cousins. She had asked herself on the spur of the moment, meeting Maggie in a shop. First she had heard a voice; then seen a hand. And it was odd, considering how little she knew them – they had lived abroad – how strongly, sitting there at the counter before Maggie saw her, simply from the sound of her voice, she had felt – she supposed it was affection? – some feeling bred of blood in common. She had got up and said May I come and see you? busy as she was, hating to break her day in the middle. She walked on. They live in Hyams Place, over the river – Hyams Place, that little crescent of old houses with the name carved in the middle which she used to pass so often when she lived down here. She used to ask herself in those far-off days Who was Hyam? But she had never solved the question to her satisfaction. She walked on, across the river.

Source: Woolf, 1937/1977, pp.124–6

EXTRACT 5.6
Virginia Woolf: 'Three guineas'

Let us then by way of a very elementary beginning lay before you a photograph – a crudely coloured photograph – of your world as it appears to us who see it from the threshold of the private house; through the shadow of the veil that St Paul still lays upon our eyes; from the bridge which connects the private house with the world of public life.

Your world, then, the world of professional, of public life, seen from this angle undoubtedly looks queer. At first sight it is enormously impressive. Within quite a small space are crowded together St Paul's, the Bank of England, the Mansion House, the massive if funereal battlements of the Law Courts; and on the other side, Westminster Abbey and the Houses of Parliament. There, we say to ourselves, pausing, in this moment of transition on the bridge, our father and brothers have spent their lives. All these hundreds of years they have been mounting those steps, passing in and out of those doors, ascending those pulpits, preaching, money-making, administering justice. It is from this world that the private house (somewhere, roughly speaking, in the West End) has derived its creeds, its laws, its clothes and carpets, its beef and mutton. And then, as is now permissible, cautiously pushing aside the swing doors of one of these temples, we enter on tiptoe and survey the scene in greater detail. The first sensation of colossal size, of majestic masonry is broken up into a myriad points of amazement mixed with interrogation. Your clothes in the first

> place make us gape with astonishment. How many, how splendid, how extremely ornate they are – the clothes worn by the educated man in his public capacity! Now you dress in violet; a jewelled crucifix swings on your breast; now your shoulders are covered with lace; now furred with ermine; now slung with many linked chains set with precious stones. Now you wear wigs on your heads; rows of graduated curls descend to your necks. Now your hats are boat-shaped, or cocked; now they mount in cones of black fur; now they are made of brass and scuttle shaped; now plumes of red, now of blue hair surmount them. Sometimes gowns cover your legs; sometimes gaiters. Tabards embroidered with lions and unicorns swing from your shoulders; metal objects cut in star shapes or in circles glitter and twinkle upon your breasts. Ribbons of all colours – blue, purple, crimson – cross from shoulder to shoulder. After the comparative simplicity of your dress at home, the splendour of your public attire is dazzling.
>
> But far stranger are two other facts that gradually reveal themselves when our eyes have recovered from their first amazement. Not only are whole bodies of men dressed alike summer and winter – a strange characteristic to a sex which changes its clothes according to the season, and for reasons of private taste and comfort – but every button, rosette and stripe seems to have some symbolical meaning. Some have the right to wear plain buttons only; others rosettes; some may wear a single stripe; others three, four, five or six. And each curl or stripe is sewn on at precisely the right distance apart; it may be one inch for one man, one inch and a quarter for another. Rules again regulate the gold wire on the shoulders, the braid on the trousers, the cockades on the hats – but no single pair of eyes can observe all these distinctions, let alone account for them accurately.
>
> Even stranger, however, than the symbolic splendour of your clothes are the ceremonies that take place when you wear them. Here you kneel; there you bow; here you advance in procession behind a man carrying a silver poker; here you mount a carved chair; here you appear to do homage to a piece of painted wood; here you abase yourselves before tables covered with richly worked tapestry. And whatever these ceremonies may mean you perform them always together, always in step, always in the uniform proper to the man and the occasion.
>
> Source: Woolf, 1938/1993, pp.133–7

As you read the extract from *The Years*, did you notice how Woolf contrasts the city as a material space with the flux of subjectivity, or to put this another way, how she contrasts public space with private affect? What is triggered for Rose in the public space of the city is her private reverie and memory. So although she is visibly occupying a public space she is mentally inhabiting the private space of her mind. At the same time, in this piece, Woolf contrasts a past with the present – the city remains fixed and immortalized in its buildings 'there's *that*' – as one becomes a different person or one remembers different parts of oneself or different places. Here, the bridge acts as a metaphor for transition and for a sense of history. As we see in the second extract, the solidity of the state, which for Woolf embodies men's laws, is symbolized there in the city as a set of formidable buildings – the Houses of Parliament.

In the extract from *Three Guineas* the bridge appears once more as a metaphor of transition, the movement from private to public space. This time, Woolf is writing of the public world of men which women are beginning to enter. Yet this is a public space which is defined by men's customs and rituals and the symbolic edifices of law and religion. So once again we see how public spaces are riddled with affect – they are experienced differently depending who you are (see also **Allen, 1999a**).

My point in discussing these two extracts is to suggest that city politics is as much about symbolic attachments and meaning as it is about the very material aspects of the city, to which Woolf also refers. It is these felt intensities, and the symbolic and cultural meanings, of city politics that are sometimes hard to define and explore, with the result that they sit uneasily with discussions of urban policy and planning.

The opportunity to represent oneself in the spaces and places of the city is another aspect of city politics. Representation takes different forms. We looked earlier at the questions of power and politics where being represented in local government or other key institutions in the city offers the opportunity to influence decisions, outcomes and the distribution of resources. But here I am interested in another form of representation which is the making visible a particular identity or culture in the spaces of the city. This also is empowering.

FIGURE 5.8 *Homeless people use graffiti to highlight their exclusion*

In the last two decades public and community art has become an important strategy for marginalized groups to make their presence felt. Sculptures, structures, paintings, graffiti representing the lives and experiences of black, Asian, Latino, gay, women – the list is endless – can be found in almost every city across the globe. Brook Andrew, who is a young Aboriginal gay man living in Sydney, has constructed artworks for the public arena with the intention of shocking, provoking and politicizing. In one work he painted billboards close to the Opera House on which were written FEAR in large letters. In another work he has placed five neon boomerangs – which cross an image of an Aboriginal icon with a suggestion of male sexuality – on a building façade above Oxford Street in the gay area of Sydney. He says this about his work:

> As a black gay man it is sometimes difficult in claiming my cultural tools in regards to being gay, being gay to many Koori people seems to be my only prominent identity, my identity is fluid, it is multiple. In representing the boomerang as a pink neon light show – I synchronistically claim ownership to both my blackness and sexuality, my cultural reality. I display this proudly.

We have seen how notions of identity have become increasingly central to city politics over the last 25 years or so. We could summarize some of the key points as follows:

- Political identities can be understood as shifting, multiple, fluid and complex.
- A politics of difference which recognizes differences of race, gender, sexuality as well as class, has increasingly come to define the political agenda, especially in cities where these relationships are intensified.
- The meaning and significance of places and spaces is an important arena for city politics. So too is the opportunity of representing our concerns in a visible way in the spaces of the city.

In the next section we shall look at two examples of another way in which new political identities have become more visible in the city.

5.2 RECLAIMING THE STREETS

An important strategy for constructing, celebrating and asserting different political identities has been to take over the streets of the city either through demonstrations or carnival. The new urban interconnections have produced new forms of street politics and carnival where sexual politics and different identities are celebrated in the public spaces of the city from New York to Amsterdam to Rio de Janeiro to Sydney. A performative politics might be one way of describing the spatial political practices which are developed in one place and often adapted by similar political groups to their own locality. These political practices have been important in constructing new political urban networks. The two examples we shall examine here are the Reclaim the Night marches of the women's liberation movement and gay and lesbian Mardi Gras carnivals.

FIGURE 5.9 *Reclaim the Night march*

Reclaim the Night demonstrations were first organized by feminists in the USA in the early 1970s. The aim was to have marches through city streets that were seen to be unsafe for women at night. Women carried candles and banners and chanted. The marches typically passed areas of sex shops and clubs. The feminist movement has always had strong international links with the result that over the following years similar marches were organized in Europe, Australia and New Zealand. Despite a similarity in form, the politics and issues thrown up by these marches in each country differed. For example, in Leeds the marches were criticized by anti-racist groups for passing through black people's areas, since by implication there was a suggestion that black men were more likely to represent a greater threat to women. While in London the marches went through Soho, which many argued was a necessary place of employment for women sex workers. So locality made a difference to a city politics which represented the concerns of women in many countries across the world. It is also worth noting here the similarity with the concerns expressed by Doreen Massey in Extract 5.2 about the meaning of local politics and the problems with jumping to too easy assumptions about what is going on in any one instance (see also **McDowell, 1999**).

Mardi Gras is a gay carnival which is celebrated in San Francisco and Sydney. The Sydney festival is globally produced in two senses: politically, it has emerged out of the growing strength worldwide of gay movements over the last 20 years; and literally, in the sense that many gay people travel to these cities to participate in and enjoy these celebrations. You will get a striking impression from Iain Bruce's extract of the vibrancy and excitement of the event. But Mardi Gras is not simply about celebration of different sexualities and the playfulness with which these are performed and represented in the public spaces of the city. It is also a serious form of city politics – it is a politics which asserts the rights of gay people to live and love publicly in all the spaces of the city without fear of harassment.

ACTIVITY 5.7 Read the extract from Iain Bruce's 'Gay sites and the pink dollar' (Reading 5B). What do you think has made the Mardi Gras such a success in Sydney? How does this parade challenge traditional ideas about masculinity? ◆

Iain Bruce's account of the 20-year history of the Sydney Mardi Gras charts the shift of the parade from its early violent confrontations with police authorities to its current status as Sydney's largest and most successful public political event. Far from preventing the parade, the New South Wales government and statutory authorities go out of their way to publicize, support and celebrate the Mardi Gras, recognizing the publicity and income – the pink dollar – that it brings to the city. What started as a contested and violent terrain has been capitalized upon by government, but it has also changed the meaning of Sydney as a city, and what it means to live in Sydney.

FIGURE 5.10 *Sydney gay and lesbian Mardi Gras parade*

Though the idea of the Mardi Gras was originally adopted from San Francisco the form it now takes is very specifically Australian. Partly this is as a result of the positive attitude by the state government, since through many of the Mardi Gras parades since 1978 there has been a gay-friendly Labor government in power. But it is also, as you may have picked up from the reading, a socio-cultural phenomenon which is particularly Australian, and that is a very strong element of self parody and irreverence in humour. The Mardi Gras is a very good illustration of this, to me at least, delightful trait.

One of the ways in which it seems to me that the Mardi Gras is important is that it provides an opportunity for gay men and lesbians to represent themselves in a public space in many different ways. Some of these are parody, while others are simply trying to challenge uniform ideas about what it means to be heterosexual in Australian society and, particularly, in Sydney. An important impact of the Mardi Gras is that it has facilitated the transformation of what was once a white newly gentrified suburb – Paddington – into a safe and fun place for gay men to live, work, run businesses and find entertainment. Oxford Street, as Bruce describes, is the centre of this gay area. Here we see an interesting connection between city politics – which started in the form of a demonstration – becoming the city politics of everyday life where gay and lesbian people can now move through the streets and live openly and in comparative safety from harassment. This story also illustrates very clearly the point that city politics do not stand still – quite the opposite in fact, they shift and change with remarkable rapidity.

6 Conclusion

In this chapter we have seen that city politics can be understood in a number of ways. Rather than focus on city politics as urban governance or administration, this chapter instead has highlighted the importance of everyday struggles over resources, spaces and places of representation in the city, and pointed out how these spring from social and economic inequalities. The intense proximity of different people and different activities competing for space in the city leads to conflicts of many kinds and forms, and in order to understand their resolution, we must also look at power in the city. Overall, as we have seen, the power to determine how resources are distributed and who has access to valued urban spaces is strongly implicated in relations of race, gender, class, sexuality and other social divisions.

Political struggle takes place at many different levels, some of which concern visible material resources or the differentiation of urban space, while others are more hidden and private. Thus the power to express or create one's identity in the city, to find new ways of inhabiting space, or to represent marginalized interests in the city is also a crucial dimension of city politics. Though these are less traditional focuses of urban analysis, they are no less significant.

We have seen also that city politics do not stand still. New issues and problems can emerge overnight or shift in significance as we saw with Sydney's Mardi Gras. Different groups come together in new strategic alliances to campaign and fight for their needs. Political struggles emerge in the everyday spaces of the city as people living side by side find themselves in conflict with one another or find shared concerns with local issues around which to organize. City politics are unpredictable with new forms emerging as cities connect into the wider global networks of power and new connections between groups and institutions are made. In this chapter we have visited a number of terrains of city politics from the politics of the community and urban social movements to the politics of meaning and representation and have paid particular attention to the ways in which everyday politics have important and visible impacts on the cities in which we live.

References

Allen, J. (1999a) 'Worlds within cities' in Massey, D., Allen, J. and Pile, S. (eds) *City Worlds*, London, Routledge/The Open University (Book 1 in this series).

Allen, J. (1999b) 'Cities of power and influence: settled formations' in Allen, J. *et al.* (eds).

Allen, J., Massey, D. and Pryke, M. (eds) (1999) *Unsettling Cities*, London, Routledge/The Open University (Book 2 in this series).

Amin, A. and Graham, S. (1999) 'Cities of connection and disconnection' in Allen, J. *et al.* (eds).

Anderson, E. (1990) *Streetwise*, Chicago, University of Chicago Press.

Bruce, I. (1997) 'Gay sites and the pink dollar' in Murphy, P. and Watson, S.

Castells, M. (1977) *The Urban Question*, London, Edward Arnold.

Castells, M. (1997) *The Power of Identity*, Oxford, Blackwell.

Dahl, R. (1961) *Who Governs? Democracy and Power in an American City*, New York, Yale University Press.

Davis, M. (1990) *City of Quartz*, London, Vintage.

Dunleavy, P. (1980) *Urban Political Analysis*, London, Macmillan.

Foucault, M. (1978) *The History of Sexuality*, New York, Pantheon.

Foucault, M. (1980) *Power/Knowledge*, Gordon, C. (ed.), New York, Pantheon.

Fraser, N. (1990) 'Rethinking the public sphere: a contribution to the critique of actually existing democracy', *Social Text*, vol.25–6, pp.56–80.

Hague, G. (1998) 'Domestic violence policy in the 1990s' in Watson, S. and Doyal, L. (eds) *Engendering Social Policy*, Buckingham, Open University Press.

Hall, S. (1991) 'Old and new identities, old and new ethnicities' in King, A. (ed.) *Culture, Globalization and the World System*, London, Macmillan, pp.273–326.

Harvey, D. (1973) *Social Justice and the City*, London, Edward Arnold.

International Journal of Urban and Regional Research (1978) vol.2.

Jakubowicz, A. (1984) 'The Green Ban Movement: urban struggle and class politics' in Halligan, J. and Paris, C. (eds) *Australian Urban Politics: Critical Perspectives*, Sydney, Longman Cheshire.

Keith, M. and Pile, S. (eds) (1993) *Place and the Politics of Identity*, London, Routledge.

Kemeny, J. (1980) *The Great Australian Nightmare*, Sydney, Allen and Unwin.

Kim, J. and Choe, S.-C. (1997) *Seoul: the Making of a Metropolis*, Chichester, Wiley.

Logan, M. and Molotch, H. (1987) *Urban Fortunes. The Political Economy of Place*, Berkeley, University of California Press.

London Edinburgh Weekend Return Group (1978) *In and Against the State*, Sydney, Pluto Press.

Lukes, S. (1975) *Power: a Radical View*, London, Macmillan.

McDowell, L. (1983) 'Towards an understanding of the gender division of urban space', *Society and Space*, vol.1, pp.59–72.

McDowell, L. (1999) 'City life and difference: negotiating diversity' in Allen, J. *et al*. (eds).

Massey, D. (1995) 'The conceptualization of place' in Massey, D. and Jess, P. (eds).

Massey, D. (1999) 'Living in Wythenshawe' in Borden, I., Kerr, J., Pivaro, A. and Rendell, J. (eds) *The Unknown City*, Cambridge, Mass., MIT Press.

Massey, D. and Jess, P. (eds) (1995) *A Place in the World? Places, Cultures and Globalization*, Oxford, Oxford University Press in association with The Open University.

Murphy, P. and Watson, S. (1997) *Surface City: Sydney at the Millennium*, Sydney, Pluto Press.

Pahl, R. (1975) *Whose City?*, London, Longman.

Rutheiser, C. (1996) *Imagineering Atlanta*, London, Verso.

Sennett, R. (1970) *The Uses of Disorder: Personal Identity and City Life*, London, Allen Lane.

Sennett, R. (1994) *Flesh and Stone: the Body and the City in Western Civilization*, London, Faber and Faber.

Sibley, D. (1995) *Geographies of Exclusion*, London, Routledge.

Stone, C. (1989) *Regime Politics*, Lawrence, University of Kansas Press.

Thompson, S. (1995) 'Suburbs of opportunity: the power of home for migrant women' in Gibson, K. and Watson, S. (eds) *Metropolis Now*, Sydney, Pluto Press, pp.33–46.

Thrift, N. (1999) 'Cities and economic change: global governance?' in Allen, J. *et al*. (eds).

Wallerstein, I. (1974) *The World System*, New York, Academic Press.

Ward, P. (1997) 'A view to the future', *Australian*, 10 October.

Watson, S. (1988) *Accommodating Inequality*, Sydney, Allen and Unwin.

Watson, S. and Gibson, K. (1995) *Postmodern Cities and Spaces*, Oxford, Blackwell.

Woolf, V. (1937/1977) *The Years*, London, Grafton Books.

Woolf, V. (1938/1993) *Three Guineas*, London, Penguin.

Young, I. (1990) *Justice and the Politics of Difference*, Princeton, NJ, Princeton University Press.

Young, M. and Willmott, P. (1962/1986) *Family and Kinship in East London*, Harmondsworth, Penguin.

READING 5A
Elijah Anderson: 'Streetwise'

Black male strangers confront problems of street navigation in similar ways. This field note illustrates some of the rules that city dwellers must internalise:

> At 3.00 Sunday morning I parked my car one street over from my home. To get to my front door I now had to walk to the corner, turn up the street to another corner, turn again and walk about fifty yards. It was a misty morning and the streets were exceptionally quiet. Before leaving the car, I found my door key. Then, sitting in the parked car with the lights out, I looked up and down the street, at the high bushes, at the shadows. After determining it was safe I got out of the car, holding the key, and walked to the first corner. As I moved down the street I heard a man's heavy footsteps behind me. I looked back and saw a dark figure in a trench coat. I slowed down, and he continued past me. I said nothing, but I consciously allowed him to get in front of me. Now I was left with the choice of walking about five feet behind the stranger or of crossing the street, going out of my way, and walking parallel to him on the other side. I chose to cross the street.

All these actions fall in line with rules of etiquette designed to deal with public encounters. First, before I left the safety of my car I did everything possible to ensure speedy entrance into my home. I turned off the car lights and looked in every direction and took my house key in hand. Second, I immediately looked back when I heard footsteps so that I could assess the person approaching. Next, I determined that the stranger could be a mugger in search of a victim – one of many possible identities, but naturally the one that concerned me the most. I knew that at night it is important to defer to strangers by giving them room, so I established distance between us by dropping back after he passed me. Further providing for the possibility that he was simply a pedestrian on his way home, I crossed the street to allow him clear and safe passage, a norm that would have been violated had I continued to follow close behind him.

> When I reached the corner, after walking parallel to the stranger for a block, I waited until he had crossed the next street and had moved ahead. Then I crossed to his side of the street; I was now about thirty yards behind him, and we were now walking away from each other at right angles. We moved farther and farther apart. He looked back. Our eyes met. I continued to look over my shoulder until I reached my door, unlocked it, and entered. We both continued to follow certain rules of the street. We did not cross the street simultaneously, which might have caused our paths to cross a second time. We both continued to 'watch our backs' until the stranger was no longer a threat.

In this situation skin colour was important. I believe the man on the street distrusted me in part because I was black and I distrusted him for the same reason. Further we were both able bodied and young. Although we were cautious toward each other, in a sense we were well matched. This is not the case when lone women meet strangers.

A woman being approached from behind by a strange man, especially a young black man, would be more likely to cross the street so that he could pass on the opposite side. If he gave any sign of following her, she might head for the middle of the street, perhaps at a slight run toward a 'safe spot'. She might call for help, or she might detour from her initial travel plan and approach a store or a well-lit porch where she might feel secure. In numerous situations like those described above, a law abiding, streetwise black man, in an attempt to put the white woman at ease, might cross the street or simply try to avoid encountering her at all. There are times when such men – any male who seems to be 'safe' will do – serve women of any colour as protective company on an otherwise lonely and forbidding street.

This quasi 'with' is initiated by the woman, usually as she closely follows the man ahead of her 'piggyback' style. Although the woman is fully aware of the nature of the relationship, the man is usually not, though he may pick up on it in the face of danger or demonstrable threat. The existence of this 'with', loose and extended as it may be, gives comfort and promises aid in case of trouble, and it thereby serves to ward off real danger. Or at least the participants believe it does.

Eye work

Many blacks perceive whites as tense or hostile to them in public. They pay attention to the amount of eye contact given. In general, black males get far less time in this regard than do white males. Whites tend not to 'hold' the eyes of a black person. It is more common for black and white strangers to meet each other's eyes for only a few seconds and then to avert their gaze abruptly. Such behaviour seems to say 'I am aware of your presence', and no more. Women especially feel that eye contact invites unwanted advances, but some white men feel the same and want to be clear about what they intend. This eye work is a way to maintain distance, mainly for safety and social purposes. Consistent with this, some blacks are very surprised to find a white person who holds their eyes longer than is normal according to the rules of the public sphere. As one middle-aged white female resident commented:

> … just this morning, I saw a (black) guy when I went over to Mr Chow's to get some milk at 7.15. You always greet people you see at 7.15, and I looked at him and smiled. And he said 'Hello' or 'Good Morning' or something. I smiled again. It was clear that he saw this as surprising.

Many people, particularly those who see themselves as more economically privileged than others in the community, are careful not to let their eyes stray, in order to avoid an uncomfortable situation. As they walk down the street they pretend not to see other pedestrians, or they look right at them without speaking, a behaviour blacks find offensive.

Moreover, whites of the Village often scowl to keep young blacks at a social and physical distance. As they venture out onto the streets of the Village and, to a lesser extent, of Northton, they may plant this look on their faces to ward off others who might mean them harm. Scowling by whites may be compared to gritting by blacks as a coping strategy. At times members of either group make such faces with little regard for circumstances, as if they were dressing for inclement weather. But on the Village streets it does not always storm, and such overcoats repel the sunshine as well as the rain, frustrating many attempts at spontaneous human communication.

Source: Anderson, 1990

READING 5B
Iain Bruce: 'Gay sites and the pink dollar'

Mardi Gras

The greatest assertion of gay and lesbian territoriality in Sydney is Mardi Gras, held every year at the end of February or the beginning of March. By occupying the streets, the parade asserts the claim of gays and lesbians to organise and be recognised. Mardi Gras is a political statement in disguise. It differs from the tradition of the political protest marches that have characterised the century: the massed marching (or shuffling) group; the display of demands on banners; the rallying point; and the speeches – a more solemn style still to be found in the London Gay Pride March and, indeed, in other gay and lesbian marches in Sydney such as the Stonewall commemoration in June (attended in 1992 by 1500 people) and the march in support of the New South Wales Anti-Discrimination (Homosexual Vilification) Bill held late in 1993. The Mardi Gras is a different sort of event: it is not obviously confronting. It involves the crowds by its flamboyance and outrageousness, by its spectacle and sound, by its boldness and inventiveness. In 1994 it attracted 600 000 spectators out of a Greater Sydney population of 3.5 million. The fact that it happens at all indicates how far the public position of gays and lesbians in Sydney has advanced since the early 1980s.

What the crowd is witnessing, however, is a subversive act, and subversive not simply because of the sexuality on display … Mardi Gras also saw off one of the *ersatz* carnival substitutes. Leo Schofield, a columnist on the *Sydney Morning Herald* recalls:

> There used to be a lethally dull Festival of Sydney Parade called the Waratah Festival Parade. It was breathtakingly parochial and uninspired. Everyone who had something to sell – from a new Toyota Corolla to whatever – put it in the parade and it was alternated with the odd band and marching girls. It was just like some hick town in the mid-West. It was horrible.

The re-emergence of the repressed in the form of participants on their floats and bikes and roller-blades mocks Christian-derived bourgeois values. The Reverend Fred Nile, member of the New South Wales Legislative Council, head of the Call to Australia Party and leading member of the Festival of Light, is right to see it as inimical to traditional values. It is precisely because those traditional values made no space for lesbians and gay men that they have muscled their way onto the streets.

Saturday, 24 June 1978 saw two parades in Sydney by gays and lesbians. The first was to commemorate the raid in 1969 by New York police on the gay Stonewall Bar in Greenwich Village and to celebrate the unexpected and unprecedented resistance by the gay customers which resulted in the riots that mark the birth of the modern gay liberation movement. This first Sydney march took place in the morning, and involved 300 to 500 people, a large number by Australian standards, and had the traditional protest form. On the same day a forum on gay issues was held at Paddington Town Hall. There already existed in Sydney a growing gay confidence fostered by CAMP (Campaign Against Moral Prejudice) NSW founded in 1971 and a more militant gay liberation organisation. The second parade was intended to be 'non-confrontational, a celebration of gay pride. Something that would link up the politics of gay freedom with the people in Oxford Street in the commercial scene' and be 'purely a street party, a walk down Oxford Street where we could be a bit frivolous'. It was held at night to encourage those who feared public exposure to join in. The name 'Mardi Gras' was not on any of the leaflets or posters advertising the event. Ken Davis remembers that it became current in the succeeding days, initially with an indefinite rather than definite article to reflect this mood. Compared with the morning's turn-out, some 1000 people started off, their numbers swollen to 1500 by bystanders and those who responded to the calls of 'out of the bars and into the streets'. Few people were in costume though Davis, who hailed from Tamworth (the country and western capital of New South Wales) was appropriately dressed in a large orange and mission-brown polka-dot dress, 'a little Loretta Lynn number', he said. In a tactical move, he was to shed this shortly when it became wise to be less conspicuous.

Twenty minutes after its start, the parade had reached its licensed destination at Hyde Park where the police turned off the loudspeakers of the truck and tried to arrest the driver. The call went up to move to Kings Cross. There, what had been a celebratory and jubilant affair had, by 11.30: 'become a two-hour spree of screaming, bashing and arrests. In one incident, police took off their identification numbers and waded into a crowd of homosexuals'. Bystanders at Kings Cross joined in to rescue those who had been arrested. The police were deeply hated by anyone who was considered a social deviant, as were many in the red-light district of Kings Cross. The police had a record of corruption and of violence which extended to bashings and rapes within Darlinghurst Police Station cells. An unlikely alliance of straights and gays was formed for the occasion, cemented by a common loathing of the police. In the ensuing melee thirty men and twenty-three women were arrested, not all necessarily gay or lesbian. The *Sydney Morning Herald* published their names, streets and suburbs of residence, ages and occupations. Of those arrested, only three were over 28; over 60 per cent were under 24 years of age. Many were in semi- or unskilled jobs and there were both students and unemployed among them. In the end, only two of those arrested were fined, the others being released without bail and the charges subsequently dropped.

It was as if the gauntlet had been thrown down and seized by the gay community. Activists targeted the New South Wales *Summary Offences Act* under which the police had charged those arrested on 24 June 1978. They argued that this Act allowed the police, in effect, to determine standards of acceptable behaviour. It was used against all minorities against whom the police held a prejudice, be they gays or Aborigines. Nor was there a right to march and demonstrate. Gay activists developed the tactic of responding to arrests at protest marches by demonstrating again and demanding that the first group arrested be freed. June 1979 saw another Mardi Gras on the theme 'Power in the Darkness' in which an impressive 5000 took part. The repeal in 1979 of the *Summary Offences Act* by the Labor government of Neville Wran was entirely the work of the gay movement …

In 1981 the parade was held in summer, leaving Stonewall day in June as a more directly political event. With the rescheduling of Mardi Gras came a declaration of its nature. It was decided that it should be: 'a celebration of coming out, with its main political goal being to demonstrate the size of the gay community, its variety of lifestyles and its right to celebrate in the streets of Sydney so as to enable a broadening of support for gay rights'. Decisions taken about the format of the 1981 Mardi Gras, unfortunately, deepened the split between gays and lesbians. Lesbians feared that the inclusion of floats by businesses would result in sexist and racist use of drag. They also felt that the focus would shift away from the political. Having led the attack at the first Mardi Gras, the lesbians increasingly absented themselves from the celebration, returning in significant numbers only towards the end of the decade.

In the years that followed, the Mardi Gras parade and subsequent party grew dramatically in terms of the number of participants and the number of spectators. In 1983, there were 44 floats in the parade which was watched by 20 000 spectators. Some 2000 people attended the post-parade party. By 1994, there were 137 floats in the parade, watched by 600 000 spectators and 19 500 people attended the post-parade party. The Australian Broadcasting Corporation (ABC) created a national audience for Mardi Gras in 1994 after it broadcast an hour of edited highlights of the parade in prime time on the next day. The program achieved top ratings for the ABC, a rare achievement. Relays on ATV International also took the Mardi Gras parade into South-East Asia.

What does it mean for the parade to occupy, so flamboyantly, prime space within the city? At its most basic level, it is to do with self-presentation. The image presented can be described by Fred Nile as 'a promiscuous orgy' and as 'offensive, obscene and blasphemous' or as 'outlandish, vibrant, attractive and innovative' in the words of Chief Inspector Kerry Beggs at Surrey Hills Police Station, in whose territory most of the parade takes place.

After centuries of being corked, there is a lot of bubble in the bottle. The images that are made public are the significant, subversive topoi and icons of the gay and lesbian community. Generally forced to conceal its identity, the community bursts out perversely in disguise, fancy dress, travesty and drag. It flaunts gay identity as if to say: 'Haven't you spotted us; we're so obvious'. It parodies its own oppression. Two T-shirts present paradoxes: 'Nobody Knows I'm Gay' and 'I'm not gay but my boyfriend is'. Each offers the norm while wittily undermining it ...

The parade displays sexual stereotyping overturned. There is dressing up and dressing down. Bodies are transformed. They are seen as the *opportunities* for display – either the severe (generally leather) or the exuberant (generally drag). The Dykes on Bikes roar by – wearing less in 1993 than in 1992, weaving and roaring, seizing the stereotyped image of the macho-male Harley Davidson bikers, parading in a phalanx, bare-breasted and buttocked, sporting shades, leathers and tattoos. Women sprout dildos and exaggerated nipples. Another group of women strip down to become human versions of the mannequins on which clothes shops hang their wares – divested of display, apparently naked but for the high heels and bouffant hairstyle. A group of men turn against male uniformity, swathing themselves in diaphanous material to become birds of display with costumes that shimmer, that move, that blow, that sparkle, that reveal, that catch the eye, that enlarge and extend and have as their cultural referents the eastern voluptuary and the harem. They embody the essence of camp proposed by Susan Sontag: 'the love of artifice and exaggeration'. She adds: 'The most refined form of sexual attractiveness (as well as the most refined form of sexual pleasure) consists in going against the grain of one's sex'. The displays in Mardi Gras combine the subjective pleasure of the exotic with the desire to shock.

There is, too, an ironising of masculinity through exaggerated worship of the muscled male body. This is a less clear-cut case than the travesties and parodies. The muscles on display are not put on in the hours before the 8.30 p.m. start of the parade. They are sweated for, with or without the help of steroids, in countless hours in the gym. They are the same muscles that serve behind the bar at the Albury or that stand, inviting admiration, at the Midnight Shift. They are real and for real. Nonetheless, in the context of the parade, they become display of another sort. The attempt to become more masculine than the queer-basher who will prey on you is subsumed into the knowing exaggeration that is the essence of this most unrepresentative night. By concentrating on spectacle, a fantasy world is temporarily created which soothes or amuses. What the parade submerges are the dangers faced by gay men and lesbians. The bashers in the crowd look at the queers on the floats: the queers know the bashers are there and know that that is another aspect of the reality of being gay in the golden mile.

The extent to which the parade is 'readable' by the majority of those who watch it is debatable. At a quite literal level, messages and banners are often indecipherable. Acronyms and abbreviations mean little to most of the spectators. There is rarely a relationship between the purpose of the organisation and the image offered. (The Gay and Lesbian Teachers and Students Alliance were an exception, going for academic gowns and school uniforms, but this 'text' may have been too dull to read with interest.) The danger is that the parade becomes pure spectacle leaving its overt political points to be understood by the few. Bombarded with glitter and light, the straights in the crowd may simply read what they see and confirm what they always thought they knew: 'queens will be queens' ...

Oxford Street

Sydney, like all major cities, has a continuing homosexual tradition. Jan Morris in her book *Sydney* notes the existence of homosexuality from 1788 but submerges it disingenuously in an account of the orgy which greeted the landing of the women convicts a fortnight after the First Fleet's men: 'Sex in Sydney ... began with a bang' ...

The development of a distinctive gay area is comparatively recent, although it is still subject to change. Between the wars, bars catering to homosexual males were exclusively in the city; by 1960 they were clustered in Kings Cross and the city, but had begun to spread to the Darlinghurst section of Surry Hills and to Bondi Junction; by 1982

Darlinghurst had the lion's share of forty-two sites of differing types. Space was claimed in Newtown, Leichhardt, Rozelle and Balmain; by 1990, Darlinghurst had forty-eight recorded gay sites, including bars, baths, bookshops, cafes, restaurants and clothes shops …

It is easy to romanticise the development of Oxford Street as a 'gay space', but the forces behind it were and are uncompromisingly commercial. There is much speculation about the degree of shady money involved. Gay premises are owned in three different ways. The first is the truly independently owned business; the second is the semi-independent business which is technically owned by an individual or group but which is financed by undisclosed parties; the third is the fronted business run by an employee but owned and controlled by a syndicate. In some respects, the 1960s tradition of dealing on Oxford Street has not changed …

Oxford Street by the early 1980s had established itself as the gay heartland. The Mardi Gras parades helped consolidate that role, but it was not a straightforward gain. If the forces of capital and organised crime were quite happy to exploit the new market, the institutions of the state – principally the police – were not so amenable. The 1979–80 parades started from the heart of hostile territory. The march was routed through Saturday-night audiences coming out of the cinemas in George Street. It was channelled down one side of the street as if to declare that it would not be allowed to disrupt city life. Its disruptive element was further lessened by moving its start to the Domain, parkland tucked away from the city centre. As the size of the parade grew and overt hostility diminished, a new relationship began to be established between the organisers and the regulating police. It began to move from being protest to spectacular, in which spontaneity was limited. Bans were put on people standing on the awnings over the pavements after male strippers from the Flinders put on a show for the crowds below after the parade had passed. Barricades were erected, which meant that people in the crowd could no longer join in the parade, which is firmly marshalled by hundreds of volunteers supplementing a force of 100 police officers. The Mardi Gras organisers pay for the clean-up of the streets and there are demands that it pay the police costs of the evening. Understandably, the growth in size of the spectacle forces negotiation with the state authorities. This marks another stage in the way which the gay and lesbian community has learnt to deal with the institutions of the state – just as they have in the party political arena.

Source: Bruce, 1997, pp.62–76

CHAPTER 6
The unsustainable city?

by Andrew Blowers and Kathy Pain

1 Introduction

In this chapter we explore the key question: can cities be sustainable? To answer this we need to understand what 'sustainability' means in an urban context. Sustainability is not merely a physical concept, concerned with the relations between cities and the natural environment. In its broadest sense it is also a social concept, concerned with the processes and patterns of development and their unequal spatial consequences. Thus we are concerned with cities as not just sources of global environmental degradation, but as sites of environmental and social injustice.

Whereas contemporary patterns of social development and the inequalities to which they give rise may result in cities being environmentally unsustainable, the idea of *sustainable development* conveys the possibility that, under different social conditions, sustainability can be achieved. This suggests that there are spatial forms and social processes that are sustainable and that these can be encouraged through methods of organization and governance which respect ecological limits.

Here we approach the issue of sustainable cities by investigating the relationships between environmental and social change within and between cities and the implications for urban governance. This will help us to contemplate the criteria and components necessary for the attainment of sustainable cities.

FIGURE 6.1 *'The ark of nature' focuses attention on cities as environmentally unsustainable*

2 *City connections*

2.1 CITIES IN DANGER?

Cities occupy only 2 per cent of the earth's land surface, yet at the beginning of the twenty-first century they contain almost half of the world's population. The world's urban population is increasing by about 55 million each year, with growth rates as high as 5 per cent per year in parts of the South. By the year 2020, urban population is likely to rise to over 75 per cent of global population. Figure 6.2 shows the percentage of the world's population that was urban in 1890 and 1995, and the predicted trend for 2025.

The influence of intense and geographically concentrated human activity and technology extends far beyond the physical or administrative boundaries of cities, impacting on people and the environment across the globe. Cities are centres of commerce, culture, power and excitement, and we have seen in earlier chapters that they can also be perceived as places of tension, risk and danger. But are cities a global hazard? Are cities themselves in danger?

Hinchliffe (1999) has shown elsewhere in this series that anti-urban representations typically portray cities negatively, as a threat to 'nature' and global environmental balance. They focus on cities as major consumers of natural resources and producers of pollution. Indeed, cities use around three-quarters of the world's resources and they generate, concentrate and disperse pollutants across the globe. They consume the products of the ecosphere and also discharge their wastes into it. Cities are ecological entities, but their relationship to nature is scarcely perceived, as people in cities seem to be cut off from their 'natural' surroundings. Yet it is the interrelationships between cities, society and the 'natural' environment that determine their sustainability.

Sustainability is usually regarded as a physical concept concerned with the impact of human activities on ecological resources. The impact of cities on the environment can be visualized in terms of 'ecological footprints' (see **Massey, 1999a**; **Hinchliffe, 1999**). The ecological footprint of a city has been defined as 'the total area of productive land and water required on a continuous basis to produce the resources consumed, and to assimilate the wastes produced, by that population, *wherever on Earth the land is located*' (Rees, 1997, p.305). Calculations of the area required suggest that, in high-income countries, as much as 3–7 hectares of ecologically productive land is needed per capita. Thus it has been estimated that Vancouver requires a land area 180 times its own size to maintain its lifestyle, and 200 times if the marine area appropriated for seafood consumption is included. Similarly, calculations suggest that the 29 cities in the Baltic Sea drainage basin require an area 200 times that of the cities themselves. London's ecological footprint has been estimated to be 125 times its surface area: with 12 per cent of Britain's population, London would require an area equivalent to the entire productive land of Britain (Jopling and Girardet, 1996).

FIGURE 6.2(a) *Percentage of the world's population urban in 1890*

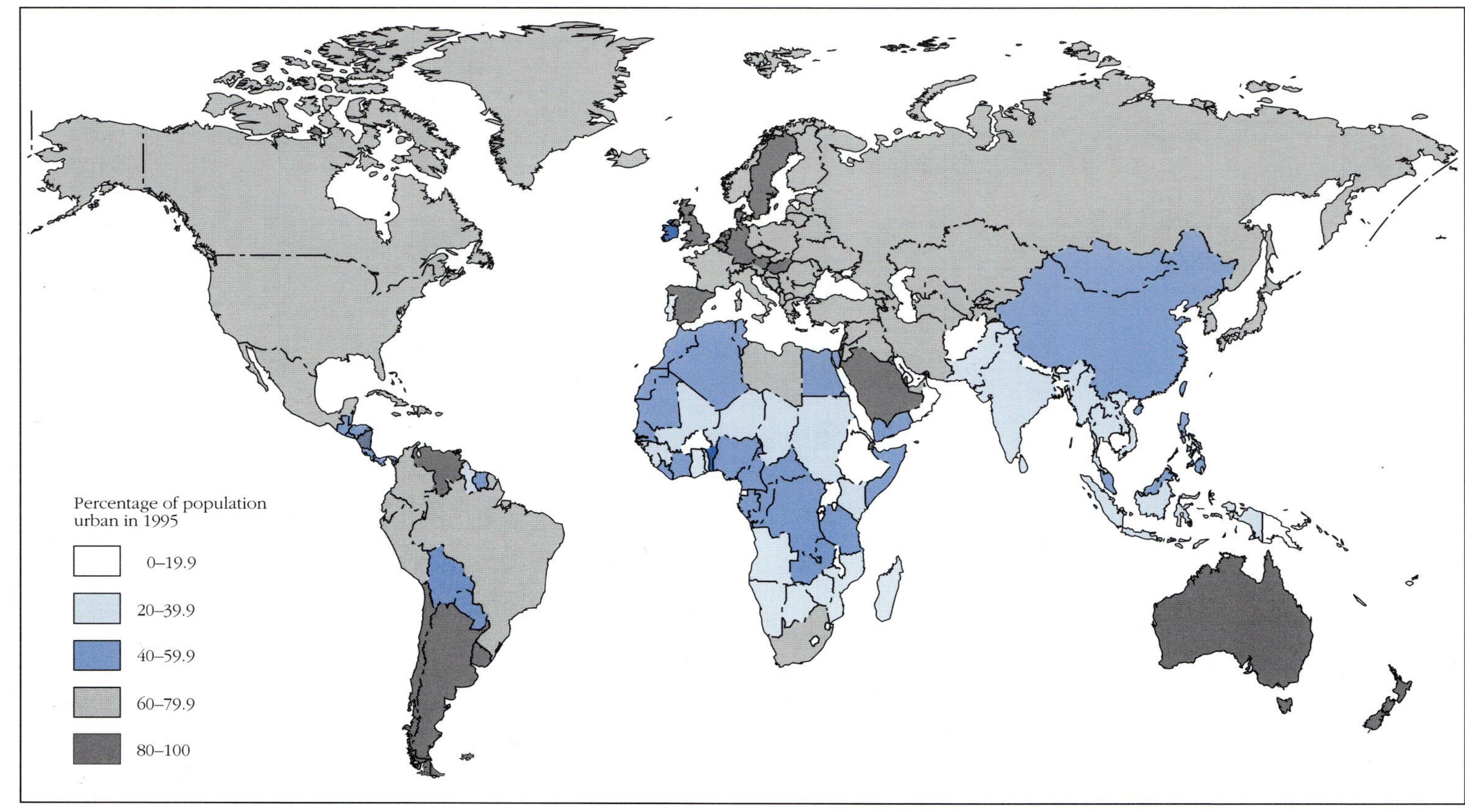

FIGURE 6.2(b) *Percentage of the world's population urban in 1995*

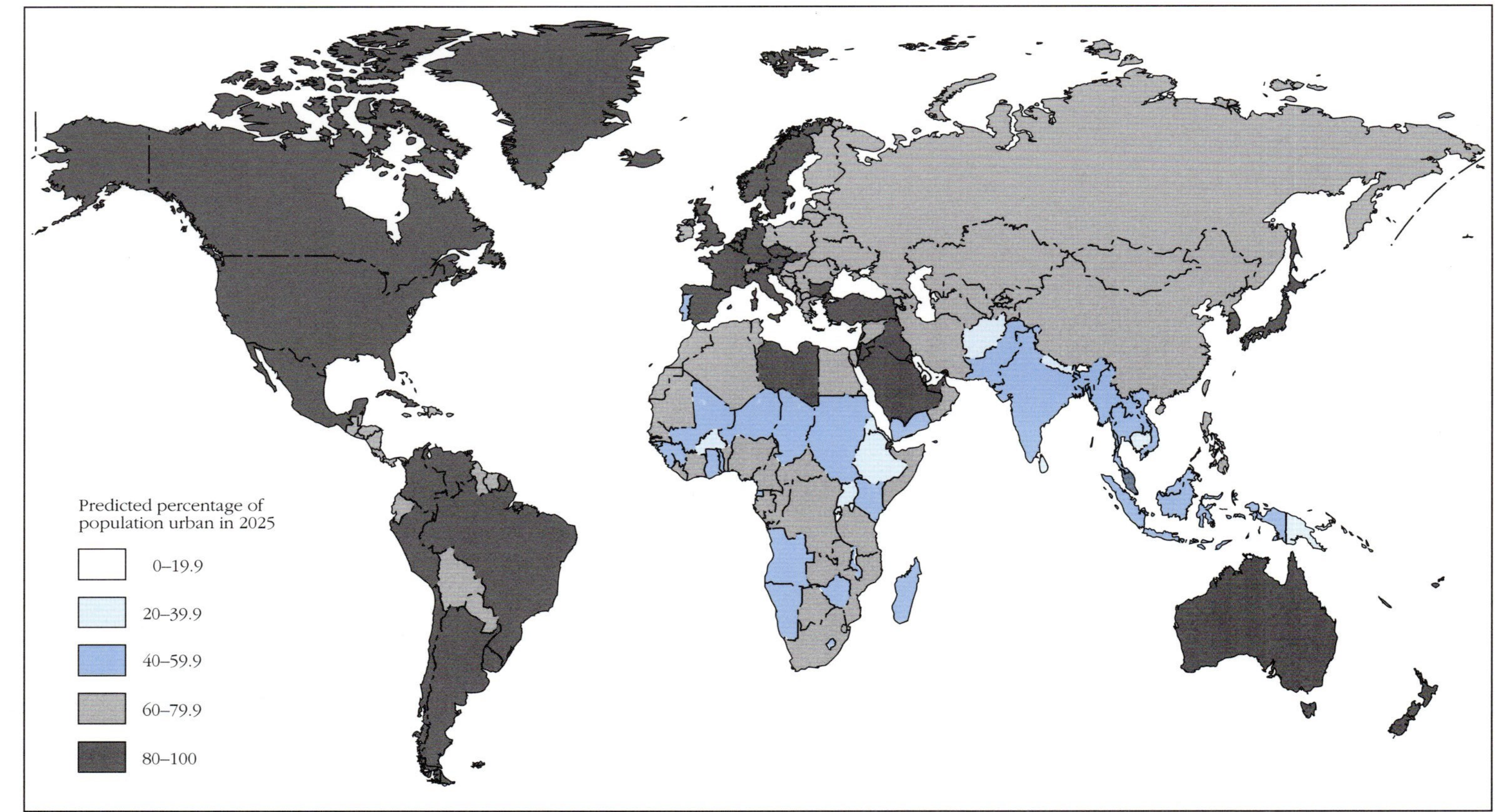

FIGURE 6.2(c) *Percentage of the world's population urban in 2025*

The notion of the ecological footprint of cities focuses attention on consumption of land and water. Of course, cities also pour wastes into the atmosphere which are rapidly dispersed and which contribute to such problems as global warming and ozone depletion. Jopling and Girardet have assessed London's input of resources and output of wastes, or its *metabolism* (see Table 6.1). Ecological footprints demonstrate how cities have become detached from their local resource base. In the past, cities lived largely off their own hinterlands, forming a localized and sustainable ecosystem. There has always been some interdependence resulting from trade, but with the development of technology and transport the sustainability of cities is now an issue of global sustainability. As networks of economic, cultural and financial relations within and between cities have stretched and intensified (see **Massey, 1999a**), so have their social and environmental impacts.

TABLE 6.1 *The metabolism of Greater London at the end of the twentieth century*

Inputs	Tonnes per year
Fuel (oil equivalent)	20,000,000
Oxygen	40,000,000
Water	1,002,000,000
Food	2,400,000
Timber	1,200,000
Paper	2,200,000
Plastics	2,100,000
Glass	360,000
Cement	1,940,000
Bricks, blocks, sand, tarmac	6,000,000
Metals	1,200,000
Wastes	**Tonnes per year**
Industry and demolition	11,400,000
Household, civic and commercial	3,900,000
Wet digested sewage sludge	7,500,000
Carbon dioxide gas	60,000,000
Sulphur dioxide gas	400,000
Nitrogen oxide gas	280,000

Source: Jopling and Girardet, 1996, p.10

Ecological footprints cannot represent all the complex networks of social and ecological interrelations concerning cities which impact on the environment. The spatial relations of production and consumption and the social processes that give rise to them cannot be modelled in this way. While ecological footprints focus attention on environmental sustainability, it is also necessary to address the questions 'who pollutes?' and 'who consumes?'.

Cities and the social processes that shape them must be understood as ecological processes that have social as well as environmental consequences. While the idea of the ecological footprint has limitations, it can help us to imagine the uneven spatial consequences of the social and economic processes that shape cities and their impacts. The gross economic disparities between cities, especially between those in the North and those in the South, exacerbate the problem of sustainability in two ways.

First, patterns of consumption in cities in the high-income countries of the North lead to continually increasing demands for resources and impose externalities in the form of pollution in areas further afield, in short expanding and intensifying their ecological footprints. Second, at the same time, the rapid growth of cities in the developing world of the South and aspirations for higher standards of living will place an increasing burden on the world's resources. In the North much attention has been given to the perceived global environmental threat that this could pose. Despite this growth, if uneven patterns of development continue within and between cities, a large proportion of the South's urban population will remain disconnected from economic prosperity and what would be regarded as basic standards of living in cities of the North. Under present conditions of development, raising the living standards of three-quarters of the world's population in line with those in the North would create insurmountable environmental problems. On the other hand, poverty results in a degraded and polluted environment. Inequality creates forms of development in both North and South that are socially and environmentally unsustainable.

In sum, 'half the people and three-quarters of the world's environmental problems reside in cities, and rich cities mainly in the developed North, impose by far the greater load on the ecosphere and global commons' (Rees, 1997, p.304). Cities are therefore a key element in what has been termed a global social and environmental crisis.

While cities can be seen as a global ecological hazard, cities are themselves at risk. Not only do cities pollute, but they are also polluted. Not only do they generate patterns of inequality, but they are characterized by inequality. Even within cities of the North, social, economic and environmental inequalities persist.

The idea that cities can be sustainable under certain social conditions and forms of governance is not new. Although sustainable development is a concept that has inspired thinking in the late twentieth century, the essential ideas were already present in the context of industrial cities a century ago.

2.2 A VISION OF THE FUTURE

At the end of the nineteenth century, Ebenezer Howard completed his influential book, *Tomorrow: A Peaceful Path to Real Reform* (1898), which was later published as the more familiar *Garden Cities of Tomorrow* (1902). Writing in reaction to the manifest problems of the world's first industrial cities of the British Victorian age, Howard commented: 'it is deeply to be deplored that the people should continue to stream into the already over-crowded cities' (Howard, 1902/1965, p.42). At this point you might wish to remind yourself of the arguments made in Chapter 1, section 3.3 (see also **Pile, 1999**, section 1.2).

During the nineteenth century, the development of the city was a commercial venture. The role of the state was seen as one of supporting private enterprise. In a context of minimal intervention, disorder and chaotic development had dire consequences for the living conditions of expanding urban populations. Howard's solution was essentially practical. He set it out in diagrammatic form in terms of the 'three magnets', as shown in Figure 6.3. Two of the magnets summarized the advantages and disadvantages of the town and the country respectively, while the third magnet, based on a fusion of town and country, was seen to have only the advantages and none of the disadvantages of either town or country alone.

Victorian responses to early industrial cities such as London and Manchester were influenced by contrasting images and perspectives. One view of Manchester in 1858, while acknowledging its environmental problems, notes that: 'not any or all of these things can prevent the image of a great city rising before us as the very symbol of civilization, foremost in the march of improvement, a grand incarnation of progress' (extract from *Chambers' Edinburgh Journal*, 1858, cited in Briggs, 1968, p.88). Others, however, saw such cities as bases for social and political action or as threats to public health, security and existing social order. There was a strand of anti-urbanism symbolized by those who made their money in the new cities choosing to live outside them.

Howard's approach was not anti-urban. He valued cities and believed that industry was integral to their life and prosperity. He proposed to combine what he saw as the advantages of the urban and the rural in a new, sustainable form of city living. Howard explained: 'There are in reality not only, as is so constantly assumed, two alternatives – town life and country life – but a third alternative, in which all the advantages of the most energetic and active town life, with all the beauty and delight of the country, may be secured in perfect combination' (Briggs, 1968, p.75). This combination would produce the garden city, which 'is not a suburb but the antithesis of a suburb: not a more rural retreat, but a more

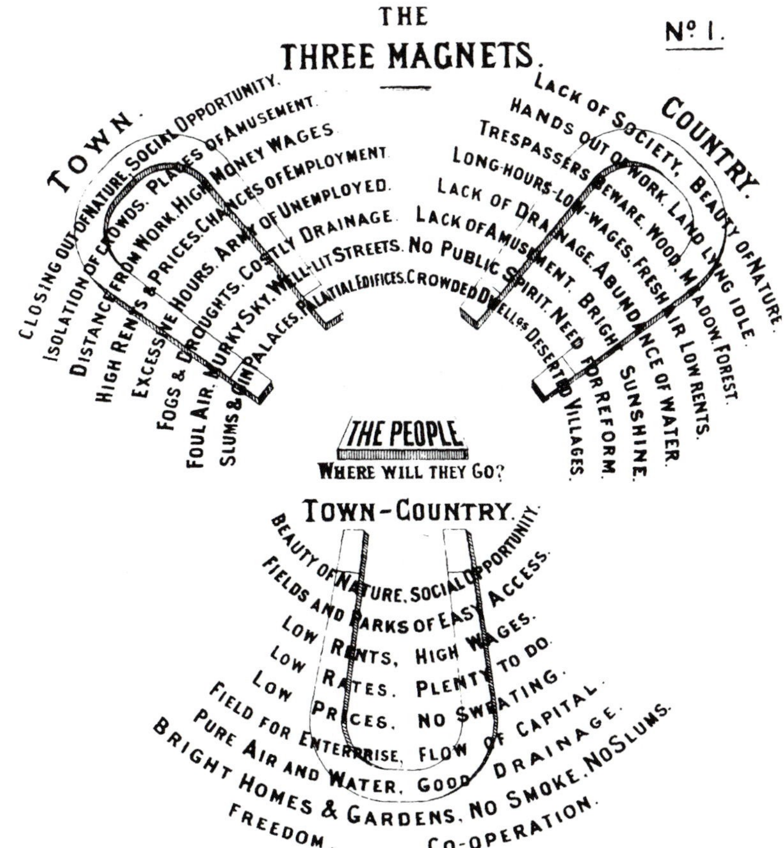

FIGURE 6.3 *Ebenezer Howard's 'three magnets'*

integrated foundation for an effective urban life' (Mumford's introductory essay to *Garden Cities of Tomorrow*, 1902/1965, p.35).

Howard's emphasis on the symbiosis between urban and rural tends to emphasize the notion of a bio-region, a city that is contained within a set of bounded interdependencies. Yet, as this course has been at pains to show, interdependencies stretch far beyond the city region in networks and flows of unbounded spatial relationships. Despite this limitation, Howard's vision was remarkably prescient in terms of many of the contemporary problems of cities. In the first place, he was concerned about the environmental context. There is an emphasis on resource conservation, with food supplied from surrounding areas and waste products used to enrich the soil. Instead of an endless sprawl, he envisaged that, once the garden city had reached a certain size (58,000 people), it would spawn satellites in a developing cluster of cities linked internally and to each other by a network of railways, thereby forming a *social city* (see Figure 6.4). Though compact, the garden city would have ample open space, a park at its centre, and tree-lined boulevards and gardens surrounding the housing. In its design the garden city demonstrates concerns about amenity, pollution and resource conservation that are at the heart of concerns about sustainability today.

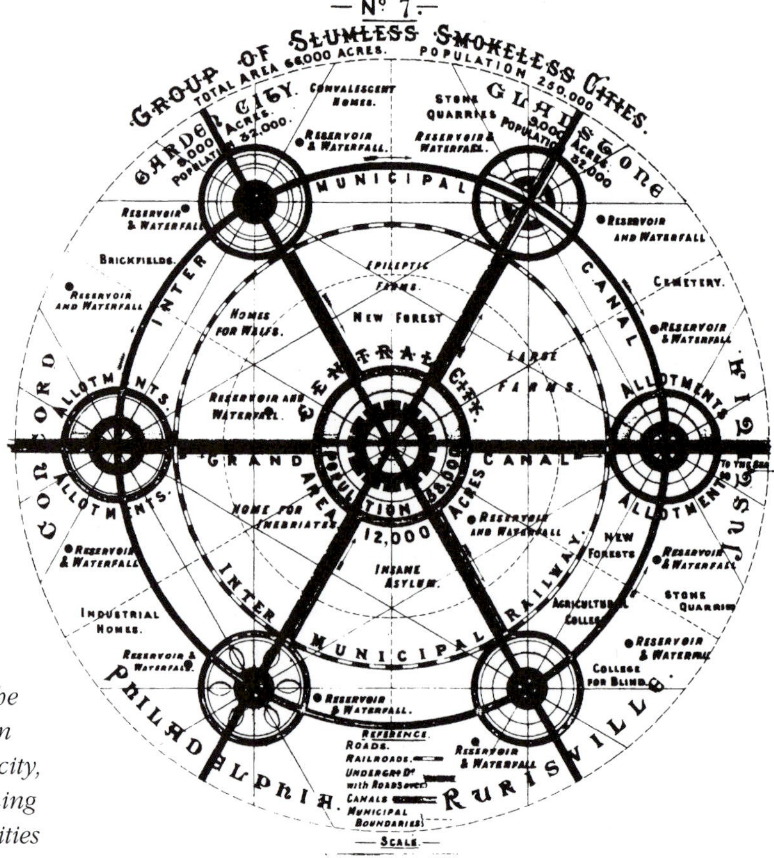

IGURE 6.4
:benezer Howard's plan of the ocial city: a cluster of garden :ities surrounding a central city, ıll interlinked and maintaining ·ural as well as urban amenities

Social and cultural life in the city are seen to be as important as economic functions, and this is reflected in the spatial form of the garden city. Cultural and leisure facilities are located centrally, and the central park provides 'ample recreation grounds within very easy access of all the people' (Howard, 1902/1965, p.53). In a foretaste of modern shopping malls, there is a 'Crystal Palace': 'here manufactured goods are exposed for sale, and here most of that class of shopping which requires the joy of deliberation and selection is done' (Howard, 1902/1965, p.54). Large public buildings were to be a focus for the whole community (see Figure 6.5).

Howard envisaged that the garden city 'should be planned as a whole, and not left to grow up in a chaotic manner'. Unity of design and purpose was not possible in cities which were the product of 'an infinite number of small, narrow, and selfish decisions' (Fishman, 1977/1996, p.42). Howard's vision was one of orderly development to combat the disorder he perceived in the cities of his day. The legacy of his ideas can be found in the green belts, protected agricultural land, low-density housing, industrial trading estates, shopping centres and neighbourhood facilities that have become integral elements of British cities. Indeed, Howard's ideas became a major influence on the development of the town and country planning system in the UK.

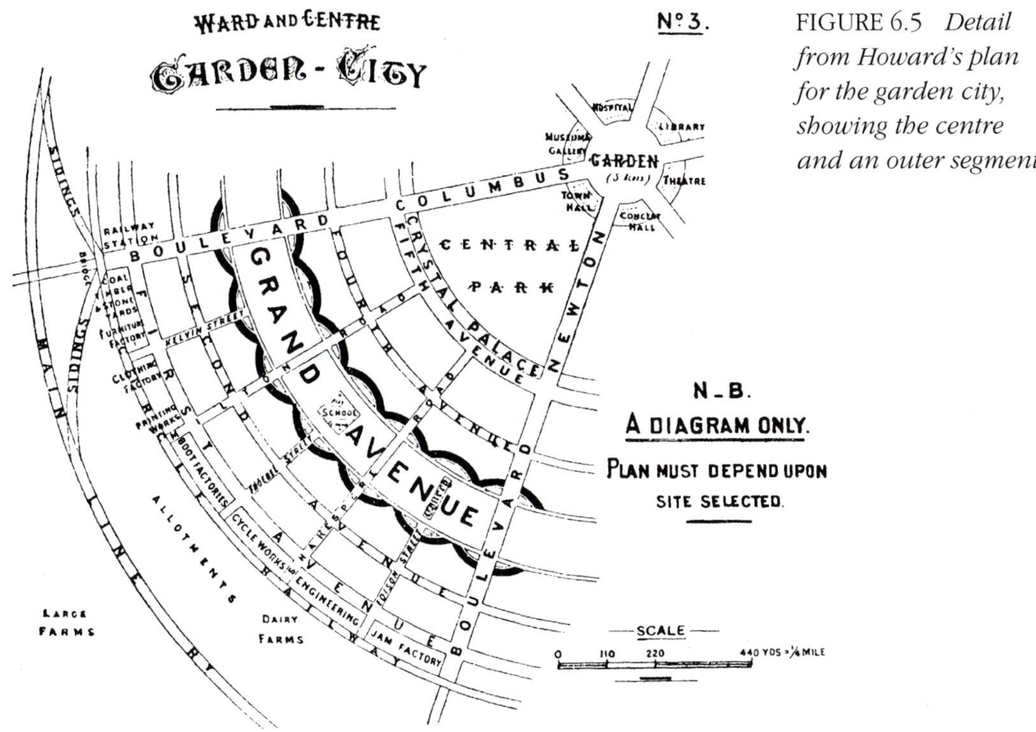

FIGURE 6.5 *Detail from Howard's plan for the garden city, showing the centre and an outer segment*

Howard recognized the relationship between the problems of urban environment, society and political economy, and the garden city was an attempt to create a better life in a better environment:

> I went into some of the crowded parts of London, and as I passed through the narrow dark streets, saw the wretched dwellings in which the majority of the people lived, observed on every hand the manifestations of a self-seeking order of society and reflected on the absolute unsoundness of our economic system, there came to me an overpowering sense of the temporary nature of all I saw, and of its entire unsuitability for the working life of the new order – the order of justice, unity and friendliness.
>
> (Howard, quoted in Fishman, 1977/1996, p.37)

The garden city was therefore not merely a physical plan; it also had a social context. Howard envisaged the reconstruction of the physical form of industrial cities as a means to address problems of social inequality and injustice. His blueprint for the garden city implied radical social change, with implications for urban economic and political relationships.

ACTIVITY 6.1 Extract 6.1 comprises several quotations from Howard's writings in *Garden Cities of Tomorrow*, and they illustrate the integration of his concerns for society and governance, as well as environment, in his vision for a new form of city and a new way of urban life. In reading the quotes in the extract, consider why Howard envisaged that garden cities would open up possibilities for greater social freedom and justice. ◆

EXTRACT 6.1
Ebenezer Howard: 'A vision for a socially balanced city'

... my proposal appeals not only to individuals but to co-operators, manufacturers, philanthropic societies, and others experienced in organization, and with the organizations under their control, to come and place themselves under conditions involving no new restraints but rather securing wider freedom. (p.116)

One cannot help wishing – so inharmonious does life seem – that the opportunity presented itself of migrating to a new planet where the 'ethical sentiments which social discipline has now produced' might be indulged in. But a new planet, or even 'a territory not yet individually portioned out', is by no means necessary if we are but in real earnest; for it has been shown that an organized, migratory movement from over-developed, high-priced land to comparatively raw and unoccupied land, will enable all who desire it to live this life of equal freedom and opportunity; and the sense of the possibility of a life on earth at once orderly and free dawns upon the heart and mind. (p.125)

> ... there is a broad path open, through a creation of new wealth forms, to a new industrial system in which the productive forces of society and of nature may be used with far greater effectiveness than at present, and in which the distribution of wealth forms so created will take place on a far juster and more equitable basis. Society may have more to divide among its members, and at the same time the greater dividend may be divided in a juster manner. (p.130)
>
> ... there is a path along which sooner or later, both the Individualist and the Socialist must inevitably travel; for I have made it abundantly clear that on a small scale, society may readily become more individualistic than now and more socialistic. (p.131)
>
> The town will grow; but it will grow in accordance with a principle which will result in this – that such growth shall not lessen or destroy, but ever add to its social opportunities, to its beauty, to its convenience. (p.140)
>
> Source: Howard, 1902/1965

Howard regarded social inequality as a major constraint on co-operation. He advocated 'social balance' as the means to achieve a 'life of equal freedom and opportunity', and he saw scale as being crucially important to the achievement of social balance. The city would be self-contained in relation to its hinterland, in its use and disposal of resources, and in the provision of employment; in other words, its ecological footprint would be more localized. The notions of balance and self-containment, equality and justice anticipate some of the debates about the social context of sustainability of the present generation.

Lewis Mumford (1966, p.590) has described Howard's vision for the garden city as follows: 'What was radically new was a rational and orderly method for dealing with complexity, through an organization capable of establishing balance and autonomy, and of maintaining order despite differentiation, and coherence and unity despite the need for growth. This was the transformative idea.'

Howard's vision amounted to a proposal for an economic, social and political experiment: 'One small Garden City must be built as a working model, and then a group of cities ... These tasks done and done well, the reconstruction of London must inevitably follow, and the power of vested interests to block the way will have been almost, if not entirely, removed' (Howard, 1902/1965, p.159). Howard's proposal was that garden cities were to be in community ownership, and much of his analysis is devoted to demonstrating the economic benefits of his scheme in terms of achieving greater equality. He argued that residents should secure the whole increase in land values due to their migration, thus putting forward the case for 'betterment' value for public gain, which later became an issue for debate. Howard's ideas of participation, with the local population having political control of the garden cities, are reflected in the continuing debates about democratic accountability in cities today. He saw the remodelling of spatial relations in cities as an opportunity to change the way in which power is held. Howard wrote that 'One of the first essential needs of Society and of the individual, is that every man, every woman, every child should have ample space in which to live, to move, and to develop' (cited in Fishman, 1977/1996, p.44) . As Fishman suggests, Howard 'added a new element to the rights of man – the right to space' (p.44).

Although Howard's thinking was remarkably perceptive, his impact has been largely confined to the physical plans of garden cities and new towns and, in a much modified form, to the layouts of suburban estates, yet his vision for the transformation of cities went far beyond physical layout. Howard saw the spatial form of the city as socially produced. His idea was to create a different urban form comprising new relationships between environment, society, economy and governance. His ambition was to achieve no less than the complete reconstruction of unequal urban spatial relationships. While Howard's thinking reflects the dominant view of society at that time in relation to gender, class and family, his observations on the relationship between the economy, society and governance of cities and their spatial and environmental implications are as applicable today as they were a century ago.

Although Howard's social programme was reformist in tone, it was utopian. While it was feasible in principle, it proved unrealistic in practice. To see why, we must look at the contrasts and continuities in the three dimensions of environment, society and governance between the cities of a century ago and those of today.

2.3 THE CONTEMPORARY PROBLEM

In Howard's day the danger to public health was both the main environmental problem for cities (de Swaan, 1988) and the stimulus to public action. Such problems persist, especially in areas of deprivation and environmental degradation. To these have been added dangers that are more widely dispersed: problems of pollution, land use and resource depletion impacting on cities, but also created by cities as the centres of production and consumption which threaten global environmental ecosystems. Howard also perceived cities as unequal and unjust social systems, occupied, used, ordered and controlled through their spatial relations. Inequalities also persist within cities, and through the interconnections and interdependencies between cities, their social as well as their environmental impacts are distributed globally. Below we look at three examples which illustrate the nature and scale of the social and environmental problems within cities of the South and the North.

ACTIVITY 6.2 As you study the examples below, make a note of the similarities and differences between environmental and social conditions in the three cities. In what ways does Howard's thinking have relevance for these? ◆

Mexico City

Massey (1999a) shows how cultural, ethnic and social diversity have given Mexico City its colour and vitality. But difference and diversity are also associated with inequality and social exclusion, which reflect connections and disconnections stretching beyond the city. In Mexico City an affluent business community is driven in air-conditioned cars, protected from dangerous levels of air pollution and street crime, to modern offices such as may be found in the business sector in cities anywhere in the world. The international business market is catered for by luxury hotels equipped with state-of-the-art faxes, computers, phones and office services available 24 hours a day. One Japanese-owned hotel has 38 floors, 750 rooms, two presidential suites, seven executive suites, two Japanese suites and four restaurants offering Japanese, French and international cuisine.

FIGURE 6.6 *The urban sprawl of Mexico City*

The insulated lifestyle of the international business community contrasts with that of the urban poor. More than half of Mexico City's housing is self-built 'shanty' development (Figure 6.6). It has been estimated that 9–10 million people live in this type of housing on land that is often unsuited to development and can be liable to floods (Gilbert, 1996). Shanty housing is generally poorly serviced, and approximately 3 million people live in housing with no sewers (Clark, 1996). Only 15 per cent of the city's housing is built on legally occupied land. This is generally located in exclusive residential areas and can only be afforded by the rich. Demand for this type of profitable exclusive development is leading to increasing land values and is further restricting housing opportunities for the poor. Traffic congestion and industry cause high concentrations of air and water pollution which particularly affect the poor, who walk or travel by bus and who cannot escape to the clubs, restaurants, high-income housing and country clubs used by the rich. Over three-quarters of the air pollution in the Valley of Mexico is produced by transportation (Gilbert, 1996). Lifestyles within Mexico City reveal sharp contrasts. Pollution, rights to space and the consumption of resources are distributed unevenly among those who use the city. Similar inequalities can be found in diverse forms in many cities of the South (see Figure 6.7).

FIGURE 6.7 *Shanty towns are a common feature of cities in the South*

Bangkok

In the Yannawa district of Bangkok, families live in high-density, wooden-structured settlements with poor access to public roads. In one community 12 per cent of residents live in structures with five or more people per room. Major fires have broken out on average once a year in the adjacent slum communities to the east, including one in 1994 that destroyed hundreds of houses. The Chea Pleung I community was established about thirty years ago on a swampy area covered with rice paddies. It grew from about ten households in 1971 to 530 with 3,183 residents in 1997. The area is bounded by a major road, an elevated expressway, railway tracks and an Esso complex. Residents are squatters and have no security of tenure, so are subject to the threat of eviction. Most residents have only primary school education and work as wage labourers. Infestation by mosquitoes, flies and rats is a serious problem. Waste water is discharged mostly through inadequate open ditches, some of which have had houses built over them, allowing rubbish to collect and form blockages beneath (Lee, 1995).

As in the case of Mexico City, Bangkok shows extreme social polarization, which is reflected in environmental conditions. Economic growth has encouraged in-migration: in 1990, 1.5 million inhabitants out of the total of 8 million in the city were unregistered immigrants. Rapid population growth has meant that land which was unsuitable for development has been built on for housing, industry and commercial uses in a haphazard free-for-all in the absence of planning. Many factories in the expanding city use hazardous chemicals which pollute the surrounding areas occupied by immigrant squatters (Lo and Yeung, 1996). The benefits of economic development are unequally shared among different groups in the city. Inadequate and poorly paid employment opportunities for migrants result in substantial numbers of the urban population living in poverty in very poor environmental conditions. Over half the population do not have access to piped fresh water (Clark, 1996).

London

Ebenezer Howard was a Londoner, and it was his hope that the garden city movement would make possible the environmental and social reconstruction of the historic heart of the city. Many of the more obvious signs of poverty and environmental degradation have disappeared since his day, but new forms of inequality have taken their place. Unequal access to space and urban resources, as well as largely invisible yet health-threatening forms of modern pollution, affect inner areas of the city such as Spitalfields.

Massey (1999a) and **Allen (1999)** demonstrate that the power which settles in the City of London is expressed through its spatial relations within the city and between cities around the world. Modern marble and steel-clad office blocks of the globally connected financial centre of the City, tower over run-down streets and terraces and the old fruit and vegetable market. In Spitalfields extreme wealth and poverty are juxtaposed. The area has an 80 per cent Bengali resident population whose lifestyles contrast strongly with those of affluent city commuters. Similar conditions of poverty juxtaposed with wealth, sweatshop labour and pollution from fumes produced by traffic congestion can be found in many cities of the South.

The City and the East End of London, with their different histories, literally blend spatially at Spitalfields, constructing new divisions, contrasts and tensions. The Brick Lane area contrasts with the carefully renovated and conserved eighteenth-century former Huguenot houses of Fournier Street or the industrial buildings converted to form bijou 'lofts'. The old fruit and vegetable market (see Figure 6.8), once the hub of the local community, is now shared between two worlds: City workers (as sandwich bars and the like have sprung up) and the local community. It therefore now represents a contested space whose future is uncertain, since the City holds the power to plan total commercial redevelopment of the site.

FIGURE 6.8 *The old fruit and vegetable market at Spitalfields is used both by City workers and the local community*

The relationships between society, economy and environment that existed in cities in Howard's day are reproduced, albeit in different forms, in contemporary cities, producing conditions which are unsustainable. While the character or scale of environmental and social problems in rich cities of the North today may appear to differ from those in cities like Mexico City or Bangkok, spatial relations within cities remain uneven and unequal. The impact of uneven spatial relations within these cities can also be seen in Nairobi, Moscow, Hong Kong, New York or any major city in the world. Different worlds for different people exist simultaneously within cities across the globe, with serious implications for the future of cities environmentally and socially and for their governance.

At the same time the scale of the problems has grown. In 1810, London was the only city in the world with a million people. At the end of the twentieth century more than 20 cities had populations exceeding 10 million, 35 had over 5 million, and 8 of the 10 biggest cities were in Asia or South America.

The prospect for urban populations in the vast cities of the South seems bleak unless radical changes in society and the economy and governance of cities takes place. The environmental and social conditions of these cities are a problem of huge proportions, but, as the example of London shows, inequality is not confined to these expanding cities. London is one of the wealthiest cities in the world, but its economic success does not produce affluence and environmental quality for all its population. This begs the question of whether sustainable cities are achievable under contemporary conditions of development – an issue that we address in section 3.

To sum up the argument so far, the analysis of Howard's ideas, together with the examples of contemporary cities, bring out the different dimensions of the problem of sustainable cities.

- The *environmental* dimension shows that the sustainability of cities continues to be threatened by problems of disease and degradation created by deterioration in their physical fabric, lack of amenities and pollution. But cities also pose a threat to global sustainability through the expansion and intensification of their ecological footprints.
- The *social* dimension emphasizes the inequalities that occur both within cities and between cities of the North and South. Social and economic inequalities lead to unsustainable patterns of production and consumption globally.
- The third dimension, *governance*, identifies the problems of order and disorder, which appear to prevent sustainable patterns of development. The problems of environmental degradation, social inequality and disorder together form the barriers to the development of sustainable cities.

Howard proposed to tackle environmental and social problems through the orderly development of planned cities which emphasized conservation and amenity. He envisaged a more equitable society in which the whole population would share in the profits created by the city and participate in its governance. While this may seem utopian in today's conditions, basic changes in social and economic organization in response to environmental needs are seen by many commentators as an essential precondition for sustainable development. By contrast, a prevailing argument contends that a sustainable environment and economic growth are compatible, not conflicting, aims. In the next section we examine these competing arguments.

3 *Sustainable development and the city*

3.1 A SOCIAL AND ENVIRONMENTAL ISSUE

Sustainable development was not a term used in Ebenezer Howard's day, yet his ideas focused on the problem of a sustained relationship between the physical environment and social well-being of cities. Today, sustainable development is a *leitmotif* of pronouncements from sources ranging from industry, through government, to environmentalists. Its rhetorical importance has increased its ambiguity.

The tendency to focus on the physical/environmental aspects of sustainability – especially in the North – and the limited interpretation of the concept 'to keep in being' or 'to keep from failing' sits comfortably with contemporary northern lifestyles. The policy focus has been on such issues as conserving land, introducing energy conservation, waste reduction and minimization, and so on. 'Sustainability indicators' have emphasized the ecological dimension of the problem, as reflected in business and government publications on sustainability.

The social implications of sustainability have been ignored or minimized. Where social criteria have been considered, they have been seen as the causes of an ecological problem, with cities as an ecological time-bomb. The question of 'sustainability for whom?' has not been seriously addressed, yet this is at the heart of the notion of *sustainable development*. Sustainable development involves issues of need both for present and future generations, a point stressed by the Brundtland Report (World Commission on Environment and Development, 1987) in its oft-quoted definition of sustainable development as 'development that meets the needs of the present without compromising the ability of future generations to meet their own needs'. Yet cities of the South are characterized by poverty and lack basic needs such as water supply, sewerage systems, decent housing and infrastructure, while those in the North, which also have problems of environmental deterioration and poverty, consume a disproportionate share of the world's resources. Economic relations within and between cities of the North and South are not seriously challenged. What may be regarded as needs in the cities of the North would be luxuries in those of the South. Both are unsustainable under current conditions of development.

3.2 ECOLOGICAL MODERNIZATION AND THE CITY

The broadly accepted approach to tackling these problems assumes that sustainable development can be secured through a process that has come to be termed 'ecological modernization'. This has four main aspects.

The first is the argument that by introducing ecological criteria and circularity into the production and consumption processes it will be possible to reduce resource consumption and pollution to sustainable levels. In terms of the development of cities this aspect is reflected in the emphasis on reducing urban traffic and congestion, on waste minimization and recycling, the reduction of industrial pollution, the improvement of energy efficiency, the conservation of green space, the planting of trees, and so on.

Second, ecological modernization provides a key role for the market economy as the most efficient and flexible way of achieving objectives of sustainability. The market will promote the economic growth that is seen as the precondition for providing the resources necessary to tackle environmental problems. Thus the UK government's White Paper on sustainable development argues: 'Sustainable development does not mean having less economic development: on the contrary a healthy economy is better able to generate the resources to meet people's needs, and new investment and environmental improvement often go hand in hand' (HMSO, 1994, p.7). Examples such as Germany and Japan, where economic prosperity has been associated with less polluting and more resource-efficient cities, are often cited as exemplars of sustainability.

A 1996 report by the Organization for Economic Co-operation and Development (OECD) on sustainable urban development claims similarly that 'first, urban economies have a capacity to generate goods and services that can alleviate urban problems, and second, that the growth of the world economy favours the expansion of the environmental sector in the urban economy' (OECD, 1996, p.153). The report goes further in suggesting not only that markets can provide solutions to environmental problems, but also that environmental problems can be seen positively as a resource that can facilitate the creation of wealth: 'Ways must be found to help businesses define urban environmental problems as opportunities for innovation. If the nature of wealth is shifting to value the ability to understand and solve problems, then urban environmental problems constitute a strategic resource' (OECD, 1996, p.153).

Where social sustainability is considered, it is related to a perceived need for economic growth to raise living standards in the South to be on a par with those of the North, and to avoid increases in the social unrest, lawlessness and fear associated with social polarization in cities of the North. Hence the Brundtland Commission argued for continued economic growth in both North and South, anticipating a five- to tenfold increase in global industrial output by the time world population stabilizes during the twenty-first century (World Commission on Environment and Development, 1987). The concept therefore supports the increasing power of neo-liberal relations within and between cities (see **Pryke, 1999**).

The third aspect of ecological modernization is the enabling role of the state, providing a regulatory framework for environmental protection. The relationship between state and market is complementary rather than conflictual. Although the primacy accorded to market forces which was especially prominent in the UK and other western countries in the 1980s has now somewhat abated, the language of 'partnership' and 'negotiation' between public and private that has followed suggests there is a mutuality in the relationship (see Chapter 5). In cities, for example, economic regeneration is now pursued through partnerships in which state support and private capital is devoted to specific areas (usually deprived) and schemes which take account of the needs for economic growth and environmental protection.

Finally, while there is a focus on the economy and the market, other interests, known sometimes as 'stakeholders', are also incorporated through participation in partnerships. The emphasis is on achieving consensus and participation by involving environmental groups and citizens in the development of ideas, projects and policies for creating sustainable cities. These are 'bottom-up' approaches that engage people in action planning, environmental forums, focus groups, citizens' juries and other innovative methods of involvement. In cities this activity has been focused around Local Agenda 21, the development of ideas and projects to promote sustainable forms of development that have emerged from the Rio Conference of 1992 and the Habitat II Conference in 1996 (see Box 6.1 overleaf).

Ecological modernization reflects some trends in modern cities. As such it is a pragmatic approach, but it has important limitations as a process for achieving sustainable cities:

- Ecological modernization is essentially a northern approach to the problem. The issues it tackles and the processes by which it works assume certain conditions such as economic prosperity, an efficient market, technological advancement, an enabling state and a plural, inclusive society. Each of these may, to an extent, be present in the advanced countries of the North. In a southern context, however, these elements may be poorly developed or altogether missing. In addition, the problems of environmental degradation and poverty in cities of the South are often simply overwhelming.
- A related point is that the approach is restricted and narrow. While improvements may be secured in many northern cities, the fundamental problems of resource depletion and pollution, which have global impacts, are not seriously tackled. Thus improvements to northern cities may be at the expense of exporting problems of pollution elsewhere. In short, the ecological footprint of these northern cities shows little sign of diminishing, and the externalities of the North continue to be imposed on the South. Meanwhile, development in the South, if it pursues a similar path to that in the North, will vastly increase the detrimental impacts on the environment. It has been estimated that if the world population stabilizes at 10–11 billion in the twenty-first century, an additional five planet Earths would be needed just to maintain the present level of ecological decline (Rees, 1996).
- The approach is optimistic, and assumes that 'All that is needed is to fast forward from the polluting industrial society of the past to the new super industrialized era of the future' (Hannigan, 1995, p.184). There is a belief

Box 6.1 Habitat II: the context

Key dates leading up to Habitat II:

1972: Only One Earth Conference (United Nations Conference on the Human Environment), Stockholm, leading to:

1976: Habitat I Forum

1980–83: Brandt Commission, leading to publications:

North-South: A Programme for Survival (1980)

Common Crisis: North-South Co-operation for World Recovery (1983)

1987: *Our Common Future* (Brundtland Report), produced by World Commission on Environment and Development

1992: Earth Summit (United Nations Conference on Environment and Development), Rio de Janeiro

1994: Signing of Agenda 21, Aalborg, Denmark

1996: Habitat II ('City Summit': United Nations Conference on Human Settlement), Istanbul

Attention was first drawn to the need for solutions to environmental problems to have regard to people and settlements at the UN Conference on the Human Environment, 1972. This resulted in the Habitat I Forum in 1976, which focused on the limited idea that the growth of cities should be slowed or reversed and that 'the urban problem' could be solved by intervention relating to their physical form.

The Brandt Commission (1980–83) was not specifically concerned with cities, but its concern with inequalities between North and South had relevance to the development of the concept of sustainable development, as defined at the 'Brundtland' conference in 1987, and to the 1992 Earth Summit in Rio de Janeiro and the adoption of Agenda 21, which was to be a comprehensive programme for sustainable development. Social equity was a key principle of Agenda 21, but in spite of its rhetoric, the summit accentuated the conflicting concerns of North and South.

The Local Agenda 21 process emerged from Rio, with the aim of encouraging local authorities to interpret the issues raised in Agenda 21 in a local context. Local Agenda 21 was the focus of Habitat II in 1996, which put the issue of cities and sustainability at centre stage. It re-emphasized the importance of the need for community participation (stakeholders and partnerships) to be involved in programmes of action for sustainable development at the local level. In cities, local government agencies, non-governmental organizations and the private sector were seen as acting in partnership with all the local interests in the community, sharing examples of best practice through global communication networks.

that present economic processes, with greater emphasis on ecological needs, will succeed in achieving sustainable cities. Ecological modernization sits comfortably with a belief in the continued liberalization of the world economy and in 'business as usual'. In practice, the impact of ecological modernization on environmental problems may be, at best, marginal.

- Ecological modernization is a consensual approach. While stressing participation, the social tensions and conflicts within and between cities are ignored. Incorporation of some groups may simply serve to legitimate the important partnership – that between government and business. Some groups may play only a marginal role, and some will be altogether excluded. Questions of 'development for whom?' and 'at whose expense?' are not dealt with.
- This leads to a further point, namely that ecological modernization, with its emphasis on the economy and technology, is largely devoid of a social dimension. Cities are composed of very divergent interests and, as this book has been at pains to show, inequalities of wealth and power are a universal condition. As we have shown, such inequalities within cities and between cities globally are a barrier to the kind of co-operation envisaged in the Agenda 21 process. Global initiatives reflect the status quo of continued economic growth in a framework of market-based 'sustainable' development and do not address the need for fundamental change.
- Finally, ecological modernization rests on the premise that the market and the state are in a compatible relationship. This is true, but only so long as the state undertakes to provide infrastructure, to assume the costs of welfare and to exercise its regulatory function lightly. Conversely, the state requires the market to generate employment, investment and taxation. It is obvious that these aims may be in conflict. As economic relations stretch and intensify between cities and beyond nation states, the relationship may become more imbalanced, since states are by definition spatially confined, while some businesses, certainly many multinationals, extend across the globe. Business can use its global reach to exercise the threat of investment withdrawal from the state's territory if the burdens imposed by government are too severe. Seen in this way, it is cities themselves that are in danger.

ACTIVITY 6.3 In view of these limitations, what do you see as the key issues that an approach that is sensitive to the social and environmental problems of cities would need to address? ◆

3.3 THE RADICAL PERSPECTIVE

If ecological modernization in cities cannot, in the end, ensure sustainable development, then what can? This rather terse question has tended to beg more questions rather than elicit satisfying answers. Commentators who take a radical environmentalist view or pursue a neo-marxist explanation of social relations are quick to draw attention to the failure of the market to promote sustainable development, even in the wealthiest cities in the world. Whereas the concept of ecological modernization emphasizes the importance of economic growth to achieve sustainable cities, the radical perspective argues that present economic development is actually a cause of environmental and social inequality and the unsustainability of cities.

The radical perspective emphasizes the role of capitalist relations of production in exploiting resources and creating patterns of uneven development, especially in shaping inequalities between North and South from colonial times. Capitalist urban spatial relations are spread between cities, the perspective argues, exploiting and creating dependencies far beyond their boundaries. Not only has ecological and social space in the South been used to support patterns of consumption and pollution in cities of the North, but capitalist social and economic relations have also effectively been exported to the developing world, with long-term consequences for sustainability. At the same time, in cities of the North, new patterns of spatial inequality have been produced. The production and reconstruction of unequal spatial relations *within* cities of the South and the North are therefore replicated in the unequal relationships *between* cities. Interconnecting urban processes work across space to create social and environmental conditions that are globally unsustainable.

Affluent lifestyles in cities of the North are dependent on the intensification of uneven and unsustainable capitalist relations between cities globally. As we have seen, the notion of sustainable development, especially in the South, has partly become a rhetorical device which avoids a challenge to existing patterns of consumption and pollution or the need for the redistribution of wealth and resources. It appears to resolve the tension between continued development, which is necessary to support global capitalist relations, and its impacts. **Hinchliffe (1999)** points out that sustainable development has been appropriated by vested interests for their own purposes. The neo-marxist or radical analyses of the causes of unsustainability suggest that fundamental changes in economic and social relations are necessary to prevent continuing environmental deterioration and social crisis.

The most gloomy of radical commentators, Ulrich Beck (1992, 1995), foresees environmental catastrophe as the product of what he terms the 'risk society'. This is a state in which we 'are confronted by the challenges of the self-created possibility, hidden at first then increasingly apparent, of the self-destruction of all life on this earth' (Beck, 1995, p.67). The 'risk society' is an outcome of high technology, which has the potential for devastating changes or destruction (and nuclear war, global warming and ozone depletion are three examples) and which is ultimately beyond our control. While Beck provides an analysis of the dangers, he is much less confident about predicting a way to avert them in the context of development patterns that include cities.

If ecological modernization cannot deliver sustainable development and we are in the grip of the 'risk society', the situation seems, literally, hopeless. Of course, social change and its environmental consequences are not inevitably disastrous. There is sufficient evidence of improvement in some cities – for instance in the decline of air pollution, in the management of traffic, in the improvement of living conditions – to suggest that change can have positive outcomes. It is obvious that British cities today are for most people vastly better places in which to live than was the case in Howard's day.

This is not to deny that the problems facing cities – especially those affecting the global environment – are deeply serious. What is needed is an analysis of the changes in the environment, society and governance which, taken together, may constitute a way of achieving sustainable cities.

As a way into considering what changes may be necessary we shall look at the work of David Harvey on socio-environmental justice (see also **Hinchliffe, 1999**). Although Harvey is not explicitly concerned with sustainability, in his paper 'The environment of justice' (1996) he nevertheless identifies processes that are essential to an understanding of the issue. We have considered some of these already. For example, Harvey emphasizes that urban social processes must also be understood as ecological processes in that human activity produces the urban environment. He goes on to underline the significance of urban inequalities of power, which result in poor and deprived people living in areas of environmental hazard. Furthermore, the interactions between people and environment within urban space are 'multilayered': 'Changing patterns of urban organization simultaneously produce configurations of uneven social and economic development at different scales coupled with multiple displacements of environmental issues to different scales' (Harvey, 1996, p.91). For instance, electric power makes for clean and healthy cities at one level but produces gases which contribute to global warming at another.

What is particularly distinctive in this analysis is, first, the primacy given to social relations as the explanation for environmental change, and second, the emphasis on social justice as the way to achieve environmental justice within cities. This is a plea for urban change that is 'conducive to the well-being of the poor'. For Harvey this means a commitment to a particular approach to ecological modernization:

> On the one hand that means embracing and subsuming the highly differentiated desire for cultural autonomy and dispersion, for the proliferation of tradition and difference within a more global politics, but on the other hand making the quest for environmental and social justice central rather than peripheral concerns.
>
> (Harvey, 1996, p.97).

Social justice in cities can only be secured if there are concomitant changes at a political and institutional level. Ecological modernization relies on the 'enabling' state as the institutional form of governance. However, ecological modernization is rather limited in this context as it is confined to relations between the state and the business sector. The role of state institutions in enabling the welfare of individuals is therefore diminished. Indeed, it is not going too far to assert that the concept of the *enabling state*, with an emphasis on partnership between government and the economy, has supplanted (or at least suppressed) the idea of the *welfare state* with its deeply embedded notion of a contractual relationship between government and society. In terms of participation, large sections of the population of cities remain excluded, and thus experience both material and political deprivation.

In certain respects the decline of state and social institutions may have been liberating, since it has given rise to the development of local spaces of power within cities through a variety of social movements. Many of these, such as environmental movements (discussed in Chapter 5) have an urban focus. Such movements often have a popular base of support and challenge the power structures based on corporatist relations between government and industry. Their engagement may be through participating in decision-making or by opposing from the outside. Environmental movements have had striking successes both in drawing attention to problems and in winning battles, especially at the local level, but they are not representative or accountable. As Harvey points out, in their concern for ecology they often marginalize or ignore issues of social justice. The question then is how it might be possible to represent politically fluid, open and intense urban interactions.

ACTIVITY 6.4 Read Extract 6.2 , which is from Harvey's paper, 'The environment of justice'. Elsewhere in this series **Massey (1999b)** posed the question: in what ways are cities 'challenges to democracy'? How do Harvey's ideas help us to consider this question in relation to sustainability? ◆

EXTRACT 6.2
David Harvey: 'Principles of justice and environments of difference'

At this point, I find myself returning to some of the basic formulations laid out in *Social Justice and the City* more than twenty years ago. There I argued that it was vital to move from 'a predisposition to regard social justice as a matter of eternal justice and morality to regard it as something contingent upon the social processes operating in society as a whole.' The practice of the environmental justice movement has its origins in the inequalities of power and the way those inequalities have distinctive environmental consequences for the marginalised and the impoverished, for those who may be freely denigrated as 'others' or as 'people out of place.' The principles of justice it enunciates are embedded in a particular experiential world and environmental objectives are coupled with a struggle for recognition, respect, and empowerment.

But as a movement embedded in multiple 'militant particularisms', it has to find a way to cross that problematic divide between action that is deeply embedded in local experience, power conditions and social relations to a much more general movement. And, like the working-class movement, it has proven, in Williams' words, 'always insufficiently aware of the quite systematic obstacles which stood in the way'. The move from tangible solidarities felt as patterns of social bonding in affective and knowable communities to a more abstract set of conceptions with universal meaning involves a move from one level of abstraction – attached to place – to quite different levels of abstraction capable of reaching across a space in which communities could not be known in the same unmediated ways. Furthermore, principles developed out of the experience of Love Canal or the fight in Warren County do not necessarily travel to places where environmental and social conditions are radically different. And in that move from the particular to the general something was bound to be lost. In comes, Williams notes, 'the politics of negation, the politics of differentiation, the politics of abstract analysis. And these, whether we liked them or not, were now necessary even to understand what was happening.'

But it is exactly here that some of the empowering rhetoric of environmental justice itself becomes a liability. Appealing to 'the sacredness of Mother Earth', for example, does not help arbitrate complex conflicts over how to organize material production and distribution in a world grown dependent upon sophisticated market interrelations and commodity production through capital accumulation. The demand to cease the production of *all* toxins, hazardous wastes and radioactive materials, if taken literally, would prove disastrous to the public health and well-being of large segments of the population, including the poor. And the right to be free of ecological destruction is posed so strongly as a negative right that it appears to preclude the positive right to transform the earth in ways conducive to the well-being of the poor, the marginalized and the oppressed. To be sure, the environmental justice movement does incorporate positive rights particularly with respect to the rights of all people to 'political, cultural and environmental self-determination', but at this point the internal contradictions within the movement become blatant.

At this conjuncture, therefore, all of those militant particularist movements around the world that loosely come together under the umbrella of environmental justice and the environmentalism of the poor are faced with a critical choice. They can either ignore the contradiction, remain within the confines of their own particularist militancies – fighting an incinerator here, a toxic waste dump there, a World Bank dam project somewhere else and commercial logging in yet another place – or they can treat the contradictions as a fecund nexus to create a more transcendent and universal politics. If they take the latter path, they have to find a discourse of universality and generality that unites the emancipatory quest for social justice with a strong recognition that social justice is impossible without environmental justice (and *vice versa*). But any such discourse has to transcend the narrow solidarities and particular affinities shaped in particular places – the preferred milieu of most grass-roots environmental activism – and adopt a politics of abstraction capable of reaching out across space, across the multiple environmental and social conditions that constitute the geography of difference in a contemporary world that capitalism has intensely shaped to its own purposes over the last two hundred years.

The abstractions cannot rest solely upon a moral politics dedicated to protecting the sanctity of Mother Earth. It has to deal in the material and institutional issues of how to organize production and distribution in general, how to confront the realities of global power politics and how to displace the hegemonic powers of capitalism not simply with dispersed, autonomous, localized and essentially communitarian solutions (apologists for which can be found on both right and left ends of the political spectrum), but with a rather more complex politics that recognizes how environmental and social justice must be sought by a rational ordering of activities at different scales. The reinsertion of the idea of 'rational ordering' indicates that such a movement will have no option, as it broadens out from its militant particularist base, but to reclaim for itself a non-coopted and non-perverted version of the theses of ecological modernization. On the one hand that means embracing and subsuming the highly geographically differentiated desire for cultural autonomy and dispersion, for the proliferation of tradition and difference within a more global politics, but on the other hand making the quest for environmental and social justice central rather than peripheral concerns.

For that to happen, the environmental justice movement has to radicalize the ecological modernization discourse itself. And that requires confronting the fundamental underlying process (and their associated power structure, social relations, institutional configurations, discourses and belief systems) that generate environmental and social injustices. Here, I revert to another key moment in the argument advanced in *Social Justice and the City* (pp.136-7): it is vital, when encountering a serious problem, not merely to try to solve the problem in itself but to confront and transform the processes that gave rise to the problem in the first place. Then, as now, the fundamental problem is that of unrelenting capital accumulation and the extraordinary asymmetries of money and political power that are embedded in that process. Alternative modes of production, consumption and distribution as well as alternative modes of environmental transformation have to be explored if the discursive spaces of the environmental justice movement and the theses of ecological modernization are to be conjoined in a programme of radical political action.

References

Harvey, D. (1973) *Social Justice and the City*, London, Arnold.

Williams, R. (1989) *Resources of Hope*, London, Verso.

Source: Harvey, 1996, pp.95–7

In this extract Harvey emphasizes the need to regard social justice as 'contingent upon the social processes operating in society as a whole'. He suggests that a discourse of universality and generality 'has to transcend the narrow solidarities and particular affinities shaped in particular places ... and adopt a politics of abstraction'. Such a politics of abstraction would have to engage with the material and institutional issues necessary to economic and political change.

From the analysis so far it should be clear that sustainable cities are as much a social and economic as an environmental concept, since issues of sustainability are inseparable from those of equity and power. Sustainable cities require different spatial arrangements, but they also require changes at a social, political and institutional level – changes that involve different ways of managing cities and developing urban society. There needs to be a concern for social justice (which includes environmental justice) if sustainable cities are to be politically attainable.

Ebenezer Howard formulated his ideas at a time when the radical movement of his day was arguing for the major transformation of capitalist society. He believed that the social city could be achieved within the capitalist system through the reconstruction of cities spatially to foster voluntary co-operation and eradicate inequality: this was to be a 'peaceful path to real reform'. The fact that the social aspects of Howard's work have not been realized in practice would seem to support the need for structural change in the spatial relations within and between cities globally. This is the challenge to democracy. Harvey suggests that the challenge will be how to develop more participative forms of democracy and governance and appropriate institutional structures within and between cities. These issues are taken up in the next chapter.

The next section of this chapter considers the nature and extent of the problems that have to be tackled by looking at specific issues and contexts. We then go on to explore the role of planning, as a form of urban governance, in attempts to tackle the problems of cities and identify the weaknesses of planning. This leads on to speculation about the processes that are necessary for sustainable cities to be achieved.

4 Unsustainable cities

In this section we use three brief case studies to demonstrate the challenges that lie ahead. The first considers a problem facing all cities: transport. The second uses the problem of managing urban land use, taking the UK as an example, to reflect upon contemporary debates about urban form in a developed country. The third explores the problem of urbanization in a context of rapid development, using the example of China, not least because its size alone has profound implications for global sustainability. The cases are illustrative and will be used to identify the spatial, social and institutional changes necessary if urban sustainability is to be achieved.

4.1 TRANSPORT AND SUSTAINABLE CITIES

While the main means of transport in cities of the South is by foot, as in African cities, or by bicycle, as in China, northern cities are dominated by motorized transport. Mass production and mass consumption have gone hand in hand to reduce the relative cost of car ownership, making this the preferred means of transport (see **Hamilton and Hoyle, 1999**).

ACTIVITY 6.5 Figure 6.9 gives some idea of the extent of dependency on motorized transport in Britain at the end of the twentieth century. Before reading on, make a note of the key points shown by the statistical information in each diagram. When you have done this, see if your points appear among the statistics we have extracted below. ◆

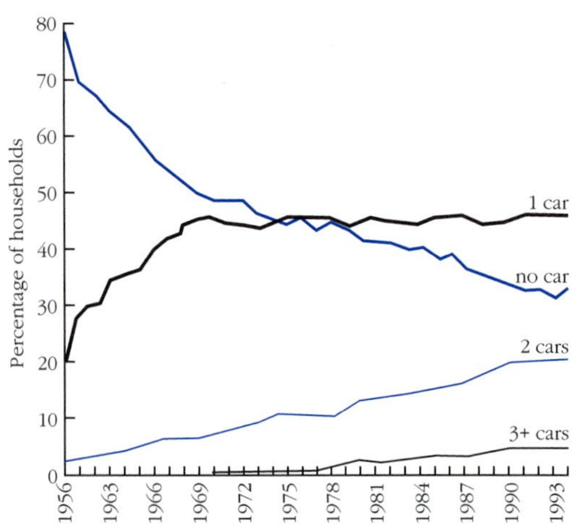

FIGURE 6.9(a) *Household car ownership in Great Britain, 1956–94*

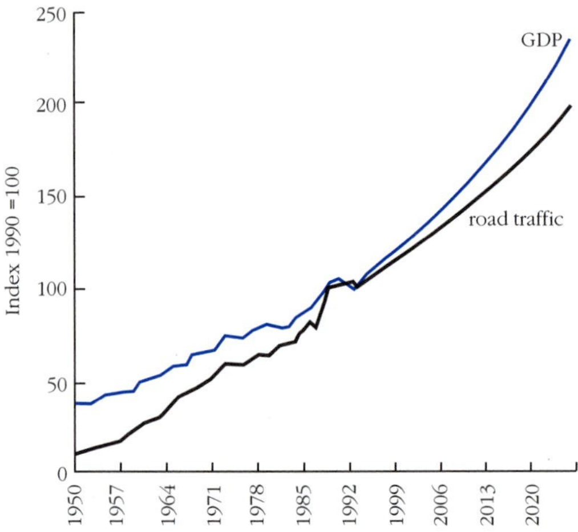

FIGURE 6.9(b) *Road traffic and GDP projections, Great Britain, 1950–2025*

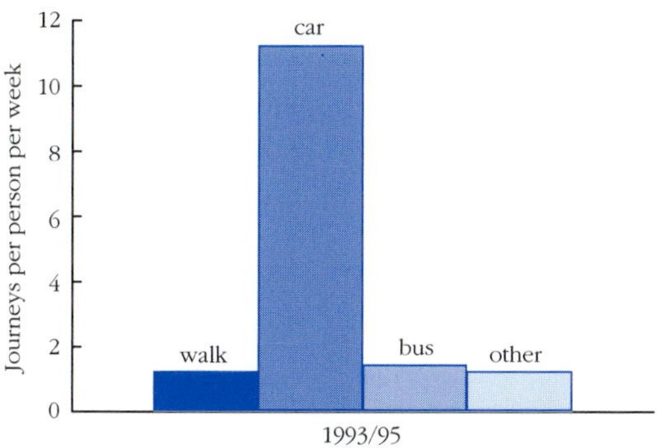

FIGURE 6.9(c) *Journeys over one mile by mode, Great Britain, 1993–95*

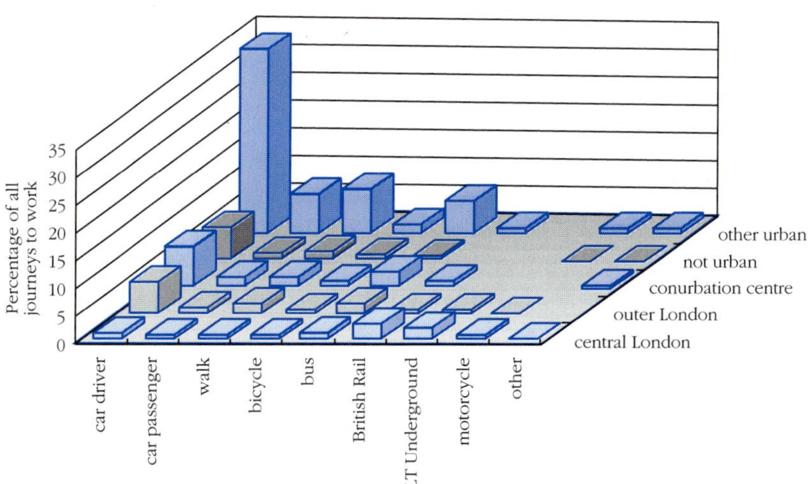

FIGURE 6.9(d) *Usual means of travel to usual place of work, Great Britain*

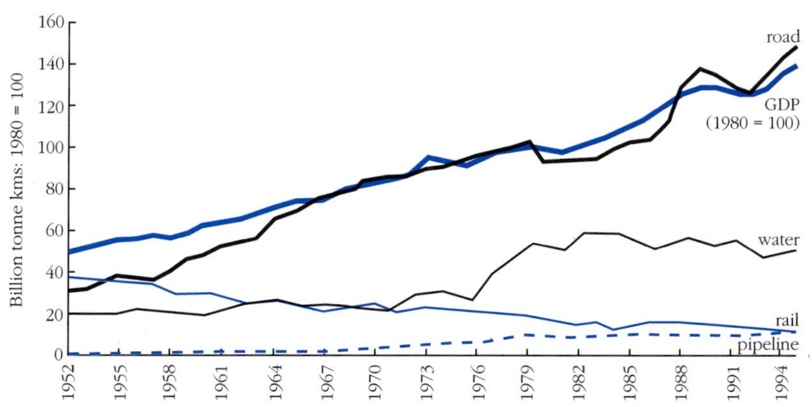

FIGURE 6.9(e) *Freight transport by mode, Great Britain, 1952–95*

- In 1956 less than one in four households owned a car; by 1994 more than two in three had at least one car.
- While around a quarter of households have two or more cars, around a third have none.
- Road traffic is expected to double between 1990 and 2025.
- The car is the main mode of transport for all journeys of over one mile; 11 journeys per week per person are made by car as against one journey by each of the other modes.
- The car is the main mode for commuting in all areas except central London.
- Road is by far the most important method of moving freight in Great Britain, accounting for two-thirds of the tonne-kilometres carried.

FIGURE 6.10 *Traffic congestion in Bangkok*

The environmental consequences of motorized transport are extensive. In terms of resource depletion, oil is the most significant: transport uses over half the UK's total consumption of this non-renewable resource (with over 40 per cent used in road transport). Transport also uses primary raw materials in the construction of vehicles and requires large volumes of aggregates and other construction materials for building roads, car-parks and other elements of infrastructure. Transport also causes amenity damage through noise, quarrying for aggregates and the construction of roads through residential areas of cities or attractive rural areas. Road projects have been the source of some of the most vigorous environmental campaigns of recent years. Traffic congestion and its associated noise, pollution and amenity damage is one of the most obvious environmental problems facing contemporary cities across the world (see Figure 6.10).

Road transport also gives rise to problems of health. Road accidents account for almost two-fifths of accidental deaths in Britain, while pollution from road transport also causes or exacerbates various respiratory diseases and increases the risk of certain cancers, especially in urban areas. Traffic produces 60 per cent of emissions of oxides of nitrogen and around a quarter of total carbon dioxide emissions, both of which contribute to the enhanced greenhouse effect. Vehicle emissions also increase ozone concentrations, intensifying pollution at ground level, and contribute heavily to the nitrogen oxides, volatile organic compounds (VOCs) and chloro-fluoro-carbons (CFCs) which cause the depletion of the tropospheric ozone layer.

Transport affects the spatial form of cities. The infrastructure associated with roads covers a fifth of urban areas and 3.3 per cent of the total UK land surface. Public transport networks facilitated the early phases of suburbanization. The greater journey length and flexibility offered by the car has encouraged suburban sprawl, which in turn has made it uneconomic to extend comprehensive public transport services. The dispersal and decentralization associated with 'spread' cities have reached their ultimate expression in the USA and Australia.

While not the cause, the car has been a necessary condition for a whole range of social changes: the dispersal of population, the decline of communities based on family and neighbourhood, commuting over long distances, the emergence of shopping malls on the urban fringe, the increasing pressure from tourism in the countryside, and so on. For those with access to a car, the experience is liberating, as it provides choice and opportunity to experience the advantages of spatial relationships created for a car-owning population. However, as **Hamilton and Hoyle (1999)** point out, lack of access is often a defining aspect of social inequality. Reade (1987) observes that the deterioration of public transport has created a new class of 'urban poor'. Moreover, those without a car often live in the very parts of cities where the problems caused by the car – noise, danger, pollution and disruption – are greatest. In a later observation, Reade (1997, p.98) concludes that the car 'is probably the chief means by which environmental inequalities are created and sustained'.

Set against the problem in developed countries, the problem in the developing world is the lack of infrastructure to accommodate increased car ownership in rapidly expanding cities. Migrants living in spatially distant shanty developments have poor access to public transport services and consequently to shops and employment, which tend to be located centrally. The combination of population growth and increased car ownership means that the costs to urban sustainability will be very high.

Motor transport is now seen as a central problem of sustainable cities in the developed world. The Royal Commission on Environmental Pollution (1994, p.1) comments: 'The unrelenting growth of transport has become possibly the greatest environmental threat facing the UK and one of the greatest obstacles to achieving sustainable development'. Not surprisingly, there is no lack of policies to reduce car dependence and boost alternatives, including integrated transport systems; an emphasis on public transport; disincentives to travel by car through taxation, road pricing and parking controls; the pedestrianization of city centres; and land-use policies to reduce the need for travel.

Such policies go far wider than simply physical and land-use changes; they involve a profound challenge to the lifestyles that have evolved around the car. As the Royal Commission puts it, 'If our lifestyles are not sustainable ... they will have to be modified sooner or later' (p.17). Transport is only a part – though a critical part – of the set of social relationships that compose the modern city. It demonstrates the interaction of environmental, social, economic and political constraints that must be tackled if sustainable cities are to be achieved (see Chapter 5). One outcome of the process of change will be a change in the spatial form of the city itself.

4.2 SUSTAINABLE URBAN FORMS

One way of reducing mobility is through changes in land use, and for this reason there has been considerable debate about sustainable urban forms. One obvious constraint on change is the high level of investment in buildings and infrastructure that has created the form of existing cities. New development only accounts for about 2 per cent per annum of the total building stock in British cities. Of course, much can be done with the existing stock to make it more sustainable: for example, the conservation of energy and other resources; waste reduction; a more intensive use of buildings; the relocation of activities within the urban area; and the reduction of pollution. In the longer term, policy could attempt to guide the spatial forms of cities, but in what way?

The European Commission's *Green Paper on the Urban Environment* (Commission of the European Communities, 1990) advocates urban 'densification' to create the 'compact city' (referred to in a different context in Chapter 4), which is seen to possess both environmental and social advantages. In particular:

> the city offers density and variety; the efficient, time-and energy-saving combination of social and economic functions ... The city's economic and social importance ultimately rests on the ease for communication offered by spatial density and the sheer variety of people and institutions which can exploit this opportunity.
>
> (Commission of the European Communities, 1990, pp.19–21)

Consequently, the Green Paper advocates high-density, mixed-use cities throughout Europe to reduce travel distances, maximize public transport use and increase energy efficiency. Spatial proximity would also encourage the economic, social, cultural and physical regeneration of declining city centres. The compact city is seen as socially and environmentally more sustainable (see also Chapter 4).

The compact city echoes the philosophy behind Le Corbusier's 'radiant city' (see Figure 6.11) and the focus on high densities associated with the 'modern

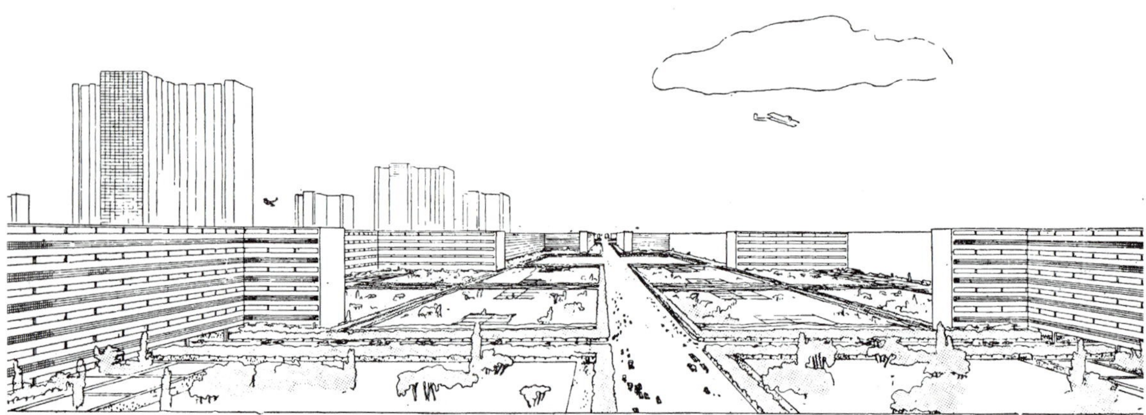

FIGURE 6.11
Le Corbusier's radiant city

movement'. However, there is much controversy about the potential advantages of high density, especially in the UK context, which mirrors the debate between advocates of the radiant and garden city of the early part of the twentieth century (see Box 6.2).

Box 6.2 Le Corbusier's 'radiant city'

Like Howard's garden city, Le Corbusier's radiant city of 1922 was planned with a strong social purpose. The radiant city was developed from the concept of a necessary integration between environment, society and governance previously identified by Howard. Le Corbusier's solution was quite different.

Le Corbusier believed that, in the 'machine age', conscious planning and control of chaotic urban development was required. He saw planning and order as a rational, technical process separated from politics. His vision was to create a system for urban living in which organization and freedom would coexist. However, unlike Howard's garden city, the radiant city was to be high-density and high-rise in form. New building technology was to be used to build vertical 'streets' or 'unités', each housing 2,700 people and set in open parkland. The unités were to be the centrepiece of the radiant city. Within them collective modes of living were to be encouraged by the inclusion of shops, cafés, restaurants, craft workshops, children's nurseries, meeting rooms, recreation facilities, and so on. Cities with high densities and proximity of people and activities were Le Corbusier's recipe for urban social co-operation.

The idea of the compact city has been attacked on the grounds that it is impracticable, undesirable and unrealistic. It is impracticable because existing trends towards decentralization cannot easily be halted; existing trends in behaviour are not easily reversed. It is undesirable because it may exacerbate inequalities as urban cramming reduces amenity, and may result in a flight to the suburbs by those who can afford more space. Much of the inner-city area experiences 'disenfranchisement by location' – a deteriorating environment of sink housing-estates and poor services (Breheny and Hall, 1996; see also Chapter 2). Meanwhile, parallel policies to restrict the growth in small towns and rural areas may also have undesirable consequences, preserving rural amenity at the expense of the less well-off. Policies of containment, supported by environmentalists, may have the social consequences of increased homelessness and overcrowding, creating 'a reservoir of unhappiness and unmet need' (Breheny and Hall, 1996, p.41).

The concept of the compact city and urban containment has been further criticized as being unrealistic since it is not self-evident that such cities are more sustainable. Alternative concepts have been put forward that incorporate the principle of minimizing the physical separation of activities and relating development to transport facilities (Owens, 1992). Cities would be linked by efficient public transport to satellites of smaller, relatively dense development. Activities would be located at public transport nodes, and combined heating and power systems could be used to reduce energy consumption. Within cities, more intensive, mixed land uses would help to revive city centres economically and socially. Such ideas combine a variety of approaches and have been described as the 'MultipliCity' approach to sustainability (Blowers, 1993, p.156). This approach proposes clusters of development combining urban infilling, the extension of the urban fringe, the creation of new settlements, and some growth in rural areas, all within the sustainable city region. Figure 6.12 provides a conceptual idea of the changes needed to ensure sustainability. Such a conception can be recognized as a revival of Howard's ideas of the 'social city region'. This is a pragmatic response to the problem – a response that sees urban development as an evolution of existing cities into more sustainable forms.

In the UK these different views of the future urban form have aroused considerable controversy. It is estimated that the population will grow by 3.6 million between 1991 and 2016, but the growth in households (due to smaller

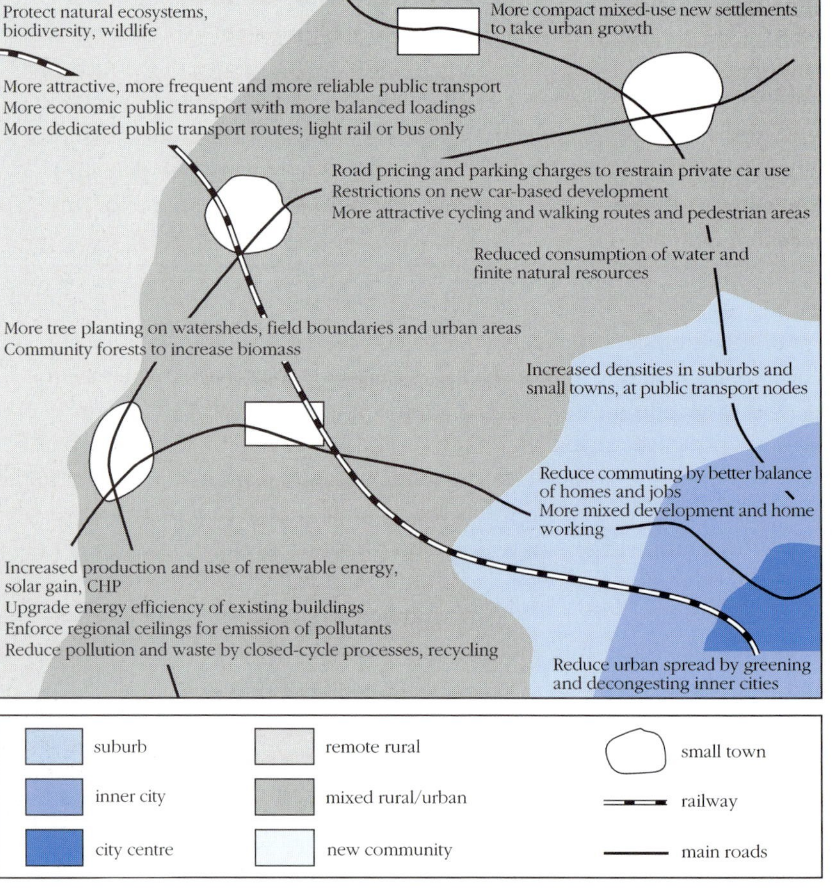

FIGURE 6.12 *Some of the changes that will need to be made to create future sustainability in the social city region*

households) will be much higher – at least 4.4 million and possibly a million more than that. The government's target is for 60 per cent of the total of new housing to be accommodated on re-used (so-called 'brownfield') sites. Advocates of urban densification have suggested that a much higher target of 75 per cent could be achieved if taxation and other measures were used to deter the use of greenfield land (UK Round Table on Sustainable Development, 1997). The opposing view is that a more realistic target is in the region of 30–40 per cent on urban land.

Behind these estimates lie profoundly different views of the meaning and prospects for sustainable urban forms. Urban form, like transport, is not simply a spatial question; it involves social processes too, and both spatial form and social process must be addressed in the search for sustainable cities. High densities such as those achieved in Hong Kong (see Box 6.3) are inconceivable in a UK context, but such densities, viewed in a wider context, are not necessarily more sustainable. In any event, a substantial shift to new urban forms involves very significant changes in lifestyles for many people.

Box 6.3 Hong Kong: a sustainable city?

Hong Kong is an example of an existing compact city which has been heralded as an example of an environmentally sustainable city form. On an area roughly the same size as Edinburgh, Hong Kong has twelve times the population. Half the people live at very high densities in high-rise state housing, and 90 per cent of commuting is by public transport. Half the area, to the north of the city, is open space, including parklands. The proximity of urban activities leads to low per capita levels of energy consumption (Stanley, 1996).

Such density would be difficult to transfer from one social context to another. The Hong Kong model would be regarded as socially unsustainable in Britain, the USA or Australia. In any case Hong Kong's economic growth depends on uneven relations within the city and beyond the city to rural areas and other cities which are drawn into its influence 'like a star cloud' (Lo and Yeung, 1996), as shown in the case of the Zhujiang Delta in the next section. The city's success depends on its ability to exploit resources through its links with other cities globally (see **Massey, 1999a**).

FIGURE 6.13 *The densely packed high-rise buildings of Hong Kong*

The problem of achieving sustainability through spatial form in the UK may be contrasted with the very different problem of meeting the challenge of sustainability in the dynamic context of China.

4.3 CHINESE CITIES: SUSTAINABILITY AND GROWTH

The detailed problems of the sustainability of cities in the UK seem to pale into insignificance when compared to the environmental and social problems arising from rapid modernization and its concomitant urbanization in China. China has 22 per cent of the global population, a forecast quadruple growth in GNP between 1980 and 2000, and rapid urbanization, from 324 cities in 1985 to 800 in 2000. There has been a massive shift from rural to urban living, with the result that Chinese cities are not only numerous, but also large. In 1995, 28 cities had a population of over two million, and 46 had a population of between one and two million (Clark, 1996, p.17).

Since 1978, economic reform has encouraged the introduction of foreign investment and the development of a free market. Nevertheless, the state has maintained its capacity to mobilize resources for major construction plants. The most controversial example is the Three Gorges project on the Yangtse River. First conceived in 1919, the project began construction in 1997 for completion in 2009. Behind a massive dam, a lake covering 253 square miles is being created to produce 18.2 thousand megawatts of electricity, the equivalent of 18 large power stations. In the process, 140 towns and 4000 villages will be drowned and 1.2 million people displaced.

However, economic liberalism has had a major impact, especially on the process of urbanization in China. (**Thrift, 1999**, discusses the networks of neo-liberal relations that stretch between and beyond Chinese cities.) The state's role in providing urban resources and controlling development has become less important, and state planning and resource allocation controls have receded. There is increasing population mobility, with urban growth dispersed among existing cities as they compete for investment. There is an urgent need to manage rapid development, particularly in the eastern coastal and south eastern regions.

The provision of social, transport and housing facilities in China is based on permanent residents, and although areas of land in the cities are being set aside for commercial housing, there are large numbers of migrants seeking work in factories or on construction sites. Indeed, in some big cities the numbers of migrants exceed the permanent residents. It is estimated that around 80 million people live in shanty settlements around China's main cities, and thus there is huge pressure on urban services and serious traffic congestion.

Urban growth has also increased the pressures on land resources through demand for higher food standards, including meat products, which require more land to produce the equivalent amount of food. Urban land uses have spread onto fertile agricultural land, causing degradation through the intensification of

FIGURE 6.14 *Construction sites in China's cities attract many migrant workers*

production. Massive forest depletion (now only 8 per cent of land cover), water shortages, increasing energy consumption (75 per cent coal but an increasing use of oil, hydro and nuclear power, each with detrimental environmental consequences) and waste management are further impacts on resources arising from urbanization. Within the cities, motorized transport is beginning to replace the bicycle: vehicle numbers are expected to increase from 28 million to 100 million by the year 2015.

The result has been a serious increase in environmental problems, with global and local impacts. These include acid rain, output of carbon dioxide (China's 11 per cent of global output is second only to the USA), methane (15 per cent of global output), urban air pollution and deteriorating water supplies and quality.

Traditionally in China, cities and their hinterlands were regarded as an integrated region, but rapid urbanization has extended the area of urban impact. While the Environmental Protection Act is helping to control pollution in the bigger cities, polluting industry has moved to smaller, growing cities where controls are less effective. China now has five of the world's ten cities with the highest levels of pollution; and river systems and rural land are contaminated. Hong Kong, a foreign investor until the handover of the territory from the UK to China in 1997, has also been a major generator of pollution on the nearby mainland. Industries such as tanning and dyeing, financed by Hong Kong-based investors, have polluted areas in the Zhujiang Delta on a massive scale.

Further growth will exacerbate environmental dangers both within China and globally. Present levels of per capita consumption are low compared with those in the developed cities, but they are rising rapidly. Cities have a much higher per capita income than rural areas, which places pressures on resource consumption and creates pollution. Rural inequalities will encourage a high rate of urbanization.

Although China has embraced Agenda 21 and put in place a programme of environmental protection, along with other developing countries, it has argued that it is unreasonable to expect a relatively poor country to slow down economic development in order to reduce global environmental impacts in the short run. Despite the relatively strong central state in China, there are limits to its ability to control behaviour, particularly as the drive for growth through economic liberalization takes root.

The example of China demonstrates the scale of the environmental problem and the conflicts between society, economy and environment in a context of rapid urbanization that need to be resolved to avert ecological and social disaster. It also illustrates the way in which the sustainability of cities is influenced by shifting networks and changing interconnections between different cities and between cities and their hinterlands.

The three case studies considered in this section, though of very different issues in the different contexts of developed and developing cities, illustrate how contemporary environmental and social processes are interacting in a spatial context to produce unsustainable cities across the world. They indicate that an unsustainable environment can be produced in conditions of both poverty and wealth; it is an outcome of processes of social inequality. But the relationship between inequality and environmental sustainability is not straightforward. In China, the process of environmental degradation has continued both under a communist regime committed to egalitarianism and in the succeeding phase of economic liberalization when inequalities have been increasing. What is perhaps clear from the examples we have discussed is that inequality is at the heart of social and spatial conflicts over resources and pollution, whether it is expressed through transport, space for housing, or rural/urban relationships. It is in this sense that inequality is a social barrier to achieving the physical outcomes and profound changes of lifestyles that are necessary if progress is to be made towards sustainable cities.

Efforts to bring about change through urban governance and planning must therefore be grounded in an understanding of social and spatial processes. In the next section we show that, in practice, planning has been rather limited in this regard.

5 *Planning the sustainable city*

5.1 THE LIMITS OF PLANNING

Planning is that aspect of urban governance which is specifically directed towards influencing the spatial outcomes of urban development. As we saw earlier, one feature of the ecological modernization thesis is that the state, acting in partnership with the market, can provide a regulatory framework which encourages more sustainable methods of production and consumption. One aspect of this regulatory framework is urban planning, a process for the spatial organization of urban land uses (which we saw in Chapter 4). This involves forms of intervention in market processes ostensibly (but not always) based on participatory and democratically accountable forms of urban governance supported by administrative institutions. In practical terms, planning involves two basic elements: one is the *development plan*, which sets out the structure and form of development; the other is the regulatory mechanism of *development control*, which prevents or enables land uses in accordance with the plan.

There is a long tradition of such intervention which in the UK goes back to Howard's day. As urbanization has proceeded elsewhere, intervention through planning has been introduced to control and order development. The styles and methods of planning have varied, but certain features of the British system have wider relevance to the issues of sustainability. Accordingly, we shall use the British example for illustration.

Howard's ideas on planning, as we have seen, incorporated social vision as well as spatial constructs. Other influential figures such as Le Corbusier, Frank Lloyd Wright and Walter Burley Griffin, also saw the connection between physical form and social purpose. (Griffin, an American architect influenced by Frank Lloyd Wright, was appointed director of design and construction for Canberra, the new federal capital of Australia, in 1913. His concept, with government buildings and a civic centre set around a large lake with surrounding green suburbs, though subsequently modified, defines the character of the city today – see Figure 6.15.) Some planners argued that social behaviour could be strongly influenced by urban form (for example Clarence Perry, who introduced the neighbourhood unit concept relating patterns of social behaviour to the location of facilities; see also Chapter 1, section 2.1). But it is the spatial and urban design aspects of their work that has been most influential, in particular their concern with the ordering of urban space. From an early stage the physical development of cities, which we may call planning, became largely separated from those aspects of urban governance dealing with social conditions, which may be called urban policy (see Chapter 7). Each became narrowly defined and each has come to rely on managing the relationship between market forces and state

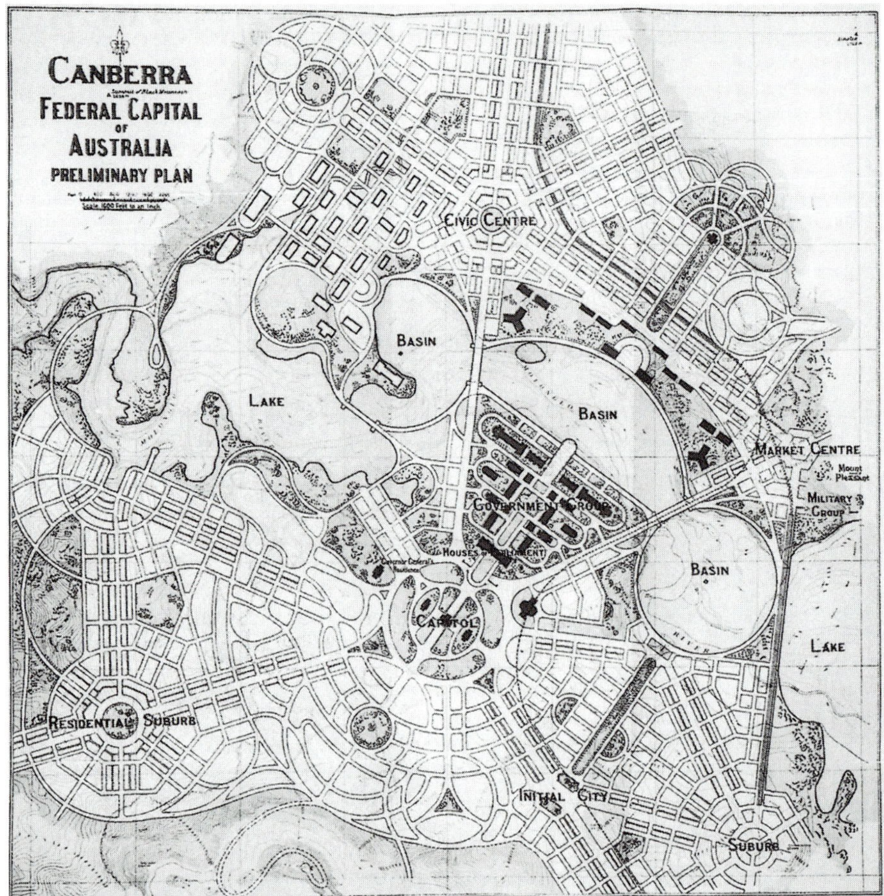

FIGURE 6.15 *Walter Burley Griffin's preliminary plan for Canberra, 1913*

intervention. As a result, planning in the UK has emerged with the following broad features:

- It focuses primarily on influencing land use in the built environment. While this can affect sustainability, as the debate on urban form has shown, other processes affecting environmental sustainability, such as waste management, air and water quality or agricultural land use, are largely beyond its remit. Although planners incorporate these issues in plans, effective control tends to be under the jurisdiction of other government agencies (for example, in England and Wales the Environment Agency is responsible for pollution control and water management).
- In so far as planning has developed any social purpose, the complexities of the relationship between social processes and the physical environment have been poorly understood, often resulting in a crude environmental determinism.
- Even within its narrow definition, planning lacks the powers to promote sustainable development. In a mixed economy power lies with

developers (public and private) and plans tend to be pragmatic and reactive. The market's short-term horizons are often in conflict with long-term sustainability.

- Planning lacks necessary resources. There have been attempts to tax the unearned increment in the value of land (often known as 'betterment') which accrues to the owner when the land use changes, for example from agricultural land to land for housing. These attempts have been rejected in favour of the operation of a free market in land, with the unearned increment going to landowners rather than the state (though owners are, of course, liable to income and wealth tax). Although the issue of betterment tax is complex, some commentators believe that such a tax would provide the planning system with the investment necessary to enable the control of development in accordance with the objectives of plans.
- Planning is a professionalized activity operating in the 'public interest'. In a context of a multiplicity of interests it is not clear whose interests are being served by whom, or what the public interest is in this context. Efforts to encourage public participation have tended to legitimate the planning process or to reveal conflicts of interest that appear to be irresolvable. The implications of a capitalist economic framework for the representation of conflicting interests in a society characterized by uneven power relations have not been articulated.
- Planning is considered to be a bureaucratic, value-free activity to which management science techniques can be applied. In practice it is a political process which tends to support powerful interests. Indeed, during the 1980s planning was explicitly linked to promoting the interests of the market.

This critique confirms that planning as it has tended to operate is an inadequate mechanism for dealing with the complexity of urban processes and spatial relationships. This complexity suggests 'that any approach to urban governance and urban planning cannot proceed on the basis of some final, formal plan, nor work with an assumption of a reachable permanent harmony or peace' (**Massey, 1999a**). The control of spatial development through urban planning has not achieved the integrated environmental and social and economic sustainability envisaged by Howard. This is not to say that planning has had little impact.

The influence of planning may be illustrated in the UK by urban containment and the protection of the countryside. Green belts, new towns and national parks are examples of planning policies which have been relatively effective. Within cities, neighbourhood planning, industrial estates and pedestrianized streets are also identifiable impacts of planning. In these and other ways it is clear that planning has left its imprint on the city.

Two points may be made here. One is that planning has made most impact when it is primarily concerned with spatial arrangements. Of course, that may be considered to be the purpose of planning. In the past, this spatial emphasis has encouraged such policies as the comprehensive redevelopment of cities, and its legacy has been displacement of part of the population and the replacement of the existing deteriorated housing by tower blocks which themselves have often degenerated into areas of social malaise.

This raises a second point – one that has been argued throughout this course – namely that space cannot be detached from its social context or consequences. Spatial organization can influence the way in which social processes interact, and it therefore plays a part in determining the sustainability of these interactions. Very often spatial planning policies have impacted on complex and interconnected urban processes without a legitimate articulation of all interests and without clear social purposes. Even where social purpose has been made explicit, unintended consequences have often resulted.

The tendency has been for planning to favour certain – often the most powerful – interests. This is hardly surprising for a process that is part of the enabling function of the state. For example, despite the rhetoric of town centre revival, planners have been powerless to resist the profusion of out-of-town retailing developments, which has led to the destabilization of many city centres and neighbourhood shopping centres and has favoured car owners at the expense of those dependent on public transport. Some commentators (for example Hall *et al*., 1971; Reade, 1987), in reviewing the overall social impact of the UK planning system, have concluded that, whatever the system's intentions, its consequence has been regressive social redistribution:

> The costs, both in terms of environmental deterioration, and in such terms as increased housing and transport costs and lower rates of capital appreciation of property values, were borne mainly by the underprivileged. The corresponding benefits, on the other hand, both in physical-environmental terms and in such terms as value for money in housing and in faster-than-average capital appreciation of house values, accrued mainly to those who were already privileged.
>
> (Reade, 1987, p.81)

The social problems of the city have tended to be tackled through 'urban policy' which has generally taken a narrowly focused, reductionist, area-based approach rather than a broader concern with interconnecting urban processes and social and economic policy as a whole. Hence, early strategies during the 1970s assumed that the 'cycle of poverty' in the cities could be broken by targeting public investment into specific areas, using such designations as 'general improvement areas' and 'housing action areas'. During the 1980s the emphasis was switched to urban regeneration promoted by the private sector. This was a period of deregulation and a reduction in local government intervention, with concepts such as 'enterprise zones' and 'urban development corporations' representing a 'shift in power away from local democratic processes, away from public involvement and an increase in decision making by central government accompanied by more freedom of action for developers' (Thornley, 1986, p.7). By the 1990s the emphasis had shifted yet again, this time to the notion of public/private partnerships competing for limited funding for regeneration. Although such policies have produced some results, large tracts of the inner cities remain socially deprived and environmentally degraded. In the absence of broad social policies directed at redistribution, fundamental inequalities are bound to persist.

5.2 PLANNING FOR SUSTAINABLE CITIES

It is relatively easy to diagnose the problems confronting cities and the limitations of urban governance and planning; far more difficult is the identification of solutions, let alone the means for achieving them. One thing, though, is clear: there is a growing recognition that sustainable forms of development for cities are not merely desirable but also necessary. Space itself does not determine the sustainability of cities, but the configuration of spatial relations does. It is not simply desirable to improve the health and quality of living for the population of cities, but sustainable development is also necessary to avoid the threat to human survival implied by the global scale of resource depletion and pollution. The quest for sustainable development has become a major theme in contemporary urban planning in the developed northern world. This is illustrated by the case of Manchester in Box 6.4.

Box 6.4 Planning for sustainable development in Manchester

Manchester has been the subject of a study in urban sustainable development published in Ravetz (1998).

Manchester was the world's first industrial city, and the dramatic changes that took place there in the 1840s have had connections with subsequent changes in cities globally. The social and environmental conditions in the city during the industrial revolution are documented in the work of well-known social commentators of the time, including Gaskell, Engels and Dickens. In spite of the great wealth produced in Manchester in the nineteenth century, the city continues to reproduce unequal spatial relations: 'it is shackled with derelict land, crumbling buildings and inefficient services. Over a million people suffer from insecurity, unemployment, deprivation and exclusion' (Ravetz, 1998, p.iv).

Greater Manchester has been calculated to contribute 1/700th of the world climate change effect. In terms of consumption of natural resources and the production of pollution, 'The city casts a huge "footprint" on the global ecology, and just to stay level, relies on continuing runaway growth in "affluence and effluence" ' (Ravetz, 1998, p.iv).

The objectives of the study for the Sustainable City-Region Working Group in 1997 focused specifically on integrated planning for long-term sustainable development of the conurbation 'to meet the needs of the present, without degrading the global and local environments and resources for future generations', 'to shift social and political structures towards greater equity and empowerment', and 'to redirect economic trends, through long term investment for eco-modernization of production and consumption' (Ravetz, 1998, p.iv). Constraints on action were seen to be the fragmentation of control, a spatial focus which is necessarily limited to the city-region, privatization of key urban services, and lack of financial resources.

Urban planning and governance are constituent parts of the spatial relations that are necessary to achieve sustainable environmental and social conditions. Sustainable development is therefore a process of which planning is a part. As we have seen, intervention through planning which does not take account of complex and interconnected urban spatial relationships, including governance, may be at best ineffective and at worst counterproductive. Planning cannot deliver sustainable development without changes in the whole realm of activities, structures and institutions that constitute and influence human existence. However, from the arguments in this chapter it is possible to delineate some of the components of change necessary to address the question, 'can cities be sustainable?'.

ACTIVITY 6.6 Before reading on, make a list of the key changes you would consider necessary for the achievement of more sustainable cities. Then compare your list to ours below. ◆

We have identified the following:

- At the physical level, it will be necessary to *change production and consumption* in the cities to reduce pollution and minimize resource depletion. This may be achieved by conserving energy, waste reduction/ minimization and recycling and pollution control. It may mean the abandonment of high-risk technologies. For example, the risks of radioactivity arising from nuclear accidents and proliferation of radioactive waste cannot be eliminated and will persist down the generations. Planning has a role in promoting such changes. More especially, planning can seek to bring about changes in urban form to achieve more sustainable relationships between land uses and activities.
- Sustainable production and consumption and accompanying urban forms must be matched by *changes in behaviour.* Sustainable lifestyles include less use of the car; changes in diet to ensure the better use of land resources; the re-use and recycling of products; and so on.
- A sustainable city must be able to cope with at times incommensurable *conflicting interests* (see Chapter 5). For example, urban social movements manifest a variety of concerns surrounding poverty, human rights and ecology which have been routinely excluded or marginalized by prevailing power structures. Yet they mobilize support for particular issues, and thus demonstrate 'recurrent patterns of collective activities which are partially institutionalized, value oriented and anti-systemic in their form and symbolism' (Pakulski *et al.*, 1991, p.xiv). The relationship between environmental movements and planners has been especially important in ensuring that ecological issues have gained a prominent place on the urban agenda.
- Related to this is the need to *broaden participation.* In most countries, and the UK is a good example, the power structure is heavily influenced by business. Business has increasingly occupied areas once the preserve of

accountable local government. Decision-making in education, the health service, urban regeneration and other spheres is increasingly carried out by people drawn from the business community.

- Consequently, *institutions need to be reshaped* to reflect the variety of interests, aspirations and needs of the wider urban community. This proposition suggests a reorientation from centralized and élitist institutions towards grassroots, participative structures of decision-making. For example, ideas such as action planning, involving all those with an interest in particular planning decisions, could become a practical rather than a theoretical exercise. Local Agenda 21 initiatives, with their emphasis on involving communities, have indicated the possibilities but so far have had little palpable influence on outcomes.
- The *process of planning itself needs to be transformed* to take account of the interrelationship between the physical and social components of change. It will need to take account of the precautionary principle that action must be taken to avoid costly damage later or the possibility of irreversible effects should action be delayed. The integrated nature of environmental processes and the consequences of action need to be clearly understood. Similarly, planning policies at various levels (local, national, international) need to be integrated to ensure that proposals are compatible and effective. A strategic approach is also necessary, by which we mean that planning needs to take a long-term view and to incorporate all those processes – land use, pollution, waste management, resource conservation – that impact on sustainability.
- Finally, planning is much more than a physical process; it is a social and political process too. The roots of urban planning lie in reform, and its practice and purposes are shaped by the social and political context within which it operates at any one time and place. Throughout this chapter we have stressed the constraints on change imposed by contemporary patterns of social inequality. Unless these are tackled, sustainable development in both its physical and social senses will prove impossible. This conclusion indicates the need for *a transformation which goes far beyond the limits of planning* as presently conceived. It indicates that planning should not simply be a form of intervention in the market or a set of regulations and controls concerned with the physical ordering of urban space. Planning in its broadest sense is a process that relates all the processes of social change to the development of physical structures (cities) capable of ensuring the continuing survival and enhancement of lifestyles for the whole population.

The prospects for such fundamental social transformation, however desirable or necessary, appear hopelessly remote in the UK, with its deeply implanted power structures, increasing social inequality and commitment to sustainable economic growth. Prevailing neo-liberal democratic philosophy in countries of the North suggests that in a pluralist society it is possible through local debate by all the stakeholders in society to define a homogeneous public good. This approach does not take account of the range of diverse and conflicting interests in contemporary urban society. The prospects for change are even more unlikely in countries of the South. In many of these countries there are authoritarian regimes where centralized power bases are coupled with weak local government and the restricted influence of environmental movements. Lack of availability of resources is a problem, as is extreme polarization between rich and poor. Raising resources to implement urban policy is problematic at both national and local levels. In this context, we may ask: is local co-operation and participation in new forms of governance possible, and is it likely to be effective in the absence of more radical changes in patterns of production, distribution and consumption? Do cities of the North or South have a social, economic and political framework that would allow such changes to occur?

ACTIVITY 6.7 Thinking back to the issues of democracy raised in section 3.3 above, in what ways would you see social and spatial relations within and between cities across the world as a constraint on their sustainability? ◆

This takes us on to an even more challenging terrain. As **Massey (1999b)** notes, it is not possible to draw a line around a city's social and spatial relations. As one city struggles to develop connections into networks of power, other cities anywhere in the world may be affected positively or negatively. At the global level capitalist social relations limit the ability and will of both local and national governments to intervene in the interests of social and environmental sustainability. While the global environmental discourse appears to have opened up space for positive action at an international level, in reality fundamental differences of interest undermine the chances of both local and global co-operation which is necessary for the achievement of sustainable development. Given the scale of present ecological and social problems, any move towards the sustainability of cities would seem to require far more radical and comprehensive forms of urban governance, including planning, stretching beyond the city and beyond the nation state itself.

6 Conclusion: what is the sustainable city?

At the outset we posed the question: can cities be sustainable? Let's reflect on the problems and prospects we have identified:

- Environmental and social problems generated by urban development, though changing over time, may be perceived as dangers, threatening the stability and even the survival of cities.
- Attention has focused primarily on the spatial and physical elements of the problem. Environmental impacts are seen to be global (beyond the city) as well as local (within the city) and to be a threat to sustainability.
- Urban sustainability has a social dimension, too. Typically the social and environmental costs of economic growth impact on the poor in cities of the South and North. Concern with poverty, inequality and development has emanated from developing countries as well as from radical urban movements in the North, including the environmental justice movement.
- The different emphases – physical/spatial and social/economic – are a source of conflict over solutions to the problems. They are integrated in the concept of sustainable development.
- Approaches to sustainable development have, nevertheless, tended to concentrate on its spatial/ecological aspects. They have generally adopted a narrow ecological modernization perspective which does not challenge existing social, economic and political frameworks.
- The interrelationships between environment, society, economy and governance create conditions in cities which are, at present, unsustainable. Urban planning can have a key influence on issues of sustainability, but so far, in spite of its traditional concern with social issues, planning has proved both limited in its focus and weak in its influence.
- Consequently, while the scale of the problem is recognized, the components of the sustainable city are only partially understood. Governance and planning interact with and shape urban spatial processes in complex ways, but there has been little understanding of the social and institutional changes necessary for urban development to shift to a more sustainable path.
- Changes at the technological level can only be secured in conjunction with changes in lifestyles and probably through different political arrangements. Greater social equality would seem to be a precondition for the social cohesion necessary to secure a transformation towards sustainability.

Taken in the context of contemporary society this last point seems absurdly utopian. The overwhelming tendency in all countries appears to be a commitment to economic growth. In so far as ecological constraints influence policy, it is in terms of ecological modernization, which presumes that environmental protection and economic growth, without any fundamental changes, are compatible objectives. Yet economic growth has done little to solve the problems of inequality or to achieve sustainability, even in cities in the rich countries of the North.

Set against the predictions of some environmentalists such as 'deep ecologists' or the glimpse of the abyss provided by Ulrich Beck and others, contemporary attitudes seem hopelessly optimistic. The 'sustainable city' can only be constructed if the processes of urban planning and governance are able to respond to the wide-ranging and competing interests that constitute the (unsustainable) spatial relations of cities. Such a response would necessarily have to confront the issue of inequality head on.

As we have seen, the interconnections and juxtapositions of poverty and wealth within cities stretch out beyond cities. What can be interpreted as dominant global forces are therefore in fact constructed within and between cities and could be changed. As presently constructed, these social and spatial relations are uneven and unsustainable, but is it possible to reconstruct positive relationships with the potential to shape the intensity of interconnecting urban processes in ways that are socially and environmentally sustainable?

There is a danger that a simplistic technological and physical determinism will continue to deflect attention from more complex and unwelcome understandings of the unsustainability of cities. Debates about how to reconstruct the spatial and social elements of cities within an institutional framework that emphasizes social equality and political participation remain as relevant in the conditions of contemporary cities as they did a century ago. It is to these debates that we turn next.

References

Allen, J. (1999) 'Cities of power and influence: settled formations', in Allen, J. *et al*. (eds).

Allen, J., Massey, D. and Pryke, M. (eds) (1999) *Unsettling Cities*, London, Routledge/The Open University (Book 2 in this series).

Beck, U. (1992) *Risk Society: Towards a New Modernity,* London, Sage.

Beck, U. (1995) *Ecological Politics in an Age of Risk,* Cambridge, Polity Press.

Blowers, A. (ed.) (1993) *Planning for a Sustainable Environment*, London, Earthscan.

Brandt, W. (1980) *North–South: A Programme for Survival*, London, Independent Commission on International Development Issues/Pan Books.

Brandt Commission (1983) *Common Crisis: North–South Co-operation for World Recovery*, London, Pan Books.

Breheny, M. and Hall, P. (1996) *The People – Where Will They Go? National Report of the TCPA Regional Inquiry into Housing Need and Provision in England,* London, Town and Country Planning Association.

Briggs, A. (1968) *Victorian Cities*, Harmondsworth, Penguin.

Clark, D. (1996) *Urban World/Global City,* London, Routledge.

Commission of the European Communities (1990) *Green Paper on the Urban Environment*, EUR 12902 EN, Brussels, CEC.

de Swaan, A. (1988) *In Care of the State,* Cambridge, Polity Press.

Fishman, R. (1977/1996) 'Urban utopias: Ebenezer Howard and Le Corbusier', reprinted in Campbell, S. and Fainstein, S. (eds) (1996) *Readings in Planning Theory*, Oxford, Blackwell.

Gilbert, A. (ed.) (1996) *The Mega-City in Latin America,* New York, United Nations University Press.

Hall, P., Thomas, R., Gracy, H. and Drewett, R. (1971) *The Containment of Urban England*, London, Allen & Unwin.

Hamilton, K. and Hoyle, S. (1999) 'Moving cities: transport connections', in Allen, J. *et al*. (eds).

Hannigan, J. (1995) *Environmental Sociology: A Social Constructionist Perspective*, London, Routledge.

Harvey, D. (1996) 'The environment of justice', in Merrifield, A. and Swyngedouw, E. (eds) *The Urbanization of Injustice*, London, Lawrence & Wishart.

Hinchliffe, S. (1999) 'Cities and natures: intimate strangers', in Allen, J. *et al*. (eds).

HMSO (1994) *Sustainable Development: the UK Strategy*, CM2426, London, HMSO.

Howard, E. (1902/1965) *Garden Cities of Tomorrow*, edited by F.J. Osborn with a preface by Lewis Mumford, London, Faber & Faber.

Jopling, J. and Girardet, H. (1996) *Creating a Sustainable London*, London, Sustainable London Trust.

Lee, Yok-shiu F. (1995) *Intermediary Institutions, Community Organisations, and Urban Environmental Management: The Case of Three Bangkok Slums*, working paper 41, UNU/IAS Working Paper Series.

Lo, Fu-chen and Yeung, Yue-man (1996) *Emerging World Cities in Pacific Asia,* New York, United Nations University Press.

Massey, D. (1999a) 'Cities in the world', in Massey, D. *et al.* (eds).

Massey, D. (1999b) 'On space and the city', in Massey, D. *et al.* (eds).

Massey, D., Allen, J. and Pile, S. (eds) *City Worlds*, London, Routledge/The Open University (Book 1 in this series).

Mumford, L. (1966) *The City in History,* Harmondsworth, Penguin.

OECD (1996) *Innovative Policies for Sustainable Urban Development – The Ecological City,* Paris, OECD.

Owens, S. (1992) 'Energy, environmental sustainability and land-use planning', in Breheny, M.J. (ed.) *Sustainable Development and Urban Form,* London, Pion Ltd.

Pakulski, J., Thomas, R., Gracey, H. and Drewett, R. (1991) *Social Movements: The Politics of Moral Protest,* Harlow, Longman.

Pile, S. (1999) 'What is a city?', in Massey, D. *et al.* (eds).

Pryke, M. (1999) 'City rhythms: neo-liberalism and the developing world', in Allen, J. *et al.* (eds).

Ravetz, J. (1998) *City-Region 2020 – Integrated Planning for Long-Term Development in a Northern Conurbation,* consultation draft, January 1998, London, Town and Country Planning Association.

Reade, E.J. (1987) *British Town and Country Planning,* Buckingham, Open University Press.

Reade, E.J. (1997) *Planning in the Future or Planning of the Future?,* in Blowers, A. and Evans, B. (eds) *Town Planning into the 21st Century,* London, Routledge.

Rees, W. (1996) 'Revisiting carrying capacity: area-based indicators of sustainability', *Population and Environment: A Journal of Interdisciplinary Studies,* vol.17, no.2, January.

Rees, W. (1997) 'Is "sustainable city" an oxymoron?', *Local Environment,* vol.2, no.3, 11 October, pp.303-10.

Royal Commission on Environmental Pollution (1994) *Report on Transport and the Environment,* London, HMSO.

Stanley, L. (1996) 'A role model for the western world', *Planning Week,* 28 November.

Thornley, A. (1986) 'Thatcherism and simplified planning zones', *Planning Practice and Research,* vol.1, no.1.

Thrift, N. (1999) 'Cities and economic change: global governance?', in Allen, J. *et al.* (eds).

UK Round Table on Sustainable Development (1997) *Housing and Urban Capacity,* London.

World Commission on Environment and Development (1987) *Our Common Future* (The Brundtland Report), Oxford, Oxford University Press.

CHAPTER 7
Administered cities

by Allan Cochrane

1 *Introduction: managing unruly places*

As **Pile (1999)** argues, cities are inherently paradoxical places. The experience of urban life is characterized by a series of individual and collective negotiations over apparently inconsistent practices which nevertheless manage to survive alongside each other. Cities are constituted through movement, fluidity and change, yet also by settlement and the search for security and stability (**Allen, 1999a**). They encourage mixing, forcing people together, yet they are also characterized by spatial differentiation and segregation (as Jenny Robinson showed in Chapter 4). They are defined by the intensification of social relations, yet, as Steve Pile notes in Chapter 1, the experience of those living in some parts of some cities is the opposite. Cities are shaped by their internal interconnections and disconnections, but they are also the products of complex and overlapping networks of interconnection to other cities and other places.

Massey (1999) captures these tensions through the notion of 'open intensity'. She suggests that cities are places that are defined through the concentration of particularly dense networks of cultural and economic interaction. The very density of these networks, as well as the density of settlement, helps to generate the intensity of social relations that is associated with living in cities. There is no simple or unitary order to be found in cities, because they are also defined through the multiple stories of those who live in them and the rhythms associated with them. However, it is possible, as Massey points out, to identify distinct space-times within the generalized intensity of urban life, which is itself the outcome of a continuous dance of space-time configurations, even if all the partners in the dance are not equal. Harvey (1997, pp.22–3) similarly argues that 'our cities are constituted not by one but by multiple spatio-temporalities, producing multiple frameworks within which conflictual social processes are worked out'.

Cities are inherently unruly and heterogeneous. At the same time as providing powerful images of utopia – or at least a better life – cities have also become carriers of dystopian visions. As Chapter 6 indicated, there are genuine fears about the sustainability of cities in social and environmental terms, yet at the same time cities are also associated with possibilities and hope for the future. They are the places where interaction between cultures encourages the development of new ideas, new developments and (as Sophie Watson noted in Chapter 5) new social movements. Cities offer people opportunities for social and economic advancement, and they promise escape from the relatively fixed social hierarchies of small towns and rural settlements. Yet they are also associated with images of poverty and decay, of dramatic differences between rich and poor, of urban disorder (see Chapter 2) and structures of surveillance and policing to manage disorder (see Chapter 3).

The attempt to capture and to manage all this complexity, diversity and unmanageable change in ways that will benefit the people who live in cities is the fundamental challenge of urban governance, or what might be called 'public administration'. This stretches from the operations of the state, through those of a range of institutions including churches, voluntary and social organizations, to the shared practices and understandings of urban life.

This chapter focuses on some of these forms of urban governance and some of the key debates around them. At the heart of these debates is a tension between those who seek to bring stable order to the apparent chaos of urban existence and those who stress that, because cities are defined by uncertainty, fluidity and change, a more differentiated approach is required. These debates and the different shapes taken by urban governance and administration reflect conflicting understandings about what makes a 'good' city, as well as alternative visions of how to build urban utopias. This chapter therefore:

- looks at those different understandings and the institutional approaches that flow from them, and
- explores the ways in which different spatial imaginations are reflected in and shaped by particular forms of urban governance.

Section 2 of the chapter focuses on approaches that seek to administer cities through forms of urban design, while section 3 develops this by looking at the grand plans of the urban utopians. Section 4 turns to forms of governance that emphasize the making of alliances between major interests as the main driving force of urban politics. Section 5 explores alternative approaches which seek to build on the complexity of overlapping relationships that exists in cities as a potential basis for diverse and democratic forms of administration.

2 *Administration by design*

Urbanization has been accompanied by a continuing series of attempts to capture the details and complexity of social relations through the practices of counting, mapping and calculating. Cities are epitomized by the density of official representations of the activities that take place across urban space – in the records of health and social services, the records of land ownership, the attempts to chart economic flows, the delineation of social problems, the identification and specification of land uses, and the identification of sources of tax income. In a sense all this data collection represents an attempt to capture or freeze processes of urban change, even if the realities it uncovers are almost immediately overtaken by the ways in which those processes work out in practice. Everything that can be pinned down is given a meaning through its place within the surveys, reports and account books of government, creating citizens – the governed – and defining the urban as manageable space.

Historically, the most powerful expression of this approach has been found in the professions of *urban design,* such as architecture and urban planning (although more recently, as Andy Blowers and Kathy Pain showed in Chapter 6, there has been a significant shift in emphasis towards more flexible models). In its strongest form, urban planning is a powerful expression of the attempt to control through recording, quantifying and mapping. The language of *rational planning* presents it as a more or less neutral set of techniques that are able to highlight the choices available to societies and governing agencies, but in practice its operation has sometimes simply offered the means of bureaucratically normalizing or reinforcing existing relations of power through the differentiation of urban space. In many colonial cities, for example, the drawing of lines on maps was used to institutionalize divisions between races, to create protected spaces for the European rulers. In 1955, for example, the East African Royal Commission reported that:

> Centres had to be established where they [the Europeans] could live free from the dangers of tropical disease and from which the surrounding countryside could be administered ... The Town was not a suitable habitat for a permanent African society ... The towns have, therefore, been regarded rather as bases for administration and commercial activities than as centres of civilizing influence, still less of permanent African population.
>
> (quoted in Doherty, 1980, p.13)

In Chapter 4, Jenny Robinson highlighted the ways in which the methods of urban planning, drawing on international 'best practice', were used to underpin the practices of racialized spatial differentiation in apartheid South

Africa. But she also noted the ways in which some of the methods adopted in South Africa themselves fed back into approaches to urban design in countries like the UK and the USA, which helped to reinforce the workings of spatial differentiation and social segregation in those countries too.

2.1 URBAN DESIGN AND POLITICAL POWER

Urban design can also operate more subtly to incorporate and reflect other relations of power, producing a landscape which, in principle at least, helps to locate residents within spatial, social and political hierarchies. In other words, it can be used to help construct particular urban orders. In the nineteenth and early twentieth centuries colonial networks ensured that government buildings in Pretoria and Belfast were almost indistinguishable. As Hall (1996, pp.183–4) notes, 'the British India Office and Colonial Office found themselves employing consultants to create instant capital cities in far corners of the globe'. These new cities were invariably planned in ways which left them located alongside, but separate from, existing developments. For example, New Delhi (designed by Lutyens and Baker) was commissioned in 1912 and completed in 1931. It was built to represent the power of the British Raj in the process of transition towards a post-imperial era, combining moves towards home rule with the continuation of a colonial economy (Frampton, 1985, p.211). In developing an acceptable Anglo-Indian style, it moderates clear statements of classical symmetry with an 'Indian style of architecture', at least in symbolic terms (Hall, 1996, p.188). The

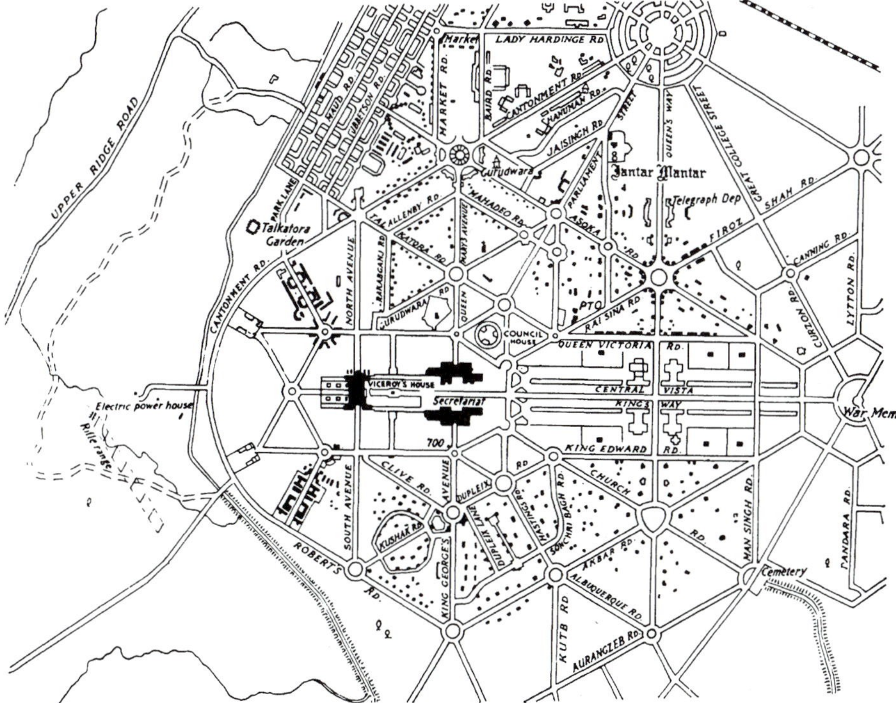

FIGURE 7.1 *The Lutyens-Baker plan of New Delhi*

FIGURE 7.2
New Delhi's Anglo-Indian architecture

offices and blocks of housing are arranged in hexagonal grids around the Government House to reflect the different social (and racial) status of their residents. In a sense, this form of urban design was intended to express and freeze social relations in built form, or at least to help shape directions of social change (in this case preparing for a post-colonial as well as a colonial world).

The Nazi attempt to redesign Berlin in the 1930s and 1940s provides a particularly powerful example of some of the ways in which architecture and urban design may be used as part of a strategy to impose political power on an urban population, as well as to provide clear messages to national and international audiences. Berlin represented a particular problem for the Nazi regime. The aim was to impose a Nazi identity on a (potentially) recalcitrant Berlin, to show that the cosmopolitanism and radicalism with which the city was popularly associated had been defeated. The Jewish population was, of course, a prime target of attention. However, Berlin was also a traditional base of the Communist and Social Democratic parties, and in order to transform Berlin into the capital of, first, a centralized Nazi Germany and then of a Nazi Europe, it was considered necessary to make extensive changes to the fabric of the city. The architecture of power expressed in the Nazis' buildings (such as the Olympiastadion built for the 1936 Olympics) and their use of boulevards such as Unter den Linden for massive parades helped to stamp the image of their rule on the physical geography of the city. Hitler and Albert Speer (his architectural adviser) had still more grandiose plans for the reconstruction of Berlin under the

FIGURE 7.3 *Albert Speer's model of New Berlin*

name of 'Germania' – 'the code word for a gigantic operation in destruction' (Sombart, 1987, p.20) – to confirm Hitler's rule and the power of his will in reshaping Germany. Two massive new roads (north-south and east-west) were to meet in the centre, which was to be covered by a huge dome. Hall (1996, p.198) describes the aims of Speer and Hitler as being to make Berlin 'a totally mechanized, totally anti-human city of parade and spectacle'. The scars of Hitler's plans, in part preserved by the later siting of the Berlin Wall, which cut across the centre of the city, are still to be seen in the extensive dereliction near the Reichstag.

The experience of the Soviet Union and of Eastern European cities after 1945 also reflected a strong belief in urban design as a means of expressing (and imposing) political power. The Stalinist vision implied that human will could overcome physical obstacles, under the leadership of state planners and the Communist Party. In the immediate post-war years of reconstruction, a series of new towns was built in the countries of Eastern Europe, and the imagery of urban reconstruction dominated political propaganda. In his film *Man of Marble*, for example, Andrej Wajda powerfully draws on the experience of Nowa Huta (near Cracow) in Poland, which was built as a new town to house the workers of a massive steel mill (Huta im Lenina) that was constructed as part of a strategy to transform the Polish economy and to create a strong urban working class, loyal to communism. The film highlights the way in which Nowa Huta was presented as the product of a new society, whose design reflected those priorities, and focuses on the experience of one of the building workers who is identified as a hero of labour for his contribution to the construction process.

FIGURE 7.4
Karl-Marx-Allee in 1997

In East Germany (the German Democratic Republic) the new town of Stalinstadt (now Eisenhüttenstadt) was another of the new towns built around new steel mills at the same time and with similar ambitions (Beier, 1997). Even the city centres were expected to play new roles. In East Berlin, for example, Karl-Marx-Allee (originally Stalinallee) was constructed as a wide boulevard stretching out from Alexanderplatz, the main square in the centre, towards the Polish border, as a powerful statement both of the 'success' of communism and of a significantly different approach to urban development in which the private market played no part. Despite its central position, Alexanderplatz incorporates housing for thousands of people and a large amount of open space, in line with the GDR's planning emphasis on symbolic public space 'with broad streets and plazas and striking monuments' (Strom, 1996, p.6). The mix of housing and other uses of land which in western cities would be dominated by commercial and retail uses appeared to highlight the dominant role of state planning in development. The symbolic space is that of bureaucratic rather than corporate power.

The experience of Soviet-style urban design also highlights some of the inherent weaknesses of attempts to centralize planning. One of the downsides of the Soviet model of urban planning was that it was driven from the top, with little consideration being given to those who lived in either the new or the old housing. The attempt to impose a unified and homogeneous model fitted uneasily with the differentiation that continued to exist and was being reproduced in those societies. The failures of economic planning also had their impact on the urban forms taken by 'actually existing' socialism. The imagery of equalization masked dramatic differences, as the older parts of cities were left to decay and the new housing developments were left incomplete and unfinished. The supply of housing was never able to grow fast enough to meet the demands of urban populations, even in the GDR where the overall population actually fell in the post-war period. Priority was given to investment in heavy industry.

The failure of the post-war experiment in Soviet-style communism has come to symbolize a wider perceived failure of planning as an approach both to urban development and running the economy.

FIGURE 7.5 *Youth workers take a break from building the future in the German Democratic Republic*

FIGURE 7.6 *Eisenhüttenstadt, formerly Stalinstadt, a post-war new town*

2.2 REINTERPRETING URBAN FORMS

One way of interpreting the workings of urban design is to see them as expressing the understandings of the social relations which predominate (or which the designers think should predominate) in particular cities.

Urban design seeks to translate these relationships into built form. In other words, it can be seen as helping to produce a sort of urban language, story or narrative within which city residents are then expected to operate. But it is important to recognize that this does not render those residents completely powerless – on the contrary, they are active participants in interpreting, re-interpreting, redefining and retranslating that language. It is through their actions that the forms of urban design are given their meanings in practice. As **Allen (1999b)** and **Massey (1999)** emphasize, the same architecture and the same urban space may be re-used and re-interpreted in different ways by those who experience and use it.

Take, for example, Chandigarh, a new city built as a capital for the government of Punjab between 1951 and 1965. The final city plan was prepared by Le Corbusier, a Swiss-born architect (Le Corbusier was an adopted name) who helped to create what we have come to understand as 'modern architecture'. Chandigarh was, ironically, the only more or less fully realized expression of Le Corbusier's vision of urban development (even if it was no longer the vision of endlessly repeated geometrical patterns of high-rise blocks which characterized the plans of the 1920s and 1930s for which Le Corbusier has become famous). The city was to be a symbol of the new India as a modern industrial state, but like the Lutyens-designed New Delhi, it translated visions of a western future into the Indian present. The master plan prepared by Le Corbusier and his colleagues built on (and largely accepted the lay-out of) an earlier decentralized grid-based plan developed by a US planner. Le Corbusier's own main architectural input was to the cluster of monumental administrative and government buildings at the centre. Chandigarh was 'a city designed for automobiles' in a country where many lacked a bicycle (Frampton, 1985, p.230). It was constructed around clear divisions of function between its different parts (residential, institutional, commercial, and open space).

However carefully the buildings and spaces were colour-coded for different functions, in practice they now have different ones – the spaces have been taken over by unplanned uses. Instead of open space, there is crowded space. The carefully constructed hierarchies of the plan have been questioned by those living with the pattern it creates. As Hall writes,

> By the 1970s, 15 per cent of the population were living in squatter or semi-squatter settlements; more than half the traders were operating informally from barrows or stalls. Since they conflicted with the Master Plan's concept of urban order, the authorities made repeated attempts to harass and break them up.
>
> (Hall, 1996, p.215)

FIGURE 7.7 *Chandigarh in the 1990s: an expression of Le Corbusier's vision of urban development*

These attempts were met with imaginative forms of resistance, which Hall describes as 'a series of public events worthy of an Indian version of an old Ealing comedy'. So, for example, the opening of 'a new illegal market' was incorporated into 'a series of sacred Sikh religious events', and on another occasion 'traders stage-managed elaborate funeral ceremonies for the Prime Minister who had just died' (Hall, 1996, p.215).

To recapitulate the discussion in this section:

- Urban design has played a central part in attempts that have been made to shape, simplify and manage the complexity and intensity of urban living. This has found a powerful expression in the maps of urban planners as well as the built form promised by the architects.
- It sometimes seems as if the ways we live our lives in cities are determined by the ways in which they are designed, but in practice it is the ways in which people imagine and use the built environment that gives those designs their lived meaning.
- Debates about urban design have also incorporated different visions of what might constitute the 'good' city and how it might be built. This has found its strongest expression in the attempts to construct urban utopias, each of which has also carried with it a negative vision of the urban dystopias from which the utopians are trying to escape.

It is on the ideas of two of these utopians that we shall focus in the next section.

3 *Building urban utopias*

There has been a strong popular, as well as architectural and professional, tradition that links utopian visions of the future with attempts to plan urban development on a grand scale (see, for example, the review of attempts to imagine the city of the future in the USA throughout the twentieth century in Corn and Horrigan, 1995). Approaches which seek to remake cities in the image of planned utopias (however modest) have been subjected to increasingly harsh criticism in recent years. The most influential examples of such approaches, on which attention has frequently been focused, are those associated with Ebenezer Howard (see Chapter 6) and Le Corbusier.

ACTIVITY 7.1 Through the creation of new urban environments, both Howard and Le Corbusier, in their different ways, promise a better future to existing urban populations and to those who wish to migrate to the new environments. Some aspects of Howard's model have already been discussed in section 2.2 of Chapter 6, and you should now remind yourself of his arguments. Then turn to Reading 7A, which comes from the writings of Le Corbusier. Try to identify the main differences and similarities in approach between Howard and Le Corbusier. It may seem easier to identify differences, but try to think about possible similarities, too. In particular, think about the visions of dystopia to which they are responding, as well as the utopias they wish to build.

How do you think they would define a 'good' city? ◆

In some respects the approaches of these two writers look dramatically different. Howard aims for low density, while Le Corbusier celebrates high density. Howard's version is of networks of relatively small garden cities linked into a social city; Le Corbusier starts from the vision of the central city and aims for a population of three million. Howard is looking for ways of bypassing the intensity of urban life; Le Corbusier is excited by it, but believes it can be managed. Although both emphasize the value of green space in the cities, Le Corbusier's approach creates that space by constructing high-rise buildings above the earth, alongside elevated roadways, while Howard incorporates green areas within his networks of settlements and in wedges which stretch into the heart of his cities. Commentators on these ideas have also been divided. Sennett (1990, p.171) notes that Le Corbusier dismisses Howard's vision for producing 'worthy, healthy, organic, and boring quasi-suburbs, dedicated to the proposition that coziness is life'. Hall (1996) dismisses Le Corbusier as 'an authoritarian centralist', arguing that Howard's vision, by contrast, was rooted in anarchist urban utopias of self-help in constructing the 'good city' (see also Ward, 1993). Howard's garden cities, stresses Hall, 'were merely the vehicles for a progressive reconstruction of capitalist society into an infinity of co-operative commonwealths' (p.87).

Others, however, have pointed to similarities, suggesting that both dismiss the messiness and inherent conflict of urban life and seek to replace them with sanitized alternatives. Both favoured starting with cleared or greenfield sites: Howard by using land outside existing cities, and Le Corbusier by levelling ground to create empty space on which to build. According to Jacobs (1961), the rather different visions of Ebenezer Howard and Le Corbusier both remove the possibility of surprise and serve to undermine the possibility of organic development within cities, limiting the scope for social innovation. Howard, she argues, 'wanted to freeze power, people and the uses and increments of money into an easily manageable and static pattern ... a pattern that was already obsolete' (Jacobs, 1961, p.303). He imagined 'the restoration of a static society, ruled – in everything that mattered – by a new aristocracy of planning experts' (p.304.). She criticizes the view – held both by Howard and Le Corbusier, she suggests – that 'sees only disorder in the life of city streets and itches to erase it, standardize it, suburbanize it' (p.460).

As Harvey points out, they appeared to share a belief that

> the proper design of *things* would solve all the problems in the social process. It was assumed that if you could just build your urban village, like Ebenezer Howard, or your Radiant City, like Le Corbusier, then the thing would have the power to keep the process forever in harmonious state.
>
> (Harvey, 1997, p.24)

Both Howard and Le Corbusier could be accused of seeking to solve (and indeed to define) what they saw as the social problems of the city in physical terms (although Hall, 1996, seeks to defend Howard against this charge). The answer for both of them was to change the layout of cities, to provide clearer and more logical forms of social segregation.

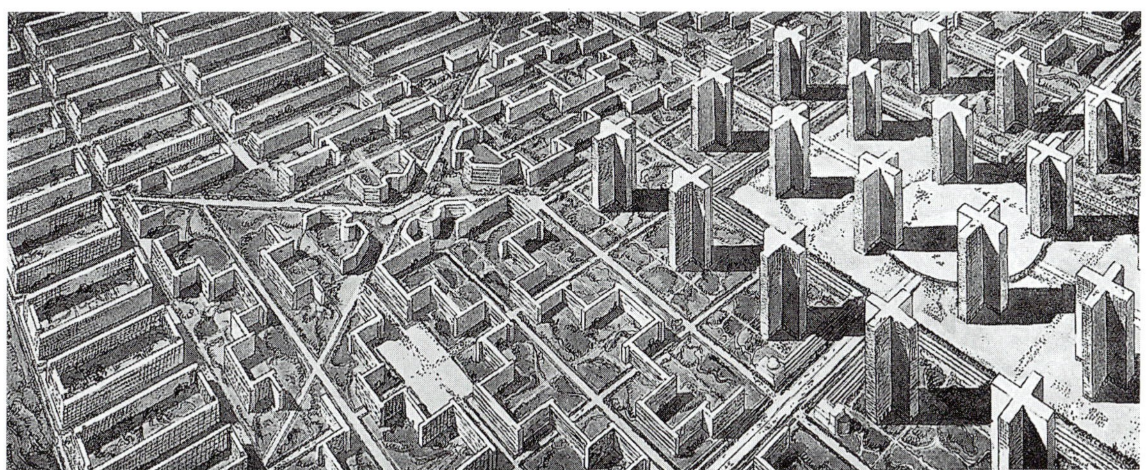

FIGURE 7.8 *Le Corbusier's vision of the contemporary city*

3.1 FROM UTOPIA TO PRACTICE

Although neither Howard nor Le Corbusier saw their own plans implemented on any scale, their approaches helped to influence the direction of state intervention in many urban areas by suggesting that the social problems of the slums might be overcome through forms of physical planning and spatial sorting. If there is any universal 'villain' identified by those writing on the management of urban space and urban development in recent years, it is Le Corbusier. That may seem strange for someone who saw only relatively few examples of his work – and none of his grand schemes – achieve built form. Neither the 'contemporary city' nor the 'radiant city' was ever built. Chandigarh turned out to be both more modest and based on rather different design principles. Yet Le Corbusier's influence on a generation of planners and architects is taken for granted. His stress on 'killing' the street (by separating roads from pedestrians) and the clear hierarchies of function and style have been blamed for the more mundane and less ambitious plans for public housing and high-rise development that characterized much urban planning in the 1950s and 1960s.

Sennett (1990, p.171) argues that Le Corbusier's approach (reflected in his plan for Paris, the *Plan Voisin*) 'was the prototype for the grade-flat-and-build urban development that has defaced cities all over the world', while Corn and Horrigan (1995, p.52) note that the comprehensive plans developed for most of the USA's major cities after the Second World War 'all had the crystalline ring of Le Corbusier about them'. Jacobs confirms that

> Le Corbusier's dream city has had an immense impact upon our cities. It was hailed deliriously by architects, and has gradually been embodied in scores of projects ranging from low-income public housing to office building projects ... The vision and its bold symbolism have been all but irresistible to planners, housers, designers, and to developers, lenders and mayors too.
>
> (Jacobs, 1961, p.33)

Nor does Howard escape from these critics. Sennett (1990, p.201) complains that 'Faced with the fact of social hostility in the city, the planner's impulse in the real world is to seal off conflicting or dissonant sides, to build internal walls rather than permeable borders'. He links this impulse back to the garden city movement, whose techniques, he argues, 'are now increasingly used in the city centre to remove the threat of classes or races touching, to create a city of secure inner walls'. In other words, the argument is that such models leave no room for alternative spatial orderings, since they seek to set existing hierarchies in steel, concrete, glass and otherwise unexceptional 'buffer zones' (see Chapter 4, section 2). Instead of celebrating and building on networks of connection within cities, these models seek to maintain existing disconnections and impose new ones through forms of social as well as functional segregation: that is, separating housing from employment and green space from traffic (see Chapter 2, section 4.1 on functional segregation in Istanbul).

It is worth emphasizing, however, that the translation of the dreams of Howard and Le Corbusier into the practices of urban administration has involved a systematic degrading of some aspects of their ideas at the expense of others. Some possibilities have been deemed to be more practical than others. So, for example, in the case of US cities in the 1940s and 1950s,

> Such plans were window dressing for the more mundane instrumentalities of the city planning process. Municipal planning departments grew into enormous cumbersome bureaucracies, fed by ever increasing aid from the new US Department of Housing and Urban Development. The sensational, dramatic gestures – the clusters of shining towers, the 'people movers', the down-town pedestrianized malls – received a great deal of publicity. But they remained, for the most part, confident façades for the perpetuation of a status quo of private development and profit making in American cities. The towers went up, but on the existing city grid, providing little in the way of urban amenity beyond a number of chilly wind-swept plazas – certainly no people movers or monorails. The knottier problems of housing the poor and providing efficient and safe public transit were left to the various branches of government which delivered those services in a perfunctory and mechanical fashion, and – especially in housing – with little success. Planners embraced the modernist vision of the future, but relied on the *image* itself to carry it out, slighting the necessary underpinnings of real social change.
>
> (Corn and Horrigan, 1995, pp.52–3)

In other words, the grand visions of the utopian urbanists were translated into urban practice in a series of grey, day-to-day piecemeal decisions, driven by the priorities of developers and exigencies of budgetary constraint.

ACTIVITY 7.2 Think about the area in which you live, or a city with which you are familiar.

- Are there any aspects of the built environment – housing, roads, parks, offices – that you think have been influenced by the ideas of the utopians?
- Do you think they work in ways that would have been expected or intended by Howard or Le Corbusier?
- Has their local legacy been more negative or positive?
- In what respects? ◆

3.2 CRITICIZING THE UTOPIANS

There is a real tension in the critiques that have been made of the utopian urbanists. On the one hand it is suggested that their aim was to remove conflict and tension and to sanitize and deaden urban life – to plan away the possibility of conflict to achieve the general de-intensification of urban life (see Chapter 1 for a discussion of de-intensification in the context of suburbanization). The role of the urban planners after 1945 has been criticized as an attempt to sanitize and to remove unevenness, when it is precisely the unevenness and variety that gives urban places their vitality (see Jacobs, 1961; Amin and Graham, 1997). Thus Donald (1997, p.200) criticizes Le Corbusier's vision of the radiant city for having 'the chill of Necropolis about it', and for 'Removing the organic and substituting the mechanical'. Jacobs (1961, p.386) claims that '*a city cannot be a work of art* ... To approach a city, or even a city neighbourhood, as if it were a larger architectural problem, capable of being given order by converting it into a disciplined work of art, is to make the mistake of attempting to substitute art for life' (emphasis in original); and Rogers (1997, p.17) argues that 'such single-minded visions of cities are no longer relevant to the diversity and complexity of modern society'.

At the same time the utopians are criticized for the failure, not so much of their visions, but their inherent unattainability in practice. From this perspective, then, the problem is

- not so much that what they produce is sanitized, standardized and suburbanized (that is, removing uncertainty and fear),
- but rather that it produces spaces which are less protected and generate more violence and more fear.

Newman (1972, p.24), for example, is quite explicit in claiming that the legacy of Le Corbusier is the building of housing which does not provide 'defensible space' for its residents. 'Defensible space', according to Newman, is space in which 'latent territoriality and sense of community in the inhabitants can be translated into responsibility for ensuring a safe, productive, and well-maintained living space. The potential criminal perceives such a space as controlled by its residents, leaving him an intruder easily recognized and dealt with' (Newman, 1972, p.3). Such space, he argues, disappears in the enclosed vision of Le Corbusier and those pursuing his legacy, in part because of the density of settlement associated with them. Similarly (as was noted in Chapter 3) Jacobs' espousal of the glories of the sidewalk and the city street (as an alternative to Le Corbusier's belief in internal streets) explicitly refers to the importance of eyes watching and overseeing what goes on – to informal processes of surveillance. In a sense, her starting point is to identify more informal methods of administration rather than those promised in grand plans, but she too wants to control and minimize the tensions of urban life. Street life seems to provide the community capable of disciplining its own members (Jacobs, 1961, chapters 2–4).

ACTIVITY 7.3 Turn now to Reading 7B, 'The generators of diversity' by Jane Jacobs. In this extract Jacobs sets out her own view of what constitutes urban living and, in particular, what makes a 'good' city. She also outlines her own strategy for administering city life.

- What are the main characteristics of cities identified by Jacobs?
- What features does she identify as making cities 'good' or successful?
- What are the main strategies she favours as means of encouraging positive development?
- Think back to your reading of Reading 7A by Le Corbusier. Try to identify the differences and similarities of approach between Jacobs and Le Corbusier. ◆

Like Steve Pile in Chapter 1, Jacobs stresses the extent to which cities are constituted through the mixture, or heterogeneity, of uses that is concentrated within them. She emphasizes the extent to which cities generate both diversity and innovation. Although she does not use these terms, she suggests that density generates an intensification of economic and social interaction which underpins these processes. But she also recognizes that not all cities are successful in these terms, and that there is also unevenness within cities. She suggests that four conditions are required for successful urban development to be achieved. They include the need for mixed uses and design, density of population, and short streets with lots of corners.

The contrasts with Le Corbusier are significant. Where he stresses speed expressed in a series of road grids separated from housing and employment, Jacobs integrates traffic into her neighbourhoods on a smaller scale. Where Le Corbusier seeks to separate functions, she seeks to mix them. Where he looks for a consistency of building styles, Jacobs calls for a mixing of styles and periods of building. Yet there are also some similarities. Both see population density as important. More important, perhaps, both also acknowledge the significance of the ways in which urban space is used in defining cities as well as the shared and divergent identities of those who live in them. Both see physical design as important. The conclusions they draw from this shared understanding are, however, markedly different. Jacobs' work is a fundamental critique of Le Corbusier's approach to urban space and its definition. Her emphasis is on the creation and defence of messy spaces which encourage interaction; his is on the creation of disciplined spaces which separate activities and functions. Her view of the city is from the street; his is from the sky, in the aeroplane arriving from Constantinople.

ACTIVITY 7.4

- Do you think Jacobs' emphasis on messy spaces will encourage orderliness or disorderliness?
- Do you think that Le Corbusier's plan for disciplined urban spaces will sustain or undermine social order?
- Whose ideas do you think are more likely to create a 'good' city?

Take some time to note down the reasons for your answers. ◆

3.3 BEYOND THE CRITICISMS

It is all too easy just to dismiss the visions of the utopians and to conclude that the answer is simply to let everything just work itself out in some more or less organic or natural way. Paradoxically, despite Jacobs' sharp criticism of the emphasis on physical approaches to social problems, she too provides her own physical solutions, which turn out to be the obverse of those she criticizes: she favours streets with small blocks, the retention of older housing, the maintenance of mixed uses in neighbourhoods. It is not enough to counterpose the 'logic' of the plan to 'natural' processes of urban evolution (as does Alexander, 1965, for example); or to the collective processes by which cities are constructed over time like a 'collage' from overlapping activity spaces (see Rowe and Koetter, 1979); or to the metaphor of weaving and reweaving social relationships across space (which Jacobs favours). All of these metaphors are highly suggestive, and help to undermine the supposed rigidities of the grand planners. They highlight some of the inherent difficulties in imposing fixed models on complex spatial processes of change and accommodation.

Yet they also seem inadequate to deal with some of the challenges that face us in seeking collectively to manage our lives within cities. Once the urban 'problem' is defined in terms of the rigidities of urban planning and design (as the dominant model of urban administration) then it becomes difficult to acknowledge that 'organic' or 'natural' change may create problems of its own (see Chapter 6). It would be easy to conclude from the criticisms that have been made of the urban utopians and their followers that it is only the intervention of the state (expressed in the form of urban planning or housing development) that has encouraged social segregation and generated insecurity and criminality. In the end, however, drawing such a conclusion would be misleading. The stress on informal arrangements is important, but informal arrangements may also generate powerful forms of exclusion and may even be underpinned by mechanisms of force. The suburbs so reviled by Jacobs were not the products of Howard or Le Corbusier, but of real estate developers and building firms (see Garreau, 1991, for a discussion of the role of developers). Even before Le Corbusier, there were dangerous places in cities, and there are dangerous places in cities which are so far relatively untouched by the power of urban planning and the orthogonal grid.

- The rejection of grand plans does not remove the problems of administration.

It merely forces us to rethink them. The rejection of any form of collective intervention implies a lack of popular control over development, but collective control does not mean that the development process ceases to generate inequality or to produce negative outcomes for some urban residents.

- Patterns of urban development reflect and express the relations of power that predominate in cities.

So, for example, the rejection of administration through state planning might also be consistent with a model of privatized control in which gated cities protect (and isolate) the overwhelmingly white middle classes from the poor, black, Hispanic and working-class people. Davis (1993) highlights the dangers of what he calls 'Ulsterization' in the Los Angeles metropolitan area, with the development of a series of protected 'edge cities' in a ring beyond the suburbs and far away from the city itself. He suggests that the very 'success' of these edge cities in providing protected space also serves to reinforce the fear of their residents. The pursuit of de-intensification (discussed in Chapter 1) leaves them feeling isolated (and disconnected) from other aspects of urban life which are then seen as inherently threatening, so that the response is to place even more emphasis on maintaining a protective and exclusionary shell. There is an underlying fear of an explosion of social chaos – of the war against the edge cities. In its most developed and dystopian form, this implies the development of a series of wealthy enclaves in a sea of disorganization and violence, with the wealthy simply emerging from armed camps from time to time in the so-called 'Blade Runner scenario' (Garreau, 1991, pp.284–5; Davis, 1992, 1993). Davis (1992) describes the emerging urban form as expressing an 'ecology of fear' (see Reading 3B; see also the discussion of fortified enclaves in Chapter 3 and by **Allen, 1999b** and **Amin and Graham, 1999**).

The absence of state planning does not mean that there is no planning. For example, Houston in Texas has frequently been identified as an unplanned city – that is, one in which the market is left to determine the appropriate form of development with little in the way of formal intervention from state professionals, and little attempt to impose a plan from on high. As Sudjic notes, however,

> Houston, supposedly the most unplanned city of all, is in fact nothing of the kind. It is perfectly true that Houston has no zoning legislation, unlike almost every other major city in the United States. The voters of Houston rejected the idea as recently as 1962, but though they might change their minds one day, the city is certainly not without planning, and it has an armoury of regulations that achieve very much the same ends. What makes Houston – the Third Coast as it presumptuously calls itself – different is the extent to which development control is in the hands of the private landowners and developers who have made fortunes out of the city's explosive growth.
>
> Right up until the 1960s, Houston was run from regular meetings at a suite in the Lamar Hotel by a clique that made the decisions, formed the committees and fired the mayors. Taxes and services were kept at survival level, with infrastructure investment put off until crisis point was reached. They used tax money to bail out their flooded suburbs and to provide water and sewers for private developments …
>
> Land use in Houston is controlled by restrictive covenants in the title deeds that cover most of the city's developed land. There are ten thousand such covenants dealing with use, size, cost and height, administered by 630 civic clubs which function exactly like private zoning boards, backed up by the city itself, which has acquired the power to enforce private covenants.
>
> (Sudjic, 1992, p.97)

To conclude, then:

- Even where it is imagined that development is the outcome of almost 'natural' market processes, it is important to recognize that the urban property market (as Jenny Robinson notes in Chapter 4, section 2.3) is itself 'a social creation', in which a range of different actors negotiate to determine outcomes.
- But, of course, that negotiation rarely takes place in a public context, and outcomes reflect the power of the participants. If there are not socialized forms of administering cities, then more privatized ones are likely to emerge.
- The dismissal of planning from on high as a strategy for ordering urban life can only be a first step towards considering other means of administration.

4 *Administration as governance*

The recognition that planning is not simply a 'rational' or a technical process, but is rather one of negotiation, argument and accommodation, opens up other possibilities for conceiving ways of managing, or administering, urban development. Models that assume an institutional structure in which a government makes decisions that can then be implemented in a relatively straightforward way are not convincing in the context of the contested orderings, fluidity and negotiation highlighted in the chapters of this book. Order cannot simply be imposed. Instead it must be recognized that the world of urban politics is made up of a series of negotiations and partnerships, linking a range of social actors, which might include elected governments and their professional staffs, non-elected statutory institutions, non-statutory and private sector agencies of various sorts, as well as communities, groups and individual citizens. Chapter 5 noted the extent to which key aspects of urban politics – particularly in conflicts over development – may represent battles over what Massey (1998) calls 'spatialized social power'.

The term 'governance' is increasingly preferred to 'government' to capture both the potential complexity of these relationships and to indicate that some more or less stable or accepted set of arrangements may arise from them, at least for significant lengths of time. In other words, just because there is a high degree of complexity, the rules, practices and political structures of urban living may still operate to maintain order in an apparently 'unruly' urban world.

The problem of urban governance can be expressed thus:

- If urban areas are characterized by differences, tensions and divisions, how is it possible for shared agendas to emerge?

Mayer (1995) identifies three main related trends in the development of urban governance at the end of the twentieth century. The first is an increased focus on the development of 'proactive development strategies' – that is, an emphasis on negotiating with business and developers to encourage growth and attract investment. The second trend she identifies is a subordination of more traditional social policies to those in support of economic development. The third is a move towards 'an expansion of the sphere of local political action to involve not merely the local authority but also a range of private and semi-public actors' (Mayer, 1995, p.232).

A shared agenda is developed out of the search for urban economic and cultural success. The argument runs roughly as follows. The emergence of wider global networks and of a more globalized economy has generated new pressures and opened up new possibilities for cities. They now have the task of positioning themselves within these networks in ways that may bring them investment and other economic advantages. Economic success and social welfare are linked, to the extent that if particular cities lose out in the urban hierarchy, their residents will also lose out. At the same time the traditionally relatively clear boundaries between state and other agencies have been eroded, with the explicit emergence of new forms of joint working and partnership between parts of the business sector and community and voluntary organizations. In this context an attempt has been made to explain the emergence of growth politics as the dominant politics of urban development, utilizing notions such as urban 'growth machine' or 'growth coalitions' (Logan and Molotch, 1987; Cox and Mair, 1989). Although there are important theoretical differences between the writers using these terms, the key shared understanding is that strategic political alliances within cities will tend to form around the priority of achieving economic growth, reflected in increased real estate development, rising prices for land and property, more business investment, and higher economic returns. Locally elected politicians and professional officers, it is believed, have little choice but to work with a growth agenda, because in most cases the alternative appears to be decline and decay.

4.1 GLOBAL PLACE-MARKETING

Shared objectives have also included the role of self-promotion in the development of urban growth. In Chapter 5 Sophie Watson noted the significance of cultural and symbolic processes in generating new images of the city. She identified some of the gains and losses for the residents of Sydney that have arisen from the 'selling of the city' through attractions such as the 2000 Olympic Games and the Gay and Lesbian Mardi Gras. The competition for prestige global events of this type provides a powerful representation of success around which it is possible to construct popular forms of politics, focused around a positive identification of the city, both externally and internally. Manchester, too, campaigned to attract the Olympics:

> The politics of the Manchester Olympics bids of the 1980s and 1990s powerfully symbolize the changes which have been taking place. Bidding to present major global events such as the Olympics has become a key aspect of the local politics of place-marketing. In the case of Manchester, it also appeared that the bids were led by a buccaneering free spirit, Bob Scott, deeply rooted in the private sector. The old images of municipal welfarist (bureaucratic) politics were replaced by those of a dynamic charismatic (entrepreneurial) business leadership. Unlike some other aspects of the restructured welfare state (for example the newly marketized and managerialized health service), Manchester's Olympic strategy was both populist and popular. Its promotional campaigns suggest that it has been able to mobilize both capital and people around glossy images of success.
>
> (Cochrane *et al.*, 1996, p.132)

The place-marketers attempt to build on existing ('locally networked') traditions and understandings to reposition each city along lines that fit with the limited range of images preferred by managers, up-market tourists and potential investors, while also hoping to make the city more attractive to local residents (Kearns and Philo, 1993). These stories are not just produced for the outside world. Instead they are also reflected back to local residents, in an active process which helps to define their understandings of the city in which they live. They help to build a shared politics around a spatial imagination which unifies the city, instead of one that emphasizes conflict and contention.

In exploring what he calls a 'space of flows', Castells (1996, pp.415–18) argues that élites are cosmopolitan. The cosmopolitan élites identified by Castells are both concerned to create protective spaces for themselves and to redefine debates about urban development in ways which reinforce their positions, for instance through global place-marketing and the imagery of world cities and global architecture. In the case of Berlin, for example, in the wake of global restructuring and the reunification of Germany, the centre of the city became a massive building site, and drew on global architects such as Richard Rogers, I.M. Pei, Aldo Rossi and Helmut Jahn, to claim a new role for the city on the world stage. The process of development has itself become a means of defining Berlin as a place of the future. Popular pamphlets have been produced on Berlin's building sites, with the Potsdamer Platz proudly described as the biggest building site in Europe (Schneider and Schubert, 1997). A bright red exhibition centre, the Info Box, placed on stilts at the centre of the building sites on the Potsdamer Platz, has become a major tourist site, attracting two million visitors between its opening in October 1995 and January 1997 (Info Box, 1996). The visitors to the Info Box are explicitly encouraged to accept the imagination of architects, planners and developers as the future reality of a city that is in the process of becoming something greater.

German reunification opened a debate about Berlin's position in the world economy. Local politicians set out to develop strategies for place-marketing, staking out their own claims to a special position for the city within a highly competitive global market-place. Like other major cities, Berlin has its own panoply of partnership marketing and development agencies which promise intensive co-operation between the public and private sectors. A key aspect of this redefinition and repositioning stresses Berlin's political-geographical location at the heart of a 'new' Europe, itself in the process of construction: 'When the wall fell in November 1989 the city moved out of the peripheral island in which it had been for many decades back into the centre of Europe' (Info Box, 1996, p.17). Berlin's own propaganda claims that the city 'is developing into a European metropolis at the heart of Europe with growing business strength' (Berlin Press and Information Office, 1995, p.40). Daimler Benz (along with Sony) is the biggest developer of the massive Potsdamer Platz site, and according to Dr Manfred Gentz, a member of Daimler Benz AG Board, the company 'has harnessed its entrepreneurial energy and vision of the future to the task of replacing this jewel in the old and new capital – with its old lustre but a new refinement'. It is 'a symbol of starting out into a common European future' (Info Box, 1996, p.111). In other words, at the heart of this redefinition is

BERLIN'S POTSDAMER PLATZ THROUGH THE TWENTIETH CENTURY

FIGURE 7.9
1907: the heart of an elegant city

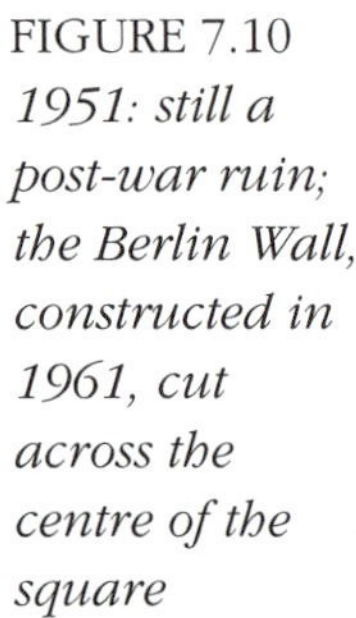

FIGURE 7.10
1951: still a post-war ruin; the Berlin Wall, constructed in 1961, cut across the centre of the square

FIGURE 7.11 *the 1990s: the square under reconstruction after German reunification*

FIGURE 7.12 *1998: the reconstruction nears completion*

FIGURE 7.13 *Berlin's Potsdamer Platz, October 1998: the opening ceremony for the new square*

an attempt to reclaim Berlin's position at the heart of a changed set of economic flows and global interconnections.

ACTIVITY 7.5 You should now turn to Reading 7C, 'Berlin Mitte' ('Berlin's central district') by Nicholas Howe, in which Howe describes the experience of living and working in the centre of Berlin during the process of physical transformation. Think about the ways in which Howe interprets the impact of the changes, and about some of the different ways in which the emergent urban spaces can be understood.

To what extent do you think the new development is succeeding in imposing a new identity on the city? ◆

Howe presents an ambivalent story of reconstruction in Berlin. He acknowledges the scale of change, and he confirms the ways in which it has disrupted some of the existing urban rhythms and undermined some of the taken-for-granted assumptions of people living in the city (**Allen, 1999b**, explores some of the ways in which similar experiences affected other East German cities in the 1990s). Howe highlights the extent to which attempts are being made to produce a new Berlin defined through the construction of what he calls the corporate palaces. But he also emphasizes the ambiguity of the interconnections and juxtapositions produced by this process, and he emphasizes the implicit and explicit contestation that is taking place across urban space. He points to some of the ways in which new forms of division and segregation are being produced by the investments that are supposed to produce unification. As well as noting the continued controversies about some symbolic sites (such as the Palace of the Republic), he shows how even those sites are being used in ways which mean they have lost some of their symbolic power (at least for 10 year-olds). Although he recognizes the attempt by major developers to 'ward off the workings of time and history' (and he might have added geography), he is highly sceptical about their ability to do so.

4.2 GOVERNANCE THROUGH REGIMES

The notion of 'urban regime' offers one means of understanding the processes of negotiation that underlie the politics of urban development, without exaggerating the power of the developers. It is a helpful notion because of its stress on the processes by which a broadly shared political agenda may be constructed, recognizing the importance of existing relations of power without assuming that any one group necessarily always dominates in the political or development process.

ACTIVITY 7.6 You should now turn to Reading 7D, 'Regime theory and urban politics', in which Gerry Stoker explores the main features of a regime-based approach to urban governance.

Stoker sets out the main features of urban regime theory and outlines the ways in which governing regimes may be constructed to build shared agendas and develop local programmes for urban development. Regime theory offers the prospect of administration through the development of these shared agendas (see also Chapter 5). Try to identify the main features that underpin urban regime theory, and also try to identify some of the main problems associated with it as a means of explaining the ways in which urban disorder may be administered. ◆

One strength of regime theory lies in the way it allows us to explore political arrangements and power beyond the boundaries of government as it is commonly understood. It brings together neo-pluralist analyses of power with a recognition of the importance of economic power. It also means we do not have to choose between believing that elected or appointed officials determine policy, or that local business simply forces local governments to follow its political agenda. Instead it points to the need for a coming together, 'blending' or 'meshing', in a set of formal and informal arrangements, to achieve a more or less shared agenda (albeit with the recognition that there is an inherent bias towards business). Regime theory recognizes the importance of fluid and changing networks of power within cities and of the need to consider ways of managing them dynamically, instead of seeking to fix them for all time. The power involved is presented as a power to act to achieve certain ends, rather than a power over others **(Allen, 1999a)**. The political logic underlying stable regimes is driven neither by pressure group influence on particular issues, nor by electoral pressure, but by the ability to mobilize resources around the (relatively fluid) policy agenda that has been developed over time. Such agendas are likely to be focused around a commitment to economic growth, since that promises long-term material gains whose benefits can also be shared locally. In principle, regimes might also be constructed around the defence of existing privileges, particularly in well-to-do suburbs, but the success of these may in turn depend on the existence of regional growth agendas or the success of growth coalitions in neighbouring jurisdictions.

There is a danger, however, that a focus on the working of regimes may encourage us to ignore the role of the marginalized and excluded in the political process. Although supporters of the regime approach explicitly recognize that regime formation may operate to exclude some sections of the urban population from the informal political networks of power, the theoretical focus of attention is necessarily on the working of the regimes rather than on those on the outside looking in (who, as Stoker acknowledges, may not even be aware of the strategies being formed through regime politics).

In Chapter 5, Sophie Watson explored the everyday struggles and conflicts over space and resources, especially those associated with urban social and political movements. Relationships like these seem to exist quite separately from the world of regimes, yet they remain central to a rounded understanding of the operations of urban politics and the governance of cities. Clarke *et al.* (1995) convincingly argue that the regime approach has worked to ensure that the role of women's organizations has been relegated to a secondary status in representations of urban politics. The process of exclusion is doubly powerful: not only are the members of such movements outside the regimes, but they are somehow necessarily excluded – permanently relegated to a politics of campaigning and resistance, and never able to be involved in an inclusive urban politics. There is a real danger that regime theory tends to divert attention from some of the tensions that characterize the management of cities. Some groups are likely to be marginalized by the working of urban regimes, but this does not mean that the agenda of the regime has been accepted. It may also serve to reinforce divisions, rather than overcoming them. In other words, the appearance of control and stability expressed in the programmatic statements about urban regimes, and in the development flowing from them, may mask continuing instability and uncertainty among those excluded by the regimes.

A further strength of regime theory is that it highlights the ways in which the identity of cities may be constructed through a process of regime formation, particularly when reinforced by a commitment to place-marketing. But it has little understanding of the spatial relations – either within urban areas or in terms of their interconnections to other places – that come together to produce cities. The existence of locally rooted élites or business interests is taken very much for granted in regime theory, and it is around these local networks that regimes are expected to form. This view is a relatively static one which concentrates attention on the ways in which governing capacity may be constructed in ways which incorporate locally embedded interests. As a result, the extent to which such élites are actually part of much wider (overlapping) networks which ensure that their members are connected to people, organizations and activity spaces far beyond the cities in which they live or work tends to be understated. In his identification of élites as 'cosmopolitan', Castells (1996, pp.415–18) emphasizes the importance of networks of communication which constitute cities through the movement of flows across space and time. For him, what matters are the points of connection to networks, rather than places. Such an understanding disrupts explanations that start from an emphasis on locally rooted relationships.

The implications of acknowledging the ways in which relations of power stretch across cities (as well as the ways in which power may be concentrated in particular cities) are explored by **Allen (1999a)**. Different kinds of power flow through different kinds of networks, with different cities, and invariably only parts of them, at their nodes. The contrast made by Castells between 'cosmopolitan élites' and 'local people' is itself difficult to sustain, however, since Allen also points to networks of interaction between 'people'. Even the most apparently marginalized and excluded of people can be seen to be part of much wider ethnic diasporas, which underpin significant lines of connection across the globe.

- The recognition that urban governance works through negotiation, the formation of partnerships and the construction of urban regimes or forms of growth coalition is important because it also highlights processes of political inclusion and exclusion which help to define the identity of cities.
- But it is also important to recognize that focusing on the building of regimes may make it more difficult to understand some of the continuing ways in which dominant definitions are contested, and the extent to which urban space remains the site of significant and everyday conflicts over forms of spatialized power.

5 Administration and democracy

At the heart of the debates about urban administration and governance is a tension between attempts to achieve fixity and to work or manage fluidity. In practice, the dominant approaches have tended to search for ways of capturing the social relations of cities in more or less unitary models of development and governance. In contrast, **Massey (1999, p.170)** highlights the possibility of an alternative approach: 'perhaps what are needed are active democratic processes for defining the public good, recognizing the conflicts over it, and allowing it to change over time'.

According to one powerful history of the governance of urban change, since the 1970s there has been a dramatic shift in approach away from models that seek to develop a coherent and holistic system of social and economic planning. This has increasingly been dismissed as undesirable as well as unfeasible. Not only, it is argued, are such approaches unable to cope with the complexity and diversity of human society, but they are likely to impose unacceptable controls on those whose lives they seek to manage or improve. Instead, it is argued, it has become necessary to focus on what makes cities successful as living environments – as 'working cities' (Jacobs, 1961), 'ordinary cities' (Amin and Graham, 1997), 'heterogeneous cities' (see Chapter 1), or 'cities of connection and disconnection' **(Amin and Graham, 1999)**.

Stress is placed on the need to move away from attempts to produce detailed master plans of urban development within which everybody has to fit, towards approaches that reflect a more organic understanding of the nature of urban development and aim at allowing (defending or creating) space in which a plurality of citizen and community voices may be heard. Instead of attempting to capture or fix social relations through the formal rules of urban living or the design of buildings, it is suggested that approaches to urban governance and urban development should concentrate on identifying and celebrating the possibilities of flexibility and fluidity. The architect Richard Rogers (1997, p.165), for example, argues explicitly that: 'We must build cities for flexibility and openness, working with and not against the now inevitable process whereby cities are subject to constant change … The aesthetics of response, change and modulation have replaced the fixed order of architecture'. Instead of top-down planning, which attempts to capture the full complexity of urban life and interaction in a series of reports, maps and statistical tables, emphasis is to be placed on decentralization and localized co-operation.

One response to concerns about the exclusion of urban residents from popular decision-making might simply be to argue for (improved) traditional representative democratic arrangements as a means of ensuring involvement in debates about urban development and the running of cities. Such arrangements generally involve the election of local councils which oversee the operation of professionals of one sort or another in the delivery of services and the operation of planning. Unfortunately, this model has itself been challenged substantially in recent years, both because its record of responsiveness and involvement has not always been impressive, and because it implies a standardized approach to an increasingly diversified set of issues and populations (see Cochrane, 1993, for a consideration of the British case). The notion of citizen that underlies the traditional western model of local democracy is a curiously passive one, in which representatives are elected to oversee the work of others, with little expectation of further involvement by electors in decision-making.

Attempts have been made, however, to identify other ways of approaching these issues. Hirst (1994) and Cohen and Rogers (1995), for example, have highlighted the possibilities of what they call 'associative democracy'. This builds on the notion that society is made up of a plurality of associations, through which individuals come together in a complex variety of ways reflecting their multiple identities (at work, in leisure, at home, as consumer, and so on). In other words, it starts from assumptions about heterogeneity and difference, rather than homogeneity. It rejects the notion of 'community' as a homogeneous basis for political activity. Although these associations are not necessarily urban, many of them are based in cities and come together in ways which help to define them in terms that recognize the importance of local diversity, and differences between places as well as between groups and individuals. The very density of city life as well as its intensity might be expected to encourage the growth of associations of one sort or another, although the disconnections expressed in some urban social relations may also help to limit them. It is important to recognize the overlapping nature of these associations, since many people are likely to belong to more than one, confirming that they define themselves in multiple ways, but also allowing for more differentiated forms of politics and, by implication, more differentiated notions of citizenship.

Hirst, Cohen and Rogers argue that instead of trying to find some overall (planning) logic to which everyone then needs to conform, it would be better to allow these groups to develop their own agendas and to implement them.

This is consistent with the arguments developed by Jane Jacobs (1961, p.16), who says that 'Cities are an immense laboratory of trial and error, failure and success, in city building and city design. This is the laboratory in which city planning should have been learning and forming and testing its theories'. 'Most city diversity', she argues, 'is the creation of incredible numbers of different people and different private organizations, with vastly differing ideas and purposes, planning outside the formal framework of public action' (Jacobs, 1961, p.256). Jacobs argues that what is needed is the prospect of gradually building from small-scale successes to a renewed urban vision, based on a carefully managed mix of uses and populations at street, neighbourhood and district level.

Jacobs' approach also highlights some of the problems with notions of associative democracy. Her emphasis on 'community-based' self-help and informal relations is based on an overwhelmingly positive view of social relations in cities, and the networks of relations she describes generate a positive spiral of social interaction and security. However, her rosy view of urban neighbourhoods fails to take into account some of the pressures of city life and, in particular, the concentration of populations with marginal economic positions whose members are excluded from employment or trapped in casualized sectors of employment in particular areas (such as Harlem –

see Chapter 1). As Berman (1992, p.325) notes, 'it was clear by the late 1960s, amid the class disparities and racial polarities that skewered American city life, no urban area anywhere, not even the liveliest and healthiest, could be free from crime, random violence, pervasive rage and fear'.

The power relations expressed in the socio-spatial patterning of cities makes some of the more optimistic hopes of 'associative democracy' look less convincing, because access to membership of the different associations is unlikely to be equal.

Who determines which people are allowed to join which groups? How can powerful agencies (such as major developers) be controlled? How, as Amin and Thrift (1995) ask, can differences between localities be resolved? And so on. One means of resolving such issues might be to recognize the continued need for elected local governments, but on a different basis – not as a direct participant but as an agency able to arbitrate and regulate, and more important to mobilize opinion and support where necessary. Donnison (1998), for example, seeks to identify a role for local politicians along these lines as 'civic leaders'.

Healey (1995) similarly develops an approach which sees the planning process – in her case land-use planning, but the points she makes are of wider relevance – not as the search for a finally agreed, frozen document, but rather as something that reflects 'management by argumentation':

> The plan becomes a record of arguments about structuring ideas, about why a particular strategy or approach has been undertaken, about reasons why specific policies and policy criteria have been selected and about impacts and who bears them ... it holds out an approach which both encourages ... alliances to expand to incorporate a broader interest base, and forces the politico-administrative nexus into a transparent and answerable relation with the rest of us.
>
> (Healey, 1995, p.269)

It must be stressed that these issues have not yet been satisfactorily resolved in theory or in practice. In particular, the danger of fragmentation in the face of powerful agendas set by urban and cosmopolitan élites is one that needs to be confronted directly. Since cities are the products of interconnections and flows, their effective management will require involvement beyond individual cities, through regional and cross-national alliances, particularly within the emergent mega-cities. There is some (albeit limited) evidence of development along these lines. In the European Union, for example, cross-border regions have been created to reflect shared agendas which cross national boundaries. However, political linkages like these remain underdeveloped.

- The key point for our purposes here is not to spell out a fully formed system of urban government, but rather to stress that this represents a different approach to the issues of achieving order, of administering cities, because it does not start from the assumption that order means uniformity and standardization.

6 Conclusion

Traditional approaches have tended to emphasize the need to control and manage urban populations and developments. Their ambition has been to maintain order in, or create order out of, a potential sea of chaos – to impose a convincing technical rationality, to capture complexity in sets of plans and data, or through processes of surveillance and supervision. In other words, they have tended to define cities as inherently threatening and disorderly. We have considered three main approaches – administration by design, the pursuit of urban utopia, and the construction of urban regimes and growth coalitions – each of which incorporates its own views of what makes a city a 'good' city. All of them are undermined by their lack of sympathy for, or inability to recognize, the dynamism and the sets of relationships, connections and disconnections that bring cities to life.

While the critique of rational planning as an overall strategy for the administration of cities is highly persuasive, the alternative visions remain unsatisfactory, too. A reliance on organic or 'natural' urban development requires us to accept the ways in which existing patterns of power and wealth play themselves out on the city streets. The critique made of the utopian urbanists like Howard and Le Corbusier sometimes seems to rely on a rather romantic vision of the urban alternative, expressed, for example, in the narratives of Jacobs. The much reviled process of suburbanization may owe something to the planners – for example where huge public sector estates have been placed on the periphery of major cities – but it also owes a great deal to a series of decentralized decisions by home-buyers and developers. Similarly, the privatized search for security within protected enclaves policed privately or publicly is as organic or 'natural' as relationships within neighbourhoods, or those discovered by individuals strolling the streets of Paris or Manhattan. The high-rise development that characterizes the centre of so many cities may owe something to the planners, but it also reflects the demands of developers to make full, intensive and profitable use of expensive land.

This suggests that some sort of administration is required in urban areas, and, indeed, the libertarian language of those who criticize the utopian urbanists soon slips into quite specific proposals for urban development. Urban policy and the emergent politics of urban governance offer some possibilities for effective administration, but they tend to fragment decision-making (making it difficult to have an overview of change) and mask the decision-making process. Criteria to explain the ways in which choices are made are rarely clearly stated. The process of urban regime formation also highlights the exclusion of popular voices from the decision-making process.

The danger is that such a shift effectively removes issues of development from the open political arena. In principle, at least, the planning process appeared to be (or had the potential to be) under some form of democratic control. In practice, of course, that was rather more difficult to guarantee. With the emergence of urban regimes and growth coalitions, the direction of change, and the nature of the political participants, are rather more explicitly restricted.

However, there are also alternative understandings of what is possible. While the process of urban planning offers a means of normalizing juxtapositions, finding ways of minimizing conflict and reducing 'fear', at the same time it has the potential to recognize their value and to mobilize resources to benefit from them. It is possible to imagine a planning process building on the plurality and heterogeneity of cities, and recognizing those features as one of its strengths. The precise structures of administration likely to emerge from such an approach are uncertain, but taking such an approach should help in rethinking what is possible – in imagining the future as a different form of urban utopia.

The implicit choice between a view from the streets – as a pedestrian or resident – and a view from the air – as a Corbusian fantasist – is not one we should have to make. These are not mutually exclusive alternatives, nor is either ultimately workable as a universal model. Instead, a more complex vision is required that is capable of incorporating both of these visions, but also others which recognize the importance of linkages and interconnections as well as disconnections and juxtapositions, within cities and beyond them.

References

Alexander, C. (1965) 'A city is not a tree', reprinted in LeGates, R. and Stout, F. (eds) (1996) *The City Reader*, London, Routledge.

Allen, J. (1999a) 'Cities of power and influence: settled formations', in Allen, J. *et al.* (eds).

Allen, J. (1999b) 'Worlds within cities', in Massey, D. *et al.* (eds).

Allen, J., Massey, D. and Pryke, M. (eds) (1999) *Unsettling Cities*, London, Routledge/The Open University (Book 2 in this series).

Amin, A. and Graham, S. (1997) 'The ordinary city', *Transactions of the Institute of British Geographers*, vol.22, pp.411–29.

Amin, A. and Graham, S. (1999) 'Cities of connection and disconnection', in Allen, J. *et al.* (eds).

Amin, A. and Thrift, N. (1995) 'Institutional issues for the European regions: from markets and plans to socioeconomics and powers of association', *Economy and Society*, vol. 24, pp. 41–66.

Beier, R. (ed.) (1997) *Aufbau West, Aufbau Ost: Die Planstädte Wolfsburg und Eisenhüttenstadt in der Nachkriegszeit*, Osfieldern-Ruit, Hatje.

Berlin Press and Information Office (1995) *Hauptstadt im Werden*, Berlin, Presse und Informationsamt des Landes Berlin.

Berman, M. (1992) *All That Is Solid Melts into Air: The Experience of Modernity*, London, Verso.

Castells, M. (1996) *The Rise of the Network Society*, Oxford, Blackwell.

Clarke, S., Staeheli, L. and Brunell, L. (1995) 'Women rethinking local politics', in Judge, D. *et al.* (eds).

Cochrane, A. (1993) *Whatever Happened to Local Government?*, Buckingham, Open University Press.

Cochrane, A., Peck, J. and Tickell, A. (1996) 'Manchester plays games: exploring the local politics of globalization', *Urban Studies*, vol.33, no.8, pp.1319–36.

Cohen, J. and Rogers, J. (1995) *Associations and Democracy*, The Real Utopias Project, vol. 1, London, Verso.

Corn, J. and Horrigan, B. (1995) *Yesterday's Tomorrows: Past Visions of the American Future*, Baltimore, Johns Hopkins University Press.

Cox, K. and Mair, A. (1989) 'Urban growth machines and the politics of local economic development', *International Journal of Urban and Regional Research*, vol.13, no.1, pp.137–46.

Davis, M. (1992) *Beyond Blade Runner: Urban Control: The Ecology of Fear*, New Jersey, Open Magazine.

Davis, M. (1993) 'Who killed Los Angeles? Part two: the verdict is given', *New Left Review*, no.100, pp.29–54.

Doherty, J. (1980) 'Ideology and town planning – the Tanzanian experience', *Antipode*, vol.11, no.3, pp.12–23.

Donald, J. (1997) 'This, here, now: imagining the modern city', in Westwood, S. and Williams, J. (eds) *Imagining Cities: Scripts, Signs, Memory*, London, Routledge.

Donnison, B. (1998) *Policies for a Just Society*, London, Macmillan.

Frampton, K. (1985) *Modern Architecture: A Critical History*, revised edition, London, Thames and Hudson.

Garreau, A. J. (1991) *Edge City: Life on the New Frontier*, New York, Doubleday.

Hall, P. (1996) *Cities of Tomorrow: An Intellectual History of Urban Planning and Design in the Twentieth Century*, Oxford, Blackwell.

Harvey, D. (1997) 'Contested cities: social process and spatial form', in Jewson, N. and MacGregor, S. (eds) *Transforming Cities: Contested Governance and New Spatial Divisions*, London, Routledge.

Healey, P. (1995) 'Discourses of integration: making frame-works for democratic urban planning', in Healey, P. *et al.* (eds).

Healey, P., Cameron, S., Davoudi, S., Graham, S. and Madani-Pour (eds) (1995) *Managing Cities: The New Urban Context*, Chichester, Wiley.

Hirst, P. (1994) *Associative Democracy: New Forms of Economic and Social Governance*, Cambridge, Polity Press.

Howe, N. (1998) 'Berlin Mitte', *Dissent,* vol. 45, no.1, pp.71–81.

Info Box (1996) *Info Box: The Catalogue*, Berlin, Dirk Nishen.

Jacobs, J. (1961) *The Death and Life of Great American Cities*, quotes from 1964 edn, Harmondsworth, Penguin.

Judge, D., Stoker, G. and Wolman, H. (eds) (1995) *Theories of Urban Politics*, London, Sage.

Kearns, G. and Philo, C. (eds) (1993) *Selling Places: The City as Cultural Capital, Past and Present*, Oxford, Pergamon Press.

Le Corbusier (1929/1987) *The City of Tomorrow*, London, The Architectural Press.

Logan, J. and Molotch, H. (1987) *Urban Fortunes: The Political Economy of Place,* Berkeley, University of California Press.

Massey, D. (1998) 'Living in Wythenshawe', in Borden, I. *et al.* (eds) *The Unknown City*, London, Routledge.

Massey, D. (1999) 'On space and the city', in Massey, D. *et al.* (eds).

Massey, D., Allen, J. and Pile, S. (eds) (1999) *City Worlds*, London, Routledge/The Open University (Book 1 in this series).

Mayer, M. (1995) 'Urban governance in the post-Fordist era', in Healey, P. *et al.* (eds).

Newman, O. (1972) *Defensible Space: People and Design in the Violent City*, London, Architectural Press.

Pile, S. (1999) 'What is a city?', in Massey, D. *et al*. (eds).

Rogers, R. (1997) *Cities for a Small Planet*, edited by Philip Gumuchdjian, London, Faber and Faber.

Rowe, C. and Koetter, F. (1979) *Collage City*, Cambridge, Mass., MIT Press.

Schneider, G. and Schubert, P. (1997) *Wer baut wo? Die grossen Bauprojekte im Überblick*, Berlin, Jaron Verlag.

Sennett, R. (1990) *The Conscience of the Eye: The Design and Social Life in Cities*, London, Faber and Faber.

Sombart, N. (1987) 'Viermal Berlin, Berliner Mythologie', in *Mythos Berlin, Zur Wahrnehmungsgeschichte einer industriellen Metropole*, exhibition catalogue, Berlin.

Stoker, G. (1995) 'Regime theory and urban politics', in Judge, D. *et al.* (eds).

Strom, E. (1996) 'The political context of real estate development: central city rebuilding in Berlin', *European Urban and Regional Studies*, vol.3, no.1, pp.3–17.

Sudjic, D. (1992) *The 100 Mile City*, San Diego, Harcourt, Brace and Company.

Ward, C. (1993) *New Town: Home Town*, London, Calouste Gulbenkian Foundation.

READING 7A
Le Corbusier: 'A contemporary city of three million inhabitants'

The desire to rebuild any great city in a modern way is to engage in a formidable battle. Can you imagine people engaging in a battle without knowing their objectives? Yet that is exactly what is happening. The authorities are compelled to do something, so they give the police white sleeves or set them on horseback, they invent sound signals and light signals, they propose to put bridges over streets or moving pavements under the streets; more garden cities are suggested, or it is decided to suppress the tramways, and so on. And these decisions are reached in a sort of frantic haste in order, as it were, to hold a wild beast at bay. That BEAST is the great city. It is infinitely more powerful than all these devices. And it is just beginning to wake. What will tomorrow bring forth to cope with it?

We must have some rule of conduct.

We must have fundamental principles for modern town planning.

Site

A level site is the ideal site. In all those places where traffic becomes over-intensified the level site gives a chance of a normal solution to the problem. Where there is less traffic, differences in level matter less …

Population

This consists of the citizens proper; of suburban dwellers; and of those of a mixed kind.

(a) Citizens are of the city: those who work and live in it.

(b) Suburban dwellers are those who work in the outer industrial zone and who do not come into the city: they live in garden cities.

(c) The mixed sort are those who work in the business parts of the city but bring up their families in garden cities.

To classify these divisions (and so make possible the transmutation of these recognized types) is to attack the most important problem in town planning, for such a classification would define the areas to be allotted to these three sections and the delimitation of their boundaries. This would enable us to formulate and resolve the following problems:

1 The *City,* as a business and residential centre.

2 The *Industrial City* in relation to the *Garden Cities* (i.e. the question of transport).

3 The *Garden Cities* and the *daily transport* of the workers.

Our first requirement will be an organ that is compact, rapid, lively and concentrated: this is the City with its well organized centre. Our second requirement will be another organ, supple, extensive and elastic; this is *the Garden City* on the periphery.

Lying between these two organs, we must *require the legal establishment* of that absolute necessity, a protective zone which allows of extension, *a reserved zone* of woods and fields, a fresh-air reserve.

Density of population

The more dense the population of a city is the less are the distances that have to be covered. The moral, therefore, is that we must *increase the density of the centres of our cities, where business affairs are carried on.*

Lungs

Work in our modern world becomes more intensified day by day, and its demands affect our nervous system in a way that grows more and more dangerous. Modern toil demands quiet and fresh air, not stale air.

The towns of today can only increase in density at the expense of the open spaces which are the lungs of a city.

We must *increase the open spaces and diminish the distances to be covered.* Therefore the centre of the city must be constructed *vertically.*

The city's residential quarters must no longer be built along 'corridor-streets', full of noise and dust and deprived of light …

The street

The street of today is still the old bare ground which has been paved over, and under which a few tube railways have been run.

The modern street in the true sense of the word is a new type of organism, a sort of stretched-out workshop, a home for many complicated and delicate organs, such as gas, water and electric mains. It is contrary to all economy, to all security, and to all sense to bury these important service mains. They ought to be accessible throughout their length. The various storeys of this stretched-out workshop will each have their own particular functions. If this type of street, which I have called a 'workshop', is to be realized, it becomes as much a matter of *construction* as are the houses with which it is customary to flank it, and the bridges which carry it over valleys and across rivers.

The modern street should be a masterpiece of civil engineering and no longer a job for navvies.

The 'corridor-street' should be tolerated no longer, for it poisons the houses that border it and leads to the construction of small internal courts or 'wells'.

Traffic

Traffic can be classified more easily than other things.

Today traffic is not classified – it is like dynamite flung at hazard into the street, killing pedestrians. Even so, *traffic does not fulfil its function.* This sacrifice of the pedestrian leads nowhere.

If we classify traffic we get:

(a) Heavy goods traffic.

(b) Lighter goods traffic, i.e. vans, etc., which make short journeys in all directions.

(c) Fast traffic, which covers a large section of the town.

Three kinds of roads are needed, and in superimposed storeys:

(a) Below-ground there would be the street for heavy traffic. This storey of the houses would consist merely of concrete piles, and between them large open spaces which would form a sort of clearing-house where heavy goods traffic could load and unload.

(b) At the ground floor level of the buildings there would be the complicated and delicate network of the ordinary streets taking traffic in every desired direction.

(c) Running north and south, and east and west, and forming the two great axes of the city, there would be great *arterial roads for fast one-way traffic* built on immense reinforced concrete bridges 120 to 180 yards in width and approached every half-mile or so by subsidiary roads from ground level. These arterial roads could therefore be jointed at any given point, so that even at the highest speeds the town can be traversed and the suburbs reached without having to negotiate any cross-roads.

The number of existing streets *should be diminished by two-thirds*. The number of crossings depends directly on the number of streets; and *cross-roads are an enemy to traffic*. The number of existing streets was fixed at a remote epoch in history. The perpetuation of the boundaries of properties has, almost without exception, preserved even the faintest tracks and footpaths of the old village and made streets of them, and sometimes even an avenue …

The result is that we have cross-roads every fifty yards, even every twenty yards or ten yards. And this leads to the ridiculous traffic congestion we all know so well.

The distance between two bus stops or two tube stations gives us the necessary unit for the distance between streets, though this unit is conditional on the speed of vehicles and the walking capacity of pedestrians. So an average measure of about 400 yards would give the normal separation between streets, and make a standard for urban distances. My city is conceived on the gridiron system with streets every 400 yards, though occasionally these distances are subdivided to give streets every 200 yards.

This triple system of superimposed levels answers every need of motor traffic (lorries, private cars, taxis, buses) because it provides for rapid and *mobile* transit …

The plan of the city

The basic principles we must follow are these:

1 We must de-congest the centres of our cities.
2 We must augment their density.
3 We must increase the means for getting about.
4 We must increase parks and open spaces.

At the very centre we have the STATION with its landing stage for aero-taxis.

Running north and south, and east and west, we have the MAIN ARTERIES for fast traffic, forming elevated roadways 120 feet wide.

At the base of the sky-scrapers and all round them we have a great open space 2,400 yards by 1,500 yards, giving an area of 3,600,000 square yards, and occupied by gardens, parks and avenues. In these parks, at the foot of and round the sky-scrapers, would be the restaurants and cafés, the luxury shops, housed in buildings with receding terraces: here too would be the theatres, halls and so on; and here the parking places or garage shelters.

The sky-scrapers are designed purely for business purposes.

On the left we have the great public buildings, the museums, the municipal and administrative offices. Still further on the left we have the 'Park' (which is available for further logical development of the heart of the city).

On the right, and traversed by one of the arms of the main arterial roads, we have the warehouses, and the industrial quarters with their goods stations.

All round the city is the *protected zone* of woods and green fields.

Further beyond are the *garden cities*, forming a wide encircling band …

The city

Here we have twenty-four sky-scrapers capable each of housing 10,000 to 50,000 employees; this is the business and hotel section, etc., and accounts for 400,000 to 600,000 inhabitants.

The residential blocks, of the two main types

already mentioned, account for a further 600,000 inhabitants.

The garden cities give us a further 2,000,000 inhabitants, or more.

In the great central open space are the cafés, restaurants, luxury shops, halls of various kinds, a magnificent forum descending by stages down to the immense parks surrounding it, the whole arrangement providing a spectacle of order and vitality …

Garden cities

A simple phrase suffices to express the necessities of tomorrow: WE MUST BUILD IN THE OPEN. The lay-out must be of a purely geometrical kind, with all its many and delicate implications …

The city of today is a dying thing because it is not geometrical. To build in the open would be to replace our present haphazard arrangements, *which are all we have today*, by a *uniform* lay-out. Unless we do this *there is no salvation.*

The result of a true geometrical lay-out is *repetition.*

The result of repetition is a *standard*, the perfect form (i.e. the creation of standard types). A geometrical lay-out means that mathematics play their part. There is no first-rate human production but has geometry at its base. It is of the very essence of Architecture. To introduce uniformity into the building of the city we must *industrialize building*. Building is the one economic activity which has so far resisted industrialization. It has thus escaped the march of progress, with the result that the cost of building is still abnormally high …

The 'interesting' or erratic site absorbs every creative faculty of the architect and wears him out. What results is equally erratic: lopsided abortions; a specialist's solution which can only please other specialists.

We must build *in the open*: both within the city and around it.

Then having worked through every necessary technical stage and using absolute ECONOMY, we shall be in a position to experience the intense joys of a creative art which is based on geometry.

The city and its aesthetic

(The plan of a city which is here presented is a direct consequence of purely geometric considerations.)

A new unit *on large scale* (400 yards) inspires everything. Though the gridiron arrangement for the streets every 400 yards (sometimes only 200) is uniform (with a consequent ease in finding one's way about), no two streets are in any way alike. This is where, in a magnificent contrapuntal symphony, the forces of geometry come into play.

Suppose we are entering the city by way of the Great Park. Our fast car takes the special elevated motor track between the majestic sky-scrapers: as we approach nearer there is seen the repetition against the sky of the twenty-four sky-scrapers; to our left and right on the outskirts of each particular area are the municipal and administrative buildings; and enclosing the space are the museums and university buildings.

Then suddenly we find ourselves at the feet of the first sky-scrapers. But here we have, not the meagre shaft of sunlight which so faintly illumines the dismal streets of New York, but an immensity of space. The whole city is a Park. The terraces stretch out over lawns and into groves. Low buildings of a horizontal kind lead the eye on to the foliage of the trees. Here is the CITY with its crowds living in peace and pure air, where noise is smothered under the foliage of green trees. The chaos of New York is overcome. Here, bathed in light, stands the modern city.

Our car has left the elevated track and has dropped its speed of sixty miles an hour to run gently through the residential quarters. The 'set-backs' permit of vast architectural perspectives. There are gardens, games and sports grounds. And sky everywhere, as far as the eye can see. The square silhouettes of the terraced roots stand clear against the sky, bordered with the verdure of the hanging gardens. The uniformity of the units that compose the picture throw into relief the firm lines on which the far-flung masses are constructed. Their outlines softened by distance, the sky-scrapers raise immense geometrical facades all of glass, and in them is reflected the blue glory of the sky. An overwhelming sensation. Immense but radiant prisms …

The traveller in his airplane, arriving from Constantinople or Pekin it may be, suddenly sees appearing through the wavering lines of rivers and patches of forests that clear imprint which marks a city which has grown in accordance with the spirit of man: the mark of the human brain at work.

As twilight falls the glass sky-scrapers seem to flame.

This is no dangerous futurism, a sort of literary dynamite flung violently at the spectator. It is a spectacle organized by an Architecture which uses plastic resources for the modulation of forms seen in light.

Source: Le Corbusier, 1929/1987, pp.164–79

READING 7B
Jane Jacobs: 'The generators of diversity'

To understand cities, we have to deal outright with combinations or mixtures of uses, not separate uses, as the essential phenomena …

A mixture of uses, if it is to be sufficiently complex to sustain city safety, public contact, and cross-use, needs an enormous diversity of ingredients. So the first question – and I think by far the most important question – about planning cities is this: how can cities generate enough mixture among uses – enough diversity – throughout enough of their territories, to sustain their own civilization?

Although it is hard to believe, while looking at dull grey areas or at housing projects or at civic centres, the fact is that big cities *are* natural generators of diversity and prolific incubators of new enterprises and ideas of all kinds. Moreover, big cities are the natural economic homes of immense numbers and ranges of small enterprises. …

The diversity, of whatever kind, that is generated by cities rests on the fact that in cities so many people are so close together, and among them contain so many different tastes, skills, needs, supplies, and bees in their bonnets.

Even quite standard, but small, operations like proprietor-and-one-clerk hardware stores, drugstores, candy stores, and bars can and do flourish in extraordinary numbers and incidence in lively districts of cities because there are enough people to support their presence at short, convenient intervals, and in turn this convenience and neighbourhood personal quality are big parts of such enterprises' stock in trade. Once they are unable to be supported at close, convenient intervals, they lose this advantage. In a given geographical territory, half as many people will not support half as many such enterprises spaced at twice the distance. When distance inconvenience sets in, the small, the various, and the personal wither away …

With urbanization, the big get bigger, but the small also get more numerous.

Smallness and diversity, to be sure, are not synonyms. The diversity of city enterprises includes all degrees of size, but great variety does mean a high proportion of small elements. A lively city scene is lively largely by virtue of its enormous collection of small elements.

… wherever we find a city district with an exuberant variety and plenty in its commerce, we are apt to find that it contains a good many other kinds of diversity also, including variety of cultural opportunities, variety of scenes, and a great variety in its population and other users. This is more than coincidence. The same physical and economic conditions that generate diverse commerce are intimately related to the production, or the presence, of other kinds of city variety.

But although cities may fairly be called natural economic generators of diversity and natural economic incubators of new enterprises, this does not mean that cities *automatically* generate diversity just by existing. They generate it because of the various efficient economic pools of use that they form. Wherever they fail to form such pools of use, they are little better, if any, at generating diversity than small settlements. And the fact that they need diversity socially, unlike small settlements, makes no difference. For our purposes here the most striking fact to note is the extraordinary uneven-ness with which cities generate diversity.

On the one hand, for example, people who live and work in Boston's North End, or New York's upper East Side or San Francisco's North Beach–Telegraph Hill, are able to use and enjoy very considerable amounts of diversity and vitality. Their visitors help immensely. But the visitors did not create the foundations of diversity in areas like these, nor in the many pockets of diversity and economic efficiency scattered here and there, sometimes most unexpectedly, in big cities. The visitors sniff out where something vigorous exists already, and come to share it, thereby further supporting it.

At the other extreme, huge city settlements of people exist without their presence generating anything much except stagnation and, ultimately, a fatal discontent with the place. It is not that they are a different kind of people, somehow duller or unappreciative of vigour and diversity. Often they include hordes of searchers, trying to sniff out these attributes somewhere, anywhere. Rather, something is wrong with their districts; something is lacking to catalyse a district population's ability to interact economically and help form effective pools of use …

So long as we are content to believe that city diversity represents accident and chaos, of course its erratic generation appears to represent a mystery.

However, the conditions that generate city diversity are quite easy to discover by observing places in which diversity flourishes and studying the economic reasons why it can flourish in these places. Although the results are intricate and the ingredients producing them may vary enormously, this complexity is based on tangible economic relationships which, in principle, are much simpler than the intricate urban mixtures they make possible.

To generate exuberant diversity in a city's streets and districts four conditions are indispensable:

1 The district, and indeed as many of its internal parts as possible, must serve more than one primary function; preferably more than two. These must ensure the presence of people who go outdoors on different schedules and are in the place for different purposes, but who are able to use many facilities in common.
2 Most blocks must be short; that is, streets and opportunities to turn corners must be frequent.
3 The district must mingle buildings that vary in age and condition, including a good proportion of old ones so that they vary in the economic yield they must produce. This mingling must be fairly close-grained.
4 There must be a sufficiently dense concentration of people, for whatever purposes they may be there. This includes dense concentration in the case of people who are there because of residence.

The necessity for these four conditions is the most important point ... In combination, these conditions create effective economic pools of use. Given these four conditions, not all city districts will produce a diversity equivalent to one another. The potentials of different districts differ for many reasons; but, given the development of these four conditions (or the best approximation to their full development that can be managed in real life), a city district should be able to realize its best potential, wherever that may lie. Obstacles to doing so will have been removed. The range may not stretch to African sculpture or schools of drama or Roumanian tea houses, but such as the possibilities are, whether for grocery stores, pottery schools, movies, candy stores, florists, art shows, immigrants' clubs, hardware stores, eating places, or whatever, they will get their best chance. And along with them, city life will get its best chances ...

All four in combination are necessary to generate city diversity; the absence of any one of the four frustrates a district's potential.

Source: Jacobs, 1961, pp.155–63

READING 7C
Nicholas Howe: 'Berlin Mitte'

When Berliners asked me last summer, usually with some apprehension, how I liked the city, they would immediately understand my response that 'like' was not the word for the place. But that easy point made, one still needed to talk about the city. At such moments, I fell into a phrase that seemed to satisfy Berliners and that allowed me to evade the force of their question. I would say that Berlin was history interrupted by construction. And if I said this in Mitte, the center of the city, there was a quick nod of recognition, for everywhere on its streets you stumble over construction and reconstruction, you feel the vibration of heavy equipment, you hear the Arabic, Polish, and English spoken by workers on scaffolds. As you walk through Mitte, your shoes pick up a thin coat of dust from its building sites. Here, construction-watching has been made into an art form: Potsdamer Platz, where once there was only the 'death strip', you can gaze out at what is called – in the inevitable formula, whether English or German – 'the largest construction site in the world'. Here the cranes have been, quite literally, choreographed to perform a ballet of mechanical movement as public performance.

For all that Berlin is a spectacle, you realize after a time that the money and energy thrown into constructing and reconstructing the city are no more than an interruption. After the equipment and workers are gone, the corporate headquarters are finished, and the art deco buildings refurbished, memories of all that happened here will still remain. The Wall is gone, but its path will be marked out again in a few years by a run of new construction from Checkpoint Charlie to Potsdamer Platz to the River Spree. The buildings of the late 1990s will divide the reunified Berlin of the future as surely as the city was divided from 1961 to 1989 by the Wall. For all their postmodern glamour, the corporate palaces of Sony, Daimler-Benz, and others will mark the map of Berlin as scar tissue. The world may close, but its trace will remain. In the meantime, it's necessary to buy, through building and rebuilding, time away from all the questions that haunt one about the city ...

Construction/reconstruction

Walking on Friedrichstrasse from Unter den Linden to the train station, a distance of three or four blocks, means weaving back and forth from one sidewalk to the other to avoid building sites. You make progress, but not in a straightforward way. Each day, it seems, the barriers have been shifted slightly, so today's route is not the same as yesterday's, nor will it be the same as tomorrow's.

These subtle shifts in daily route force shifts in the way you look at the same site. Crossing at midstreet instead of at a corner means looking at the flow of buildings in ways not possible the day before. Changes in the urban landscape appear: as one building comes down, the view of another opens and what was hidden emerges, though it may soon disappear from view as a new building appears. The history of the city gets played out in its architectural styles. Nineteenth- and early twentieth-century buildings that survived the bombing are being restored, and new glass boxes are being erected in the open spaces. From a bench on Unter den Linden you can watch a steady flow of flatbed trucks loaded with plasterboard, dump trucks filled with torn-up masonry or paving, and cement mixers churning away. The sign on one truck could serve as a motto of this new Berlin: *für Bau und Aufbau* (for construction and reconstruction). When you look up from street level, the horizon line is filled with construction cranes swinging their loads. From one spot on the edge of Potsdamer Platz I counted twenty-nine cranes before the light turned green and I moved on.

There are disconcerting, even shocking juxtapositions in Mitte between the most elegant art deco hotel covered with sexy nymphs in stone and the shabbiest, poured concrete, pre-fab building left over from the German Democratic Republic (GDR), one of the ubiquitous *Plattenbauten*. It would be hard, if you did not know the narrative of Berlin, to comprehend how it happened that such buildings came to stand next to each other. For how could any one place have such impossibly different senses of how to build and thus of how to live?

Perhaps the most troubling site of construction and reconstruction in Mitte is the Palace of the Republic, which sits on the far end of Unter den Linden from the Brandenburg Gate. Occupying the site of what had been the imperial place until it was razed because of World War II bomb damage, the Palace of the Republic today stands empty and abandoned because it is poisoned with asbestos. Surrounded by a chain-link fence, it is enormous, evocative, too eloquent a relic from the communist era to be treated merely as a building. And so, overcharged with symbolism, it remains while Berliners argue about whether it should be torn down – because it is ugly and contaminated and thus a reminder of the communist past – or be restored – because it was a place where people enjoyed themselves at restaurants and bowling alleys and thus a reminder of the communist past. Or, and this is no cheap paradox, that it should be preserved for exactly the reasons some would tear it down, or torn down for exactly the reasons some would preserve it.

Today, the bronze mirror-glass facade of the palace reflects little except the tour buses parked in what was once Marx-Engels Place. They pile in there, their drivers happy to escape the snarl of Berlin traffic. Wisecracks about the locomotive of history being replaced by tour buses forever parked in front of the abandoned Palace of the Republic are easy, almost too easy, to make. More eloquent was the scene I witnessed one afternoon of a group of ten-year-olds on an excursion with their teachers. As they walked the length of the building, almost two hundred meters along the fence, not one of them stopped to look up at the Palace. Even ten years ago that would have been a destination for their school trip; now it's only the edge of a dusty square filled with buses, piles of building material, dumpsters.

At certain sites, this process of rebuilding seeks some architectural and thus political adjustment between old and new. The Deutsche Bank on the corner of Unter den Linden and Charlottenstrasse has been impeccably restored to its pre-World War II condition; its immaculate stonework and polished metal proclaim it to be the headquarters of a long-established concern, though one absent from the neighbourhood for some time due to unavoidable circumstances. But it's back now, and the parade of black Mercedes-Benzes is there in front to assure us that important business is being conducted within. To complete the restoration, a wall of plate glass has been erected across the façade of the bank's upper four stories. The effect is striking because the glass seals the building off from the city as if it were on display behind a museum case. This glass suggests a desire to ward off the working of time and history, to exist hermetically apart from the grime and pollution that fill the Berlin air, once famously pure and now redolent with complicated odors. Looking at the wall – no less impermeable for being transparent – reminds you that banks are well practiced in the art of historical accommodation.

Source: Howe, 1998, pp. 71-3

READING 7D
Gerry Stoker: 'Regime theory and urban politics'

Regime theory provides a new perspective on the issue of power. It directs attention away from a narrow focus on power as an issue of social control towards an understanding of power expressed through social production. In a complex, fragmented urban world the paradigmatic form of power is that which enables certain interests to blend their capacities to achieve common purposes. Regime analysis directs attention to the conditions under which such effective long-term coalitions emerge in order to accomplish public purposes ...

Regime theory, pluralism and neo-pluralism

... Regime theory takes as given a set of government institutions subject to some degree of popular control and an economy guided mainly but not exclusively by privately controlled investment decisions. A regime is a set of 'arrangements by which this division of labour [between popularly controlled government and privately controlled economy] is bridged' (Stone, 1993, p.3).

... regime theorists have taken on board the central thrust of much Marxist inspired work of the 1970s. Namely that business control over investment decisions and resources central to societal welfare give it a privileged position in relation to government decision making. In the words of Stone 'we must take into account these contextual forces – the facet of community decision-making I label "systemic power"'. He continues: 'public officials form their alliances, make their decisions and plan their futures in a context in which strategically important resources are hierarchically arranged ... Systemic power therefore has to do with *the impact of the larger socioeconomic system on the predispositions of public officials*' (1980, p.979, original emphasis).

Regime theorists argue that 'politics matters' ... The founding premise of regime theory is that urban decision makers have a relative autonomy. Systemic power is constraining but scope for the influence of political forces and activity remain. Regime theorists argue that the organization of politics leads to very inadequate forms of popular control and makes government less responsive to socioeconomically disadvantaged groups. The organization of politics does not facilitate large-scale popular participation and involvement in an effective way ...

In the policy debate within cities one solution and one view about how to proceed tends to dominate. Problem solving in these circumstances is not likely to be to the benefit of citizens because desirable alternatives go unexplored.

All this suggests that regime theory stands on different ground to the 'classical urban pluralism, the reigning wisdom of 30 years ago and earlier' (Stone, 1993, p.1). Yet David Judge (1995) is right to suggest that regime theorists do share common ground with the revised statements of pluralists such as Dahl and Lindblom. In many respects regime theory takes as its starting point many of the concerns of 'neo-pluralists' (as defined by Dunleavy and O'Leary, 1987, ch. 6). It accepts the privileged position of business. It is concerned about the limits to effective democratic politics. It also shares the 'neo-pluralist' concern with the fragmentation and complexity of governmental decision making. Stone comments: 'we have a special need to think about what it means politically to live under conditions of social complexity ... about the special character of power relationships in complex social systems' (1986, p.77) ...

The contribution of regime theory

What is attractive about regime theory is that it begins to address the questions which flow from the common ground it shares with 'neo-pluralists'.

- What are the implications of social complexity for politics?
- What does the systemic advantage of certain interests imply for the nature of urban politics?
- What forms of power dominate modern systems of urban governance?
- What role is there for democratic politics and the role of disadvantaged groups?

Regime theory moves beyond 'neo-pluralism' by offering a series of distinctive answers to these questions and provides a broad framework for analysis to examine the variety of politics within cities.

In this section ... the discussion concentrates on the work of Clarence Stone. His work represents the most advanced application of regime analysis ...

Complexity is central to the regime perspective (Stone, 1986). Institutions and actors are involved in an extremely complex web of relationships. Diverse and extensive patterns of interdependence characterize the modern urban system. Lines of causation cannot be easily traced and the policy world is full of unpredicted spillover effects and unintended consequences. Fragmentation and lack of consensus also characterize the system. 'Many activities are autonomous and middle-range accommodations are worked out. In some ways the ... world is chaotic;

certainly it is loosely coupled, and most processes continue without active intervention by a leadership group' (Stone, 1989, p.227).

This kind of society 'does not lend itself to the establishment of direct and intense control over a large domain in a wide scope of activity' (Stone, 1986, p.89). Such command or social control power is limited to particular aspects or segments of society. The study of regime politics focuses on how these limited segments or domains of command power combine forces and resources for 'a publicly significant result' – a policy initiative or development.

Complexity and fragmentation limits the capacity of state as an agency of authority or control. Nor can the state simply be seen as an arbiter or judge of competing societal claims. Rather 'as complexity asserts itself government becomes ... more visible as a mobilizer and co-ordinator of resources' (Stone, 1986, p.7). It is this third type of governmental activity which is particularly the target of regime analysis ...

The state can on occasion still impose its will. It can also mediate between parties. Yet authoritative action and pluralist bargaining capture 'only a small part of political life in socially complex systems' (Stone, 1986, p.88). Politics in complex urban systems is about establishing overarching priorities and 'the issue is how to bring about enough co-operation among disparate community elements to get things done' (Stone, 1989, p.227.) Politics is about government working with and alongside other institutions and interests and about how in that process certain ideas and interests prevail.

The point is that 'to be effective, governments must *blend* their capacities with those of various non-governmental actors' (Stone, 1993, p.6, original emphasis). In responding to social change and conflict governmental and non-governmental actors are encouraged to form regimes to facilitate action and empower themselves. Thus following Stone (1989, p.4) a regime can be defined as 'an informal yet relatively stable group *with access to institutional resources* that enable it to have a sustained role in making governing decisions' (original emphasis). Participants are likely to have an institutional base – that is, they are likely to have a domain of command power. The regime, however, is formed as an informal basis for coordination and without an all encompassing structure of command.

Regimes operate not on the basis of formal hierarchy. There is no single focus of direction and control. But neither is regime politics governed by the open-ended competitive bargaining characteristic of some pluralist visions of politics. Regimes analysts point to a third mode of coodinating social life: the network. The network approach, like regime analysis, sees effective action as flowing from the cooperative efforts of different interests and organizations. Cooperation is obtained, and subsequently sustained, through the establishment of relations premised on solidarity, loyalty, trust and mutual support rather than through hierarchy or bargaining. Under the network model organizations learn to cooperate by recognizing their mutual dependency ...

Regime partners are trying to assemble long-running relationships rather than secure for themselves access to immediate spoils ... 'Politics is about the production rather than distribution of benefits ... Once formed, a relationship of cooperation becomes something of value to be protected by all of the participants' (Stone, 1993, pp.8–9). Politics is not then the fluid coalition building characteristic of many versions of pluralism. Regime theory focuses on efforts to build more stable and intense relationships in order that governmental and non-governmental actors can accomplish difficult and non-routine goals.

Politics is about achieving governing capacity which has to be created and maintained. Stone (1989) refers to power being a matter of social production rather than social control. In contrast to the old debate between pluralists and elitists which focused on the issue of 'Who Governs?' the social production perspective is concerned with a capacity to act: 'What is at issue is not so much domination and subordination as a capacity to act and accomplish goals. The power struggle concerns, not control and resistance, but gaining and fusing a capacity to act – power to, not power over' (Stone, 1989, p.229).

Unlike elite theorists regime theory recognizes that any group is unlikely to be able to exercise comprehensive control in a complex world. Regime analysts, however, do not regard governments as likely to respond to groups on the basis of their electoral power or the intensity of their preferences as some pluralists do. Rather, governments are driven to cooperate with those who hold resources essential to achieving a range of policy goals. As Stone argues: 'Instead of the power to govern being something that can be captured by an electoral victory, it is something created by bringing cooperating actors together, not as equal claimants, but often as unequal contributors to a shared set of purposes' (1993, p.8). Regime theorists emphasize how the structure of society privileges the participation of certain interests in coalitions. As Stone (1986, p.91) comments, for actors to be effective regime partners two characteristics seem especially appropriate: first, possession of strategic knowledge of social transactions and a capacity to act on the basis of that knowledge; and second, control of resources that make one an attractive coalition partner.

The US-based regime literature ... sees two groups as the key participants in most localities: elected officials and business. Beyond this, however, there is recognition that a variety of other community interests may be drawn in based on minorities, neighbourhoods or even organized labour. Writing with a comparative perspective means that it is useful to add a fourth broad category: technical/professional officials. Such officials may well be influential participants in some US coalitions but in other western democracies, especially in Europe, their leading role is difficult to deny. These officials ... may be employed by elected local government, work for various non-elected local agencies or be local agents of various central or regional government departments. Knowledge joins economic position as a key resource that gives groups privileged access to decision making.

Regime theory is concerned more with the process of government-interest group mediation than with the wider relationship between government and its citizens. Regime theory views power as structured to gain certain kinds of outcomes within particular fields of governmental endeavour. The key driving force is 'the internal politics of coalition building' (Stone, 1989, p.178). If capacity to govern is achieved, if things get done, then power has been successfully exercised and to a degree it is irrelevant whether the mass of the public agreed with, or even knew about, the policy initiative.

Yet regime theory gives some recognition to the role of popular politics, elections and public participation in liberal democratic politics. Opposition to policy agendas that are being pursued can be mobilized and disrupt established policy regimes ... Established regimes, however, can be expected to seek to incorporate certain marginal groups, to make them part of their project. People are brought in, it is suggested, less by being sold 'big ideas' or 'world views' and more by small-scale material incentives (Stone, 1989, pp.186–91). Regimes may also practise a politics of exclusion, seeking to ensure that certain interests are not provided with access to decision making. Generally all regimes have to develop strategies for coping with the wider political environment ...

The task of regime formation is about gaining a shared sense of purpose and direction. This in turn is influenced by an understanding of what is feasible and what is not. Feasibility favours linking with resource-rich actors. It also favours some goals over others whose achievement may be more intractable and problematic. The 'iron law', as it were, governing regime formation is that 'in order for [a] governing coalition to be viable, it must be able to mobilize resources commensurate with its main policy agenda' (Stone, 1993, p.21).

References

Dunleavy, P. and O'Leary, B. (1987) *Theories of the State,* London, Macmillan.

Judge, D. (1995) 'Pluralism', in Judge, D., Stoker, G. and Wolman, H. (eds) *Theories of Urban Politics,* London, Sage.

Stone, C. (1980) 'Systemic power in community decision-making: a restatement of stratification theory', *American Political Science Review,* vol.74, no.4, pp.978–90.

Stone, C. (1986) 'Power and social complexity', in Waste, R. (ed.) *Community Power: Directions for Future Research,* Newbury Park, Sage.

Stone, C. (1989) *Regime Politics: Governing Atlanta, 1946–1988,* Lawrence, University of Kansas Press.

Stone, C. (1993) 'Urban regimes and the capacity to govern', *Journal of Urban Affairs,* vol.15, no.1, pp.1–28.

Source: Stoker, 1995, pp.54-61

CHAPTER 8

On orderings and the city

by Gerry Mooney, Steve Pile and Christopher Brook

1 *The jumbled orderings of the city*

At the start of this book it was argued that cities are characterized by heterogeneity. This heterogeneity is not, however, a simple 'given' in the analysis of this book. It arises out of the ways in which cities bring people, things and activities together, enabling them to mix and meet (or not). But it isn't as simple as that either. For cities are also characterized by the relative homogeneity of certain spaces within them in comparison with other spaces. For this reason, the book has made much of the differentiation of spaces within the city, from the inner city to the suburb, whether in Istanbul, São Paulo or New York. Cities seem to have contradictory attributes: on the one hand their *heterogeneity* and, on the other, the apparent *homogeneity* of specific places or areas within them. Instead of thinking of this contradiction as an insoluble problem or a logical flaw, however, we have tended to see this tension between heterogeneity (or difference) and homogeneity (or community) as a product of the ways in which cities bring different things, people and activities together, or keep them apart.

We have argued that the mixing and meetings, the patternings and orderings of the city are best understood spatially. The spatiality of cities has been addressed from many different directions: sometimes by tracing wider connections into the city, seeing how they constitute places (as in Berlin, Manchester or Sydney); and sometimes by looking at the ways in which people move around (or not) within cities (as in Paris, Johannesburg or Glasgow). There has also been a focus on the 'within-ness' of cities: the ways in which urban spaces are produced to have certain social effects, the ways in which urban spaces have effects on social relationships, and the ways in which cityness is lived and experienced.

In Chapter 3, for example, we saw that Paris was consciously made into a more beautiful and more orderly city by intervening in the built form to improve the circulation of people and goods. It is for this reason that Paris was seen as 'the capital of the nineteenth century': the place where new-ness was encapsulated and embodied, and why Paris even today persists as a centre of fashion (see **Allen, 1999a**). But these interventions in Paris's built form had other social aims. They enabled the authorities to control 'unruly' elements in other quarters by permitting troops to move faster around the city, while the wide streets were harder to barricade. In this vignette, we can see that the city is an assemblage of bricks and mortar, cross-cutting social (class, 'race', gender, political) relationships and technical expertise, whether such expertise relates to governance, administration or policing, or to the arts of living, 'getting by' or acting politically. In Paris, as elsewhere, the outcomes have been less certain. Indeed, they have been continually renegotiated, and it is through these renegotiations, tied as they are to unequal relations of power, that the heterogeneities and homogeneities of the city and its spaces are produced, maintained, challenged and transformed.

In Chapters 6 and 7 we also saw that designs for the city, with their attendant visions of how urban society should be, have been produced by urban planners with the intention of improving the cityness of cities. These actual and conceptual attempts to plan the orderly city have produced particular kinds of urban form. In Paris and Istanbul, in Chandigarh and Milton Keynes, different urban spaces are clearly demarcated: each space has its own function, its own role to play in the living machine that is the city. In Chandigarh, for example, Le Corbusier planned the city along the lines of a human body. At the 'top' or 'head', housed in typically modernist office blocks, are the administrative functions of the state (Chandigarh was designed to be the state capital of the Punjab). Below that, in the 'lungs' of the city, is a series of rib-like housing blocks, built for the office workers. These lungs were placed at sufficient distance to allow air to circulate through the city, to allow it to breathe. In this way, Le Corbusier designed his perfect city along the lines of the perfect body, with all its functions perfectly integrated and working in harmony. But some things have not worked out as Le Corbusier imagined. Today, a shanty town is developing somewhere below the waistline! Furthermore, the tower blocks are used for purposes other than those imagined by Le Corbusier, and other kinds of houses have sprung up, filling the spaces left blank on Le Corbusier's plan. Whether seen in plan form or in sketches illustrating 'what it will be like' when it is built, the careful and conscious production of urban space gives a sense that the city can be rationally planned and socially ordered. However, this narrative has been undermined by the analyses of the city that have been presented in this book.

FIGURE 8.1 *What should the metropolis be like? A scene from the 1930 film* Just Imagine

Broadly, the sense that there can be, or should be, an ordered and orderly city is complicated by the following points:

- The derivations and perceptions of 'order' in the city are many, and the various 'orderings' of the city can often be in conflict with one another. We have seen that pressures to integrate Istanbul into the global economy have been accompanied by attempts to order it in particular ways, yet this conflicts with other kinds of order within the city, by disrupting pre-existing social hierarchies and their spatial patterns.
- Orderings also involve disorderings, either because they are an attempt to 'cure' or 'solve' disorders, or because the imposition of an ordering disorders other aspects of city life. Thus, while moving people from slums into tower blocks has been an attempt to 'cure' the ills of the inner city, it has also created even worse conditions, by disconnecting or excluding people from pre-existing social networks based on home and work.
- While disorder is often seen as a threat, it may also be viewed as enriching a city's character. So, for Jane Jacobs, in Boston and New York there is a vibrancy and creativity that flows from the encounters between different people. Indeed, it is this that produces new 'hybridities', new experiences, new kinds of subjectivities (as **Massey, 1999** argues).

As this set of bullet points builds up, a fuller picture emerges of the internal differentiation of cities. This differentiation is produced

- not by one order or a reaction against one disorder,
- but by *the jumbled orderings of the city* – orderings that may be characterized by some as orderly, but by others as disorderly.

It is the task of this final chapter to tease out these jumbled orderings of the city: that is, to think about how they emerge out of, or are imposed on, city life and how it is that people intervene to produce 'orders' and 'disorders' of various kinds. We will develop this analysis in sections 2 and 3 below. Before we get to these, however, let's think some more about the 'jumbled orderings' of the city.

ACTIVITY 8.1 Turn now to Reading 8A, which broaches these entangled issues about the intensity and diversity of city life, order/disorder and community/difference, and the internal spatial relations of the city. Before you start to read the extract, however, some context setting is appropriate. In this extract from his book *The Uses of Disorder* (1971/1996), Richard Sennett makes an impassioned plea for an 'anarchist' city – not an anarchy that suggests that there are 'no laws' (as anarchy is often read), but an anarchy where laws are unnecessary. As you read the extract, rather than agree or disagree with Sennett's political philosophy, it is important that you think about his understanding of:

- the relationship between 'disorder' and 'city life' and to think about the relationship between these;
- what city life is like;

- how urban life should be reorganized and how this reorganization involves transforming the city spatially. For example: What kind of built environment does he envisage? How are people to come together within cities? How are city spaces to be differentiated? ◆

First, you will have picked up Sennett's sense that city life should be vibrant and intense. More specifically, in his account we can see that the kind of vibrancy he wishes to promote emerges out of the juxtaposition of difference. Sennett is keen to intervene in city life to 'mix up' different people by bringing them into close proximity (with one another). But what are the social outcomes of juxtaposing differences within cities? It has been a significant argument in this book that it matters *how* people are brought together, or are segregated from one another. This argument has raised questions about the power relations within cities, and this in turn raises questions about urban politics, the politics both of city administration and of people's everyday lives. As was stated in Chapters 5 and 7, there is a politics of difference in the city, and it is important to note that this politics doesn't necessarily mean that cities are more tolerant, or more indifferent, since difference can equally induce bigotry and hostility (as we have seen). What is important for our purposes is that in order to produce a creative, mature city life, Sennett wishes

- to intervene in the spatiality of the social relationships that constitute that city life, creating neighbourhoods where different groupings are juxtaposed.

In fact, what appears to matter are the inequalities that underlie these mixings and orderings, together with an 'openness' to movement and settlement within and beyond the city. We will follow up these issues in section 3.

Second, we can see (though somewhat implicitly) that the physical form of the city intervenes in, and affects, city life. In the early part of this book we saw many examples of how the structure of neighbourhoods and districts in the city makes a difference to how the orderliness and unruliness of city life are viewed. Chapters 1 and 3, for example, drew on the work of Jane Jacobs to show that the seeming chaos of the street is underlain by other, more tacit kinds of ordering, while in Chapters 2 and 4 we saw how urban planners have sought to improve the life of the poor by building housing estates.

How the city is built makes a difference to the kinds of social relations that can take place. Thus empty streets and impoverished housing stock mean that certain kinds of social interactions are more – or less – likely. From Sennett's perspective, the city has to be built in a form that fosters the kinds of social relationships he wants to see. We have seen that producing particular built forms – whether this is intended to 'intensify' or 'de-intensify' cityness – is fraught with difficulties and ambiguities (see, for example, Chapter 2 on peripheral housing estates in Glasgow). In part, these difficulties arise because interventions in the built environment are very often underpinned by a singular ideal (perhaps a disentangled, de-intensified way of life, or the thoroughly entangled, disorderly city envisaged by Sennett). However, this narrowing of ideals ignores the multiplicity of social relations (and the various and variable forms of exclusion and marginalization) that exist in cities. Nevertheless, it seems necessary to make some kind of intervention in the physical form of the city. We will pick up these questions in section 2.

Taking these two points together, we can see that Sennett is attempting to re-invent the city. In his imagined (anarchist) utopia, he wishes to intervene in two interrelated aspects of the city:

- the physical form of the city in which social relationships play out, and
- the spatiality of these social relationships, through which the city is built and re-built.

The important point to recognize is that changing social relationships are accompanied by a changing spatiality. However, this cannot be understood as if specific interventions would inevitably produce particular effects, for good or ill. We need to understand something else about Sennett's argument.

Thus, finally, we can see in Sennett's argument that it matters who is deciding what is orderly and disorderly (an issue that was raised in Chapter 2), who is ordering and disordering, and how. In the final section of this chapter we will raise questions – as we have begun to do in the latter half of this book – about the city, citizenship and democracy. Significantly (and this point radically alters the politics of the city that Sennett envisages), we begin to see that city politics isn't simply about the 'within-ness' of cities, but also about the 'openness' of cities – to wider relationships, to alternative perspectives, both from within and from outside.

The built form of the city has often taken the foreground in attempts to envisage the 'good city'. By intervening in the built environment, it is hoped that the utopian elements of cities will be identified and fostered, while the dystopian elements are eliminated. We have seen that the utopian city has been designated as an orderly or a disorderly one. However, in this book we have suggested that orders and disorders, and indeed their kaleidoscopic character, are not so easily disentangled. If this is so, then we must raise questions about what interventions into the built environmental are seeking to achieve, and on whose behalf. It is to these issues that we turn next.

2 Interventions in urban space

Interventions in urban space result from ways of imaging city life, from particular understandings of urban order and disorder. They are also generally premised on specific knowledges of what constitutes 'urban problems'. Particularly in western discourses of city life, there are a number of recurrent themes that characterize the ways in which such 'problems' are constructed or defined, and how the policy 'solutions' are shaped and influenced. These interventions in urban space do not only stem from particular understandings of social or urban problems, but also from particular visions of what city life 'should be'. In other words, urban problems are defined against a particular urban ideal or utopia. Just as understandings of urban disorder stem from a sense of what order 'looks like', and vice versa, the depiction of city life as dystopian is also derived from a utopian vision of the city (see Chapter 7).

Throughout history, utopian and dystopian images have characterized discussions of the city. All utopias have their antithesis in dystopian visions of the city as a place 'out of control' or where social life has in some ways 'broken down': the orderly city is contrasted with the unruly city or with urban disorder; the sustainable city with cities 'in danger'. The nightmarish image and vision of city life as dystopian – images which featured in Chapters 2, 3 and 6 in particular – are

FIGURE 8.2 *A vision of the city: running from justice? A scene from the film* Bladerunner

not difficult to detect in stories of the city. Not only do they feature in numerous films, novels and media reportage of city life, but they have also been central to the ways in which the city has been theorized and to attempts to plan and order the city in particular ways. Let us consider three examples.

One of the most influential ways of thinking about the city is the notion of an *engineered city*. Imagining the city in this way has underpinned numerous urban policy interventions. Perhaps the most notable exponent of this notion of the city as something that could be engineered was Le Corbusier (see Chapter 7), whose ideas and visions of city life were strongly based on an understanding of the city as a 'machine for living'. Architecture, design and engineering for Le Corbusier were the key to transforming the urban landscape, thereby eradicating urban problems. Rationally conceived 'master' plans favoured geometric harmony, the zoning of land use and the segregation of different social groups (Gold, 1997, p.21). These grandiose plans either favoured centralized high-density agglomeration (for example Le Corbusier's plans for 'a city of towers' in the early 1920s), or decentralized schemes that relied on the development of mass transportation.

Major arterial highways, ring roads, mass housing estates, high-rise and deck-access forms of construction represented the material manifestation of such visions in cities throughout the world. As we saw in Chapter 7, for Le Corbusier the intensity of city life was not a problem, as long as this was planned and ordered in particular ways. But Le Corbusier's views of the role of 'modernist' and 'rational' planning have important historical precursors. From classical Rome through to the geometric city plans of Renaissance Europe, the planning of urban space has been central to attempts to impose certain forms of regulation and uniformity on urban life. 'Urban redevelopment' was as much a part of nineteenth-century life in many European cities as it was to be in the aftermath of the Second World War.

In the large cities of nineteenth-century Britain, for example, slum clearance and the dispersal of 'problem' populations, such as 'criminals' and 'the disreputable poor' of the rookeries of inner London, represented a major effort to 'plan out' and remove 'urban problems'. In much the same way, Haussmann's boulevards imposed a rational spatial order as part of an attempt to control the poor, the 'criminal' and the 'subversive', who were congregated in specific areas of Paris.

Le Corbusier and other proponents of the idea that cities could be 'made to work' through planning believed that urban life could be spatially created in ways that would generate a particular urban social order. Architecture and planning could bring about social change without transforming fundamental social and economic relations.

The idea of the engineered city stands in marked contrast to the idea of a *heavenly city*. Unlike the engineered city, the heavenly city and its dystopian counter, the *unheavenly city*, are underpinned by a strong anti-urbanism. In this imagery urban problems are interpreted largely as a consequence of anomie and the absence of community attachments in the modern city. In Banfield's (1968) study, for example, urban social unrest in US cities during the 1960s was attributed in the main to 'family breakdown' and to the creation of 'ghetto subcultures' which were characterized by a lack of attachment to 'mainstream' social values and institutions. There are direct links between this perspective and the underclass arguments that we met in Chapter 2, and with the depiction of certain housing estates, shanty towns and inner cities as disorganized locales (see Chapters 1, 2 and 4).

The utopian notion of a heavenly city, with its conservative emphasis on authority, community, hierarchy and order, is also reflected in some of the ideas of the garden city movement, and in the writings of its most vociferous proponents, such as Ebenezer Howard (see Chapters 6 and 7). Not only is a pervasive anti-urbanism evident in the nostalgia for idealized stable, rural communities, but there is an implicit assumption that cities destroy community life, through fragmentation and hyper-individualism, allied to their intensity. Such views were, of course, not uncommon among urban theorists such as Georg Simmel and Louis Wirth (**Pile, 1999**; **Allen, 1999b**).

The Chicago School has been credited with developing one of the most enduring notions of the city: the idea of the *city as ecology*. This vision of the city relies on the use of ecological metaphors, with the development, spread and processes of urban spatial differentiation seen as the outcome of ecological processes. For theorists of the Chicago School such as Park, Burgess and McKenzie, the human world, like the plant world, could best be understood as being characterized by competition for the basic resources of life. This competition gives rise to the differentiation of the division of labour and of functions. This leads to an orderly spatial distribution of competing classes and functions to the areas of the city for which they are best suited. Thus the city develops a series of 'natural areas' which are characterized by particular locations, housing types and social compositions.

Despite the weight of the criticisms directed against it, an ecological view of the city has survived in many forms. For example, Mike Davis (see Reading 3B) suggests that Los Angeles is in the grip of an 'ecology of fear' – an 'ecology' that stems, we would argue, from Los Angeles' particular city-nature formation, which combines the threats of earthquakes, pollution, pestilence, crime, heat and so on (see **Hinchliffe, 1999,** on city-nature formations). This 'ecological thinking' is further evident in the identification of particular problem areas where crime and other social ills are perceived to be prevalent. In Chapters 1 and 2 we saw that the 'inner city', the 'ghetto' and the 'shanty town' occupy a distinctive position within such stories of urban problems. Such places are constructed as centres of 'disorder' and residual 'social disorganization', where (only) the socially excluded reside. The argument that such places and their populations require to be managed and regulated is often premised on a vision of an ecological patterning of the city into good/bad areas, with bad areas requiring targeted policy interventions to cure the pathologies located only within them.

Despite the significant differences that exist between these competing notions and images of city life, they share similar concerns with the (dis)ordering(s) of the city and with the *intensity* of urban life. In diverse ways they offer physical solutions to the challenges created by the intensification of social relations in cities. There is a strong belief, particularly in the vision of the engineered city, that perceived social problems, from crime to urban sprawl, could be overcome through planning, zoning and segregation. This is reflected not only in mass housing estate developments and in many new towns, but also in the racialized segregation of the apartheid city (see Chapter 4).

We have already highlighted the dominant anti-urbanism in the heavenly city imagery. Such sentiments are reflected in arguments that cities need to be physically contained, with their growth and spread controlled through 'green belts', as in post-war London and Clydeside. It is also a feature of the imagery of the idealized semi-rural suburb, in Istanbul as much as North America and England, as a 'retreat' from the intensity of the city.

There are many other utopian/dystopian visions of urban life, which either conflict with or share some of the assumptions in the three visions discussed above. The vision and imagery of ideal or good cities has been significant in generating the language and ideas through which urban problems have been defined and policies determined. How urban problems are defined and experienced, then, provides the bases for interventions in urban space, and is informed by assumptions of what makes for a 'good life' in the city – indeed, of what makes for a 'good city'.

There is a recurring theme running through this discussion, namely that social problems can be resolved through physical means. One of the most notable examples of this was explored in Chapter 3. In Oscar Newman's notion of 'defensible space' there are echoes of the engineered, ecological and heavenly city. Newman's argument was premised on the assumption that crime could be 'designed out'. Architecture, design and the layout of urban space could greatly reduce the opportunities for crime to be committed. As Eugene McLaughlin and John Muncie point out, there are conflicting views as to whether such schemes represent a 'good thing', but the influence of this way of thinking should not be underestimated.

However, there is no consistent vision of what makes for a 'good life' in the city, or what constitutes 'the good city'. Indeed, we can often detect several visions operating simultaneously, conflicting and competing with each other, underpinned as they are by different values, political outlooks and senses of urban order/disorder. Visions of the good city involve some sense and understanding of the ideal city form – its shape and organization (you might remind yourself of Activity 2.1 at this point). Should cities build more parks, roads and suburban shopping centres, and install more CCTV cameras? Are gated communities offering a solution to the problems of social disorganization, or do they intensify that disorganization by cutting off juxtaposed groups from one another? Do we require more residential segregation, or should urban differentiation be spatially ordered in different ways? (You might want to think back to the discussion of Sennett in section 1 here.) Do 'we' want suburban estates, high-rise housing developments, skyscrapers, manicured lawns, etc.?

Such questions involve specific understandings of urban orders and disorders, and these extrapolate into particular visions of the city. But, as we saw in

Chapter 3, attempts to reorder and de-intensify urban space in certain ways can further exacerbate processes of social marginalization and polarization. Nineteenth-century attempts to disperse 'problem populations' geographically in cities around the world, and the concern to curtail the spread of the ghetto and shanty town today represent physical 'solutions' to urban differentiation, but work only to further differentiate and demarcate groups and places in the city. These experiments continue, whether it is in apartheid South Africa, or in the Paris described by Maspero (see **Massey *et al.*, 1999**).

Interventions into the physical form of cities, be they the siting of hospitals or of cinemas, factories or homes, are continually faced with a clash of orderings, reflecting the uneven distribution of power, inequality and conflict in the city, and in relations with the wider world. So what are the implications for the notion of a 'good city'? If we are saying that questions of good and bad depend on who you are, where in the world you live, and what your experiences are, then is it possible to identify anything that might tell us what a good city is like?

ACTIVITY 8.2 Before proceeding any further, reflect on the issues that we have raised above: you will have encountered them in different ways throughout this book. What do you think makes for a good city?

When you have noted a few of your own ideas, turn to Reading 3B, which is a short extract from Richard Rogers' book *Cities for a Small Planet*. In the extract Rogers provides a list of 'key features' of the good city.

- What do you make of this list?
- What, if anything, has Rogers missed, or included unnecessarily?
- Do you think such a city is feasible? ◆

Rogers stresses the importance of social justice, ecology, diversity, easy contact, compactness, beauty and creativity. He argues that the city 'needs to meet our … objectives', but what does he mean by 'our' social, environmental, political, cultural, economic and physical objectives? And what are these 'objectives' exactly? Little is said about the economic organization that would underpin such a sustainable city. How is social justice to be achieved? Will it address inequalities of power and wealth? Bearing in mind that different technologies are at work, affecting different groups in different ways, will 'new technology' simply lead to a growing divide between the 'information rich' and the 'information poor', instead of providing cities with 'a new lease of life'?

It might well be that within any one city, there will be a range of different answers to these questions. Once more, to think spatially about these principles suggests that we need to consider how these differences are to be accommodated and perceived within the city. For example, it may be that the principles are to be unevenly applied through the city, with certain areas providing easy contact, while others provide an ecological balance. It seems that we need to think about the spatialities *within* the city to see how the principles underlying the 'good' city are to be applied.

3 *Intervening in the spatialities of the city*

Throughout this book there have been many ways in which the spatiality of the 'within-ness' of cities has been discussed. We have used spatial terms such as proximity, juxtaposition and crossing to evoke something of the ways in which people and things are brought together in cities, and we have used others, such as segregation, borders and walls, to imply that the 'within-ness' of cities is characterized by more than mixing and meeting. In fact, the 'within-ness' of cities is better understood through its *unevenness*, not only in terms of its various felt intensities and intensifications of social relations, but of power relations and people's location within networks of power relations. But thinking about this unevenness must also be tempered by a sense of the *openness* within the city to movements and connections across apparently disconnected spaces, whether those movements are within or beyond the city. When we begin to think about improving city life, we almost inevitably need to consider intervening in these aspects of the 'within-ness' of cities, namely unevenness and openness. Let's look at each in turn.

3.1 ON UNEVENNESS ...

At the start of this book it was argued that urban space is differentiated. This sense of differentiation does not imply, however, that there are essential differences (between people, activities and things) that somehow get 'ordered' and 'placed' in the city – as if different people, different things and different activities have proper, exclusive places in the city. Instead, we have demonstrated that the differentiation of urban space arises from the spatiality of social relations within, and stretching beyond, the city. This is where terms like juxtaposition and proximity began to become important analytical tools. It was argued that social relations within and between cities produce both social differences and socially differentiated urban spaces, through bringing people, things and activities together and/or keeping them apart. This was most readily seen in the ways in which certain areas of cities have become associated with certain groups, like Harlem in New York. We saw that Harlem is 'juxtaposed' to the rest of New York in a way that exaggerates its disconnection from the financial centre. Meanwhile, we also saw how other juxtapositions, this time within Harlem, are more fortuitous, producing an intensely creative atmosphere, both culturally and politically.

In this book we have frequently come across the idea of 'the stranger', and the urban politics that stem from the idea of the stranger (remember 'nimbyism' in Chapter 5?) Threading through the book is the argument that 'the stranger' is produced out of the social production of urban space: the stranger is only a stranger in certain places in the city; in others he or she may be at home, or a neighbour. From this perspective we can suggest that the differentiation of urban space and the social distinctions between people are interrelated. Let's think about this a bit more.

ACTIVITY 8.3 Look back at section 2.1 of Chapter 1, section 2 of Chapter 3 and section 4.2 of Chapter 5, and think about the arguments that were made about 'the stranger'.

- What do these arguments tell you about the juxtaposition of difference within cities?
- What kinds of juxtaposition are there in cities? Try listing some of the juxtapositions found in cities.
- How do these juxtapositions sharpen differences within cities? ◆

We have argued that the differentiation of urban space is produced through social relations, and that these social relations also produce social differences among people in the city. On the other hand, we have also indicated that urban space is constitutive of both social relations and social difference. There seems to be a confusion here: what is producing what? In fact this is not a good question, for we are moving towards a position in which social relations, the production of urban space and social differences are seen to be inextricably interwoven. But we have also argued strongly that the differentiation of urban space is neither arbitrary nor random. Instead, we have demonstrated that power relations lie at the heart of understanding the differentiation of urban space. Indeed, it is understanding these power relations, and how they are spatially constituted, that offers a way into understanding the particular 'within-ness' of the city.

ACTIVITY 8.4 This is a fairly abstract point. To help exemplify this argument, let's examine a particular feature of many cities: suburbs. Turn to the index of this book and look up 'suburbs' (you will find the discussions in Chapters 1, 2 and 5 of most interest). Thinking about the different ways in which suburbs have been produced as a differentiated urban space, what strikes you about the way in which power relations and the differentiation of urban space are interrelated? For example, won't this differentiation intensify further the way power relations are played out in the suburbs? ◆

Allen (1999a) has argued that power is not a fixed resource, like a cake, whose ingredients are somehow sited in and unevenly distributed between cities, but that power is an aspect of social relationships. Taking this argument into our thinking about the 'within-ness' of cities, we could argue that power relationships within cities are less about the distribution of the rich and poor through the city, but rather about how social relationships lead to the *differentiation* of the city, an aspect of which would be about how the rich and poor come to live in markedly different places. Notions of juxtaposition illustrate that these differentiations can in themselves be marked, as in Chicago in the 1920s, where the gold coast and the slum existed side by side. Power relationships can also work to produce what are described in Chapter 3 as technologies of spatial regulation, such as zoning, planning, surveillance and the like. With suburbs in mind, we can see that economic and political power have underlain many suburban developments. Thus suburbs have often been produced by the rich in order to produce (distinctive) territories within cities where they feel safe, where they are in control, and which maintain their capacity to order their (urban) world. However, it can also be seen that suburbs are not quite so orderly, safe or rich. Again, one's view of suburbs will be strongly influenced by 'who you are' and 'where you live'. The impacts on women's lives have been variable, while other suburbs are better described as peripheral estates that only serve to marginalize and socially exclude people further.

We cannot therefore state that 'the powerful' live in suburbs, while 'the powerless' live in inner cities, as a *universal* observation about the internal differentiation of urban space, whether we are talking about London, Istanbul or Sydney. In fact, such an observation is more than simply inaccurate, for it suggests a misunderstanding of the ways in which the differentiation of urban space is bound up with power relationships. Instead, it would be better to argue that power relationships are constitutive of differences between people and places within the city. Thus, while it would not be accurate to say that the rich live in the suburbs, we do know that suburbs are produced through power relations that make the spaces of the city in specific ways, sometimes making suburbs for the rich to escape from the city, sometimes making suburbs where the poor can be (decently?) housed, sometimes making suburbs in which new migrants to the city can live.

Indeed, we know more than this, for these power relationships are also reacting to something that is distinctive about cityness: its felt intensities, and its intensification of social relationships. It is not simply that people and things are differentiated and distributed in particular ways, for it matters how they are brought together and kept apart. Thus, as we have seen in Harlem and Johannesburg for example, it matters how people are juxtaposed, brought into close proximity, or whether they are kept apart – whether they meet and mix, or not.

ACTIVITY 8.5 We can again exemplify this abstract point about the spatial relationships within the city by returning to Chapter 2's analysis of disorder (section 3) and Chapter 4's discussion of living in differentiated cities (section 3), focusing particularly on the issue of segregation and borders. Look back over these sections, but think specifically about

- how power relations are bound up with the juxtaposition of different groups;
- whether segregated areas are characterized by their openness or closedness; and by openness and closedness to what; and
- whether segregation is a good or bad thing, and a good/bad thing for whom? ◆

You will have noted in these sections that the authors are keen to point out that seemingly closed-off areas of the city – such as those built by the urban 'apartheids' of gated communities or of townships – are produced through power relations in which the powerful attempt to close (themselves) off (from) 'unruly' elements through various spatial technologies, such as walls, gates, CCTV, zoning and so on. Although we need to appreciate that CCTV can benefit other groups in the city, in many ways the walling-off of communities can be judged to have deleterious effects. However, we have also seen that segregation can have positive effects in terms of community formation, through which people form social ties (through common circumstances and needs) that enable them to survive and (even) thrive. The outcomes of these (economic, social) connections and (economic, social) disconnections is therefore ambiguous. Arguing about whether the uneven spatialities of these social relations are good or bad would involve a sense of whether the power relations producing these forms of differentiated urban space are good or bad. In many ways this is an ethical, moral and political decision – a decision, moreover, that is rarely clear-cut and never final, since power relations are constantly shifting, and constantly being renegotiated, as Chapters 4 and 5 demonstrated (see also **Massey, 1999**).

3.2 ON OPENNESS ...

It is important to note that these examples show that the unevenness of cities is also related to *openness within* them. This is not a question of places within cities being open or closed. Instead, the term 'openness' should be thought of as referring to the porosity and permeability of urban spaces, since even gated communities are both 'open' and 'closed', as are ghettos. People are constantly moving into, through and out of urban spaces, yet not everyone is free (nor able, nor wants) to travel or communicate everywhere. We might think again here about the stranger, and this is our second point. If we are thinking about improving city life, we not only have to consider the unevenness of the city (the differentiation of urban spaces) and whether these are good or bad, but also the openness (the porosity, the permeability) of urban spaces: who and/or what is able to move into, move through, settle in the city; who can move out of, made to move on, and so on. Once again, it is not so simple to intervene appropriately or successfully.

The city must necessarily be open to many things, for example to essential services such as water supply. Yet the openness of the city to water begs two questions: where does that water come from, and how is it to be distributed within the city? For many people in 'First World' cities, running hot and cold water is taken for granted. But Chapter 6 has raised other questions about running water, and its environmental consequences. Furthermore, we must consider how scarce

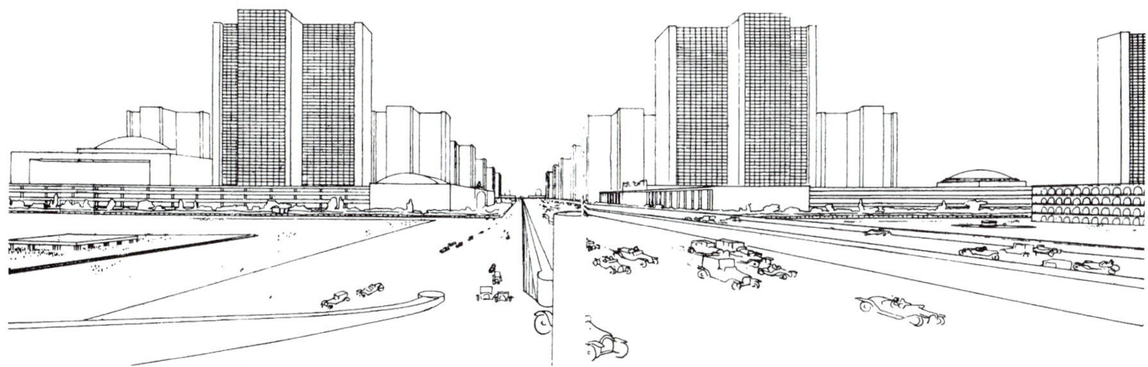

FIGURE 8.3 *The open, even and just city? Le Corbusier's vision of the city, open to traffic, open to public services, open to light and air*

resources are to be distributed equitably within the city. We could proliferate these kinds of environmental questions, and we could add economic issues, such as the openness of the city to flows of money. It may be a good thing to open up the city to international flows of money: as the financial markets get richer, there is supposed to be a 'trickle down' of money to poorer sections within the city. On the other hand there will be a leakage of this money outside the city. This openness leads to sections of the city's economy becoming more vulnerable to the vagaries of international economics, and it can also exaggerate and intensify social differences within cities between the rich and the poor. Now, are these good things or not? Should the city be opened up or closed down? Should some parts of the city be more porous than others?

To conclude this section, let's return once more to Sennett's argument about the uses of disorder in the anarchic city. You will have noted that Sennett is keen to bring the conflicts of the city out into the open, so that they can be made transparent and thereby negotiated, making the disorderliness of the city its prevailing social order. Our analysis of the social relations of the city agrees that there is an internal relation between order and disorder, but that such an understanding must be integrated with an appreciation of the unevenness and openness of the city. It is uncertain, then, how the 'disorderly orders' of the city will produce an integrated and coherent outcome, across space and time: for example, how would the conflicting interest of one neighbourhood be assessed against the demands of another? Would there be restrictions on the flow of people or things (like rubbish, water, waste, and so on) through the city? Would polluting factories be closed down, or kept open to provide jobs, and would it matter where in the city these factories were sited? Such questions demonstrate that the differentiation of urban spaces produces conflicts and struggles of resources (as we saw in Chapter 5).

Furthermore, it is important to recognize that one way to deal with conflicts between people is to allow them to have different spaces: cities are 'homes' to people who have irreconcilable views, so should they be brought together to fight it out, or kept well apart? The suggestion that the city should be comprised of segregated spaces has an unsettling side, since it can serve as an excuse to marginalize specific groups of people, and even exclude them from the city altogether. What is at issue, then, is more than the question of whether it is a good or a bad thing to bring people, things or activities into close proximity or to blur the sharpness of certain juxtapositions (say between rich and poor areas). These are significant questions, but the central issue is about how the unevenness and openness of the city are to be negotiated. It is your understanding of these issues that will determine whether you believe that walls, CCTV and segregations are good or not, and inevitable or not.

We seem to be close to drawing the conclusion that cities are unruly, but we are not sure whether or not this is a good thing – or, more precisely, under what circumstances unruliness and order might be good (or bad) things. Perhaps we need finally to ask questions about power relationships, in all their orderings and disorderings, their mixings and patternings, and about the ways in which urban dwellers might be able to intervene in city life.

4 *Unruly cities?*

Let us return to the central theme of this book, 'unruly cities?', and to some of the questions and issues that we raised in the Introduction. The 'unruly' character of city life has been evident in many of the cities discussed in this book. Dystopian images and visions of cities in 'crisis' are never far away from accounts that offer a one-sided interpretation and understanding of city life, but throughout this book the authors have been concerned to question claims of disorder and unruliness, to problematize them and to throw them into a new light. This is not to say that cities are free from enormous economic, social and environmental problems. The economic crisis that gripped many of the economies of the Pacific Rim during the late 1990s was played out on the streets of Jakarta, Seoul and numerous other cities in Asia and is testimony to the misery and poverty faced by millions (Fermont, 1998). This also highlights the centrality of conflict within cities, but to say that cities are essentially chaotic and unruly is very different, and only ignores the complexity of city life. What is being played out in all cities around the world, albeit in diverse ways, are the consequences of the uneven, porous differentiation of urban space. How this differentiated space is ordered and regulated is central to visions of the good city. The question of *how* conflict is managed and regulated is the key issue here.

Above all, such visions do not emerge in an economic and political vacuum. They represent particular ways of conceiving of social relations, social problems and their solutions. They represent conflicting visions of order and disorder in the city. Let us return to Rogers for a moment. One of the key facets of his module of the sustainable city is 'a just city where justice, food, shelter, education, health and hope are fairly distributed and where all people participate in government'. Few people reading this would disagree with such sentiments, yet they leave us with more questions to answer: What does 'fairly distributed' mean? Who decides what form participation in urban politics should take? Such questions are relevant both within and beyond the city, but they are particularly difficult when cities combine so many arenas of political debate and so many perspectives. This is their challenge, but also their opportunity. Arguably, however, it is within cities outside the West that such questions are being asked with renewed force and vigour, amidst increasing levels of poverty, inequality, polarization, exclusion and pollution, and against growing, if uneven, levels of openness.

What, then, does it take to produce a 'better city'? Clearly this is a complex and very difficult question to answer. There are issues of democracy and citizenship, as was noted in Chapter 7. But how are these to be understood and contextualized? What forms of participation are necessary to sustain the economic vitality of a city? Can this be organized in a way that benefits all the population? Different kinds of orderings will involve different kinds of participation, and recall that different forms of order continually 'clash' and come into conflict in the city. The organization and form of the urban environment is both constitutive and generative of social relations, and this is embedded in unequal power relations. But in addressing the issue of what makes for a good/better city, we need to address issues of both social and spatial relations – relations which stretch well beyond particular cities.

In posing the question, 'what makes for a better city, a sustainable city, a good city?', one need not privilege a particular sense of urban life and order. Nevertheless, the answer to such a question will look very different from the street than it would from the corporate boardroom, or from the shanty town and the gated enclave.

References

Allen, J. (1999a) 'Cities of power and influence: settled formations', in Allen, J. *et al.* (eds).

Allen, J. (1999b) 'Worlds within cities', in Massey, D. *et al.* (eds).

Allen, J., Massey, D. and Pryke, M. (eds) *Unsettling Cities*, London, Routledge/The Open University (Book 2 in this series).

Banfield, E.C. (1968) *The Unheavenly City*, Boston, Little Brown.

Fermont, C. (1998) 'Indonesia: the inferno of revolution', *International Socialism*, no.80, autumn, pp.3–33.

Gold, J.R. (1997) *The Experience of Modernism*, London, E. and F.N. Spon.

Hinchliffe, S. (1999) 'Cities and natures', in Allen, J. *et al.* (eds).

Massey, D. (1999) 'On space and the city', in Massey, D. *et al.* (eds).

Massey, D., Allen, J. and Pile, S. (eds) (1999) *City Worlds*, London, Routledge/The Open University (Book 1 in this series).

Pile, S. (1999) 'What is a city?', in Massey, D. *et al.* (eds).

Rogers, R. (1997) *Cities for a Small Planet*, London, Faber and Faber.

Sennett, R. (1971/1996) *The Uses of Disorder*, London, Faber and Faber.

READING 8A
Richard Sennett: 'Ordinary lives in disorder'

This book contrasts a society that is with a society that could be. On the one hand, there exists a life in which the institutions of the affluent city are used to lock men into adolescence even when physically adult. On the other hand, there is the possibility that affluence and the structures of a dense, disorganized city could encourage men to become more sensitive to each other as they become fully grown. I believe the society that could be is not a utopian ideal; it is a better arrangement of social materials, which as organized today are suffocating people.

Yet the feel, the quality, of a social change is difficult to envision. People have only the sense of what they already have experienced, and that makes talk of social change seem abstract and unreal. To convey an impression of how these anarchic cities would affect ordinary life and everyday problems seems to me a fitting way to close this book.

Let us try to imagine what it would be like for one intelligent young girl who grows up in an anarchic urban milieu. She lives, perhaps, on a city square, with restaurants and stores mixed among the homes of her neighbors. When she and the other children go out to play, they do not go to clean and empty lawns; they go into the midst of people who are working, shopping, or are in the neighborhood for other reasons that have nothing to do with her. Her parents, too, are involved with their neighbors in ways that do not directly center on her and the other children of the neighborhood. There are neighborhood meetings where disruptive issues, like a noisy bar people want controlled, have to be fought out. Since the neighborhood is a densely packed place, thus permitting personal styles or deviations to be expressed, and since the personnel of the neighborhood is constantly shifting, her parents are out a great deal merely to find out who their neighbors are and see what kind of accommodations can be reached where conflicts arise. A black couple down the street may feel the little girl is cruel to their children, or some days she may feel they are cruel to her; the families cannot ignore each other. They are physically thrown together without impersonal resources, like attendance at homogeneous school districts, for separation.

In fact, the schools of the neighborhood are a kind of focus of conflict and conciliation for the parents. They are controlled by the community, but the community is so diverse that the schools cannot be pushed in any one direction. Families of the school district may have the right to set moral and religious policy for the school, for example, but since in these cities Catholics, Protestants, and Jews are mixed together, accommodations concerning ethical instruction and Bible training have to be arrived at. Indeed, the rules of the school shift constantly as new people in the district, with special backgrounds and interests, assert their right to have a hand in the shaping of their children's education.

But this little girl sees, every day, that the tensions and friendships in the community or school, so transitory and unstable, do not create chaos. She is made conscious of a kind of equilibrium of disorder in the lives of adults around her and in her own circle of friends. People are not sheltered from each other, but their contacts are more explorations of a constantly shifting environment than an acting out of unchanging routines.

Therefore this little girl grows up in a neighborhood that does not permit her family or her circle of friends to be intensive and inward-turning. This fact has a liberating power for her as someone who is exceptionally bright. For at school, the complex weave of friendship and casual acquaintance makes it very difficult for the other children to exercise pressure against her for being 'different' because she is bright. In the suburbs, where social and economic backgrounds are ironed out, that pressure frequently and, in terms of the development of children like this, tragically arises. But in the city school this little girl attends, everyone is in some way different; there is a jumble of many backgrounds, and it becomes harder to shame someone who is unusual. Were this little girl exceptionally unintelligent, the same would be true. The children do not play and learn in packs; their backgrounds and their social contacts are too complex and too shifting for the brutal baiting that suburban children practice against kids who are 'different' to work.

Let us now take a glimpse of this intelligent young girl later in her life, when she is a woman. Sociologists know something of what lies ahead for her in city culture as it exists now. Beyond all the clichés – made clichés by being so often true – of the inequities she will find in her work and the fear intelligent men show of treating her as an equal, her city life as an adult is today constricted in several less obvious ways. The forums for being with men as friends are usually limited to work; it is difficult outside this sphere to get to know people who aren't also after her. If and when she marries, there is usually an enormous amount of guilt about giving up her work and becoming just a housewife, since housewifely tasks, including her relations with other people in the community where she lives, offer little scope for exercising her intelligence. From what is known of young women like this, her life in a city faces two equally unacceptable alternatives of isolation: either a professional life where the opportunities of social encounters are limited to colleagues who feel competitive and men who want possession, or the more usual housewifely and community routines, which offer no field for intellect. But in a city where men and women are forced into all sorts of contact for accommodation and mutual survival, these poles of isolation can be greatly diminished. A single woman's work wouldn't define the sole society of peers in which she had to exist. It would be possible to meet a great number of people, in a variety of situations of mutual interest and curiosity, out of the necessity of dealing with her home neighborhood, her work neighborhood, and the city-wide political and social problems. Working in voluntary organizations or political clubs wouldn't be something an intelligent girl forced herself to do in order to meet men, as so often now occurs. These skeins of association would be a natural social life that arose out of the necessity for common action. Were a woman such as this to marry, have children and leave off her career, the same skein of necessary community relations would offer her a field to use her talents in ways that mattered. It is a commonplace that large numbers of middle-aged, intelligent housewives are left in suburbs with the time and desire to work in the communities where they live, but little room is given them in which to work, save as helpers or assistant to 'real' professionals in schools or hospitals. By increasing the complexity and loosening the rules of routine in the community settings of these women, they would have a chance to be creative and have a forceful social life, even though they had opted out of professional careers.

A change in the kind of communities intelligent women can live in will not obviously change the whole complex of discrimination and fearfulness with which intelligent men and other women regard them. But for a young woman like the one I have described, the opprobrium of being different would be muted in childhood and adolescence; in adulthood the new anarchic communities would offer a means out of isolation whether the woman pursued a career or not.

Let us then try to envision the impact of anarchic cities on an ordinary group of city men: people from the 'working class' who have become relatively affluent. A popular stereotype has it that such affluent industrial laborers and service personnel have become conservative and a force for the maintenance of repressive 'law and order.' It might seem, then, that they would be most resistant to social changes that introduced greater disorder into the city.

What researchers are beginning to glimpse about affluent working-class communities is that the cries for law and order are greatest when the communities are most isolated from other people in the city. In Boston, for example, the fear of deviance and conflict is much greater in an Irish area called South Boston, which is cut off geographically from contact with the city at large, than in another Irish area, North Cambridge, which is stuck in the midst of the city and is to some extent penetrated by blacks and college students. Cities in America during the past two decades have grown in such a way that ethnic areas have become relatively homogeneous; it appears no accident that the fear of the outsider has also grown to the extent that these ethnic communities have been cut off.

If the permeability of cities' neighborhoods were increased, through zoning changes and the need to share power across comfortable ethnic lines, I believe that working-class families would become more comfortable with people unlike themselves. The cries for law and order are enormously more complicated than the effect of community setting, and no one can pretend that a different kind of neighborhood would of itself transform the feelings of status insecurity of frustration in work that are involved in the desire for law and order. But the experience of living within diverse groups has its power. The enemies lose their clear image, because every day one sees so many people who are alien but who are not all alien in the same way.

Let us imagine a family where the father is an industrial worker moving into a disorganized community that forced family members into contact with others, with black families becoming affluent themselves, with managerial and professional people, with the young as well as the middle-aged. The image and demands of the neighborhood would work against whatever desire this family had for exclusion of the immoral, unpatriotic 'them'. But these communities would have a further, positive value to such a family as well.

In a community like this, the bureaucratic manipulation of conflict is gone. A factory man can confront those around him on a more equal basis than in a situation where middle-class bureaucracies prevail. The reason is that the willingness to use impersonal bureaucracy and faceless power has become the great weapon of the middle classes today over those who do routine labor. This weapon is purposely made weak in anarchic cities. It is patterns of personal influence and personal alliance that shape the balance of disorder in the new cities; politics and less formal community relations on these terms are historically how working-class people have evolved institutions in which they feel a stake. I am convinced that in such a milieu a man who does humble work can feel more of a man in dealing with others than in circumstances where the present weapons of power prevail. He can use himself as a human being, make himself heard, rather than be muffled by those who are different and more skilled in the arts of bureaucratic management. Instead of establishing common dignity by ensuring the sameness of all who live in a community, as happens in so many affluent working-class areas now, this laboring family could establish its dignity in a more satisfying way, by having a community forum for conflict and reconciliation in which it faces other people as concrete beings who have to talk with each other.

In this way, a disordered city that forced men to deal with each other would work to tone down feelings of shame about status and helplessness in the face of large bureaucracies. Participation of this sort could mute in affluent workers that sad desire for repressive law and order.

The final effect of the anarchic cities on the feelings of ordinary people facing day-to-day problems concerns the functional efficiency of the city itself. There is a Homeric catalogue of complaints today about the quality of the services and the environmental health of cities. Transportation is jammed, the air is polluted, the streets are dirty; inadequate fire, police, and sanitation staffs strike for more pay from cities that already operate at a loss; most city schools are out of date and poorly equipped, with inadequate provisions for teachers and staff.

These problems depend on more money, and this book purposely provides no answers about new ways to make more tax dollars. Getting more money for cities in America is a brutally simple affair: the priorities of the economy have to be changed from massive military spending to a more just distribution of public resources. In the face of the military dominance of public finance, all other revenue-raising operations are 'Bandaids', publicity devices that will have little real effect. The American economy is certainly able to finance the cities; after all, much less affluent countries whose budgets are not all-absorbed in military affairs have maintained their massive cities much more adequately than has the United States. As urbanists like myself have said over and over during the last few years, the urban financial problem in this country is the problem of military spending. Converting the economy from a military-industrial to an urban-industrial base is the only real solution.

But once the money is available, what is the best social use to which it can be put? As I have tried to show in this book, many of the seemingly routine aspects of city administration, like police, housing construction, and school administration, need not be routines, but opportunities for community life, thereby revitalizing the people directly concerned. Furthermore, I have sought to illustrate how the peculiar model of routing that has guided the planning of these services – a model based on the way machines produce goods – becomes dysfunctional and, in the words of the economists, counterproductive, when applied to the management of the social affairs of city men. Once the financial base for city services is expanded by a reduction in expenses for war, these services could become more responsive to the desires of city dwellers, if men could only accustom themselves to look at conflict over city services as a necessary, desirable product of people seeking to govern themselves. The social breakdown of city services – a strike by teachers, a strike by hospital staffs, etc. – is not an immoral threat to the public; these strikes are expressions of human need, voiced by people who want to be heard and are now thwarted by the central bureaucracies. Take conflict in the public arena away, and you revert to the idea that a broad swatch of urban society can have its best interests 'managed' for it by impersonal bureaucratic means. As has been shown, this godlike presumption about other people's lives on the part of planners only builds up steam for violent disruption.

When conflict is permitted in the public sphere, when the bureaucratic routines become socialized, the product of the disorder will be a greater sensitivity in public life to the problems of connecting public services to the urban clientele. The financial crisis in city services caused by militarism has only served to reinforce the idea that 'good' public service is one in which some measure of routine can function. Once the cash is available, the threats to routine will take on an entirely new character. The threats will be a focus of 'sensitizing' public service bureaucracies to the public and to the public issues.

The fruit of this conflict – a paradox which is the essence of this book – is that in extricating the city from preplanned control, men will become more in control of themselves and more aware of each other. That is the promise, and the justification, of disorder.

Source: Sennett, 1996, pp.189–98

READING 8B
Richard Rogers: 'Cities for a small planet'

Given the will to create them, cities of the future could provide the foundation for a society in which everyone participates in health, security, inspiration and justice. New technology could give our cities a new lease of life: a more sociable, more beautiful and more exciting life, above all a life determined by its citizenry.

The concept of the sustainable city recognises that the city needs to meet our social, environmental, political and cultural objectives as well as our economic and physical ones. It is a dynamic organism as complex as the society itself and responsive enough to react swiftly to its changes. The sustainable city is a city of many facets:

– A Just City where justice, food, shelter, education, health and hope are fairly distributed and where all people participate in government;

– A Beautiful City, where art, architecture and landscape spark the imagination and move the spirit;

– A Creative City, where open-mindedness and experimentation mobilises the full potential of its human resources and allows a fast response to change;

– An Ecological City, which minimises its ecological impact, where landscape and built form are balanced and where buildings and infrastructures are safe and resource-efficient;

– A City of Easy Contact, where the public realm encourages community and mobility and where information is exchanged both face-to-face and electronically;

– A Compact and Polycentric City, which protects the countryside, focuses and integrates communities within neighbourhoods and maximises proximity;

– A Diverse City, where a broad range of overlapping activities create animation, inspiration and foster a vital public life.

The sustainable city could be the agent for delivering environmental rights (basic rights to clean water, clean air, fertile land) to our new, dominantly urban, global civilisation. There are millions now and there will soon be billions who enjoy no such rights. Commitment to environmental rights demands the emergence of the sustainable city – in fact the two are interdependent. Take the issue of clean air or health. Currently millions of shanty dwellers are burning solid fuels to cook and warm themselves; this creates dangerous levels of air pollution from which they cannot escape. A city committed to human rights must provide clean energy, it must carry out policies that increase its efficiency and reduce its pollution. This in turn alleviates the global environmental crisis for all.

The world-wide environmental and social crisis of our cities has focused minds. The call for sustainability has revived the need for considered urban planning and has demanded a rethink of its basic principles and objectives. The crisis of modern civilisation demands that governments plan for sustainable cities.

Source: Rogers, 1997, pp.167–8

Acknowledgements

Grateful acknowledgement is made to the following sources for permission to reproduce material in this book:

CHAPTER 1

Text

Etcherelli, C. (1993) 'Elise or the real life', in Wilson, J.P. and Michaels, B. (trans.) in Heron, L. (ed.) *Streets of Desire: Women's Fictions of the Twentieth-Century City*, Little Brown and Co (UK). © the author; Zukin, S. (1995) *The Cultures of Cities*, Blackwell Publishers Ltd. Copyright © Sharon Zukin, 1995; Silverstone, R. (1997) *Visions of Suburbia*, Routledge; Chambers, D. (1997) 'A stake in the country: women's experiences of suburban development', in Silverstone, R. (ed.) *Visions of Suburbia*, Routledge.

Figures

Figure 1.1: Getty Images; *Figure 1.2:* Park, R.E. and Burgess, E.W., with McKenzie, R.D. and Wirth, L. (1925/1984) *The City: Suggestions for Investigation of Human Behaviour in the Urban Environment*, p.51, University of Chicago Press; *Figure 1.3:* adapted from Smith, N. (1996) *The New Urban Frontier: Gentrification and the Revanchist City,* Routledge; *Figure 1.4: The Ascent of Ethiopia*, 1932, by Lois Mailou Jones. Milwaukee Art Museum, Purchase, African-American Acquisition Fund. Matching funds from Suzanne and Richard Pieper, with additional support from Arthur and Dorothy Nelle Sanders; *Figure 1.5: Saturday Night Street Scene* by Archibald J. Motley, jnr. Courtesy of Archie Motley; *Figure 1.6:* Clarke, A.J. (1997) 'Tupperware, suburbia, sociality and mass consumption', in Silverstone, R. (ed.) *Visions of Suburbia*, Routledge.

Photographs

p.45: Photographs and Prints Division, Schomburg Center for Research in Black Culture, The New York Public Library, Astor, Lenox and Tilden Foundations; *pp.46, 47 (right), 48, 49:* Alex Vitale; *p.47 (left):* Wide World Photos.

CHAPTER 2

Text

Branford, S. and Kucinski, B. (1995) *Brazil: Carnival of the Oppressed,* The Latin American Bureau, © Sue Branford and Bernardo Kucinski; Jellinek, L. (1997) 'Displaced by modernity: the saga of a Jakarta street-trader's family from the 1940s to the 1990s', in Gugler, J. (ed.) *Cities in the Developing World: Issues,*

Theory and Policy, Oxford University Press. Reprinted by permission of Oxford University Press; Bromley, R. (1982) 'Working in the streets of Cali, Colombia: survival strategy, necessity or unavoidable evil?', in Gilbert, A., Hardoy, J. and Ramirez, R. (eds) *Urbanization in Contemporary Latin America*, John Wiley and Sons Ltd. Reproduced by permission of John Wiley and Sons Ltd; Lomnitz, L. (1974) 'The social and economic organization of a Mexican shanty-town', in Cornelius, W.A. and Trueblood, F.M. (eds) *Anthropological Perspectives on Latin American Urbanization*, Latin American Urban Research, 4, pp.135-55. Copyright © 1974 Larissa Lomnitz. Reprinted by permission of Sage Publications, Inc.

Figures

Figure 2.2 (top p.58): Hulton Getty; *Figure 2.2 (bottom p.58):* Aerofilms; *Figure 2.2 (top p.59):* John Maier/Still Pictures; *Figure 2.2 (bottom p.59):* Architects Journal; *Figure 2.4:* Oktay Ekinci; *Figure 2.5:* Photo: Ted Ditchburn/North News, from *The Independent on Sunday,* 23 August 1992; *Figure 2.6:* Bob Collier/ Popperfoto; *Figure 2.7:* Photo: Ged Murray, from *Observer Magazine,* 21 February 1993; *Figure 2.8:* Photo: Ged Murray, from *The Observer*, 12 July 1992; *Figure 2.9:* Gustav Dobrzynski, Department of Geography, University of Salford; *Figure 2.10:* © Factory Records, Manchester; *Figure 2.13: The Sunday Times Magazine*, 26 November 1989, © Times Newspapers Ltd, 1989.

Cartoons/Ilustrations

p.57: from: *The Herblock Gallery*, Simon and Schuster, New York, 1968. © Herbert Block; *pp.66, 67:* Tan Oral; *p.68:* © Peter Till; *p.79:* reproduced by permission of *Punch*.

CHAPTER 3

Text

From *The Death and Life of Great American Cities* by Jane Jacobs. Copyright © 1961 by Jane Jacobs. Reprinted by permission of Random House, Inc.; Davis, M. (1994) *Beyond Bladerunner: Urban Control: The Ecology of Fear*, Pamphlet 23, Open Magazine Pamphlet Series. © Mike Davis, 1992, 1994; Beckett, A. (1994) 'Take a walk on the safe side', *The Independent on Sunday,* 27 February, 1994.

Figures

Figure 3.1: Denis Thorpe/*The Guardian*; *Figure 3.2:* David H. Pinkney, *Napoleon III and the Rebuilding of Paris*, 1958. Reprinted by permission of Princeton University Press; *Figure 3.3:* Gustave Caillebotte, *Paris Street; Rainy Day*, 1876/77, oil on canvas, 212.2 by 276.2cm. Photograph © 1998, The Art Institute of Chicago. All rights reserved; *Figure 3.4:* Bibliothèque Historique de la Ville de Paris; *Figure 3.5:* Jeffrey D. Needell, *A Tropical Belle Epoque*, Cambridge University Press, 1987; *Figure 3.6:* Photograph by Augusto Malta, Jennings Hoffenburg Collection. Courtesy of Professor Teresa Meade; *Figure 3.7:* Bradley

Arden; *Figure 3.8:* Martin Parr/Magnum; *Figure 3.10:* Don McPhee/*The Guardian*; *Figure 3.11:* David Rose/*The Independent*.

CHAPTER 4

Text

Anderson, K. J. (1988) 'Cultural hegemony and the race-definition process in Chinatown, Vancouver, 1880-1980', *Environment and Planning D: Society and Space*, 6 (2), pp.134-5, Pion Ltd, London; Jellinek, L. (1997) 'Displaced by modernity: the saga of a Jakarta street-trader's family from the 1940s to the 1990s', in Gugler, J. (ed.) *Cities in the Developing World: Issues, Theory and Policy*, Oxford University Press. Reprinted by permission of Oxford University Press; Goldberg, D.T. (1993) 'Polluting the body politic: racist discourse and the urban location', in Cross, M. and Keith, M. (eds) *Racism, the City and the State*, Routledge; Mayekiso, M. (1996) *Township Politics: Civic Struggles for a New South Africa*, Monthly Review Press. Copyright © 1996 by Mzwanele Mayekiso. Reprinted by permission of Monthly Review Foundation; Naga, R. (1997) 'Communal places and the politics of multiple identities', *Ecumene*, 4 (1), Arnold, Hodder Headline plc.

Figures

Figures 4a-h: Starkey, M. (1994) *London's Ethnic Communities*, London Research Centre and Census data reproduced with the permission of the Controller of Her Majesty's Stationery Office. © Crown Copyright; *Figure 4.3:* adapted from Bassett, K. and Short, J.R. (1980) *Housing and Residential Structure: Alternative Approaches*, p.30, Routledge; *Figure 4.4*: Western, J. (1981) *Outcast Cape Town*, Allen and Unwin. Photo: Terence McNally, courtesy of John Western; *Figure 4.6:* Copyright © 1993. From *Reading Rodney King*, edited by Robert Gooding-Williams. Reproduced by permission of Routledge, Inc. Graphic © J. H. Johnson 1992, Chase Langford Cartography.

Table

Table 4.1: Denton, N.A. and Massey, D.S. (1988) 'Residential segregation of Blacks, Hispanics and Asians by socioeconomic status and generation', *Social Science Quarterly*, 69, University of Texas Press.

CHAPTER 5

Text

Ward, P. (1997) 'A view to the future', *The Australian*, 10 October 1997, © News Limited 1997; Massey, D. (1999) 'Living in Wythenshawe', in Borden, I., Kerr, J., Pivaro, A. and Rendell, J. (eds) *The Unknown City*, Routledge; Murphy, P. and Watson, S. (1997) *Surface City: Sydney at the Millennium*, Pluto Press; Woolf, V.

(1937) *The Years*, Grafton Books; Woolf, V. (1938) *Three Guineas,* Penguin Books Ltd; Anderson, E. (1990) *Streetwise,* University of Chicago Press; Bruce, I. (1997) 'Gay sites and the pink dollar', in Murphy, P. and Watson, S. (eds) *Surface City: Sydney at the Millennium,* Pluto Press.

Figures

Figure 5.2: Michael Ann Mullen/Format; *Figure 5.3:* Judy Harrison/Format; *Figure 5.4:* Val Wilmer/Format; *Figures 5.5, 5.7:* Jenny Matthews/Format; *Figures 5.6, 5.9:* Maggie Murray/Format; *Figure 5.8:* Pam Isherwood/Format.

CHAPTER 6

Text

Harvey, D. (1995) 'Principles of justice and environments of difference', in Merriefield, A. and Swyngedouw, E. (eds) *The Urbanisation of Injustice,* Lawrence and Wishart Ltd, © David Harvey.

Figures

Figure 6.1: Horst Haitzinger; *Figures 6.2(a),(b),(c):* Clark, D. (1996) *Urban World/Global City,* Routledge; *Figure 6.6:* Stuart Franklin/Magnum; *Figure 6.7:* Jenny Matthews/Network; *Figure 6.8:* Mike Levers/The Open University; *Figures 6.9 (a)–(e)*: Government Statistical Service (1994) Developing and Integrated Transport Policy, August 1997, Department of Transport and Department of the Environment, © Crown Copyright is reproduced with the permission of the Controller of Her Majesty's Stationery Office; *Figure 6.10:* Mark Edwards/Still Pictures; *Figure 6.11:* Le Corbusier (1967) *The Radiant City,* Faber and Faber Ltd. © FLC/ADAGP, Paris and DACS, London, 1999; *Figure 6.12:* Breheny, M. and Rookwood, R. (1993), in Blowers, A. (ed.) *Planning for a Sustainable City Region,* Earthscan Publications Ltd.; *Figure 6.13:* Catherine Platt/Still Pictures; *Figure 6.14:* Donovan Wylie/Magnum; *Figure 6.15:* From: Peter Harrison, *Walter Burley Griffin, Landscape Architect,* 1995, Canberra, National Library of Australia, p.36. © National Library of Australia.

CHAPTER 7

Text

Le Corbusier (1929/1987) *The City of Tomorrow,* The Architectural Press, © FLC/ADAGP, Paris and DACS, London; From *The Death and Life of Great American Cities,* by Jane Jacobs, Copyright © by Jane Jacobs. Reprinted by permission of Random House Inc.; Howe, N. (1998) 'Berlin mitte', *Dissent,* 45 (1); Stoker, G. (1995) 'Regime theory and urban politics', in Judge, D., Stoker, G. and Wolman, H. (eds) *Theories of Urban Politics,* Sage Publications Ltd.

Figures

Figure 7.1: Hall, P. (1997) *Cities of Tomorrow*, Blackwell Publishers, © FLC/ADAGP, Paris and DACS, London, 1999; *Figure 7.2:* Photo: Mavis Swenerton/Architectural Association; *Figure 7.3:* Photo: Courtesy of Private Collection, Wiener Library Ltd; *Figure 7.4*: © Axel Boesten; *Figure 7.5:* Bundesarchiv Koblenz; *Figure 7.6:* Stadtarchiv Eisenhüttenstadt; *Figure 7.7:* Clara Kraft/Architectural Association; *Figure 7.8*: Le Corbusier (1987) *The City of Tomorrow*, Butterworth-Heinemann Ltd. © FLC/ADAGP, Paris and DACS, London, 1999; *Figures 7.9, 7.10:* Popperfoto; *Figure 7.11:* Gunter Schneider; *Figure 7.12*: Courtesy of Daimler Benz/debis; *Figure 7.13:* PA News.

CHAPTER 8

Text

Sennett, R. (1971) *The Uses of Disorder: Personal Identity and City Life*, Faber and Faber Ltd; Rogers, R. (1997) *Cities for a Small Planet,* Faber and Faber Ltd, © Richard Rogers, 1997.

Figures

Figure 8.1: From: David Butler's *Just Imagine*, 1930. Photo: Fox Films (Courtesy Kobal); *Figure 8.2*: From: Ridley Scott's *Bladerunner*, 1982. Photo: Ladd Company/Warner Bros (Courtesy Kobal); *Figure 8.3:* From Le Corbusier, *The City of Tomorrow and its Planning*, 1929, reprinted 1978, London, The Architectural Press Ltd, pp. 242–3. © FLC/ADAGP, Paris and DACS, London, 1999.

COVER

Photo: Paul Harris/Tony Stone Images.

Every effort has been made to trace all copyright owners. If any has been inadvertently overlooked, the publishers will be pleased to make the necessary arrangements at the first opportunity.

Index